Lecture Notes in Computer Science

Lecture Notes in Computer Science

Edited by G. Goos and J. Hartmanis

248

Networking
in Open Systems

International Seminar
Oberlech, Austria, August 18–22, 1986
Proceedings

Edited by Günter Müller and Robert P. Blanc

Springer-Verlag
Berlin Heidelberg New York London Paris Tokyo

Editors

Günter Müller
IBM European Networking Center
Tiergartenstraße 15
D-6900 Heidelberg, Federal Republic of Germany

Robert P. Blanc
Institute for Computer Sciences and Technology
National Bureau of Standards
Gaithersburg, MD 20899, USA

CR Subject Classification (1987): C.2.0, C.2.2-4

ISBN 3-540-17707-8 Springer-Verlag Berlin Heidelberg New York
ISBN 0-387-17707-8 Springer-Verlag New York Berlin Heidelberg

© Springer-Verlag Berlin Heidelberg 1987
Printed in Germany

Printing and binding: Druckhaus Beltz, Hemsbach/Bergstr.
2145/3140-543210

Preface

The contributions to this volume were prepared for and presented at the IBM Europe Institute 1986 in Oberlech, Austria. The seminar "Networking in Open Systems", which was held from August 18 to August 22, was organized jointly by the IBM European Networking Center (ENC) Heidelberg and the National Bureau of Standards (NBS), Gaithersburg, USA, under the guidance of the Director of Scientific Programs, Dr. H. Budd (IBM), and the organization of Dr. L. Hyvarinen (IBM). The IBM Europe Institute provides in a series of weekly seminars on various topics the opportunity for leading researchers to reflect on the current advances and future directions of their field.

Networking is one of the most challenging areas of computer science. The challenge is to develop techniques to take advantage of new communication technologies such as local area networks, fiber optics, or ISDN in order to provide students, researchers or office workers with the means to communicate via electronic mail, computer conferencing systems, or the ability to retrieve a variety of documents from remote data bases, etc..

Standards, architectures, and implementations of networks following the Open Systems Interconnection model are receiving wide technical, economical, and political attention. The need to interconnect heterogeneous networks and computers for the purpose of transparent access to information via standardized protocols is accepted by many. However, many technical and strategic issues remain controversial and are open to continued research efforts:

1. There is a growing need for increased availability of advanced applications requiring access to remote services in networks of heterogeneous hard- and software. Neutral parties are going to play a crucial role to define the success of OSI to assure a balance of interests of all involved groups ranging from endusers via network providers to manufacturers.

2. OSI is designed as an architecture to interconnect autonomous systems. OSI protocols will require modifications and additions to achieve high performance and network management functions.

3. Standardization and implementation are slow by nature. Ambiguities in standards and necessary compromises cause misunderstandings and create the need for techniques to produce implementations at least parallel to evolving standards. Formal descriptions and program generators will need a closer relationship to prove helpful to real systems.

4. The dramatic advance in technology and the huge market in communication and computer networks, the progressing digitalization of information demand through OSI a consistent use of protocols between private and public applications in networks and the incorporation of ISDN into OSI. Incompatibilities in requirements and proposals from CCITT, e.g. X.400, and industry, e.g. MAP and TOP, are already evident.

The purpose of this book is to provide an overview of the state-of-the-art and to explore the directions of future research in telecommunications and distributed systems. Firstly, the main part concentrates on the evaluation of the CCITT recommendations and ISO standards with respect to OSI and ISDN and user services. Secondly, the book reviews in a top-down fashion new applications that are becoming available in modern offices, manufacturing, and the public domain. Thirdly, different approaches are discussed as to how these applications can be built more easily with advanced programming tools, and what support is needed for network operation in OSI by contrasting OSI to proprietary network architectures. The following main topic deals with operating systems support for communication systems. Many of the currently available operating systems are inadequate with respect to their communication primitives. Very often, today's implementations of OSI do not meet efficiency requirements. These critical deficiencies must be further investigated in order to achieve an open exchange of data. The final part of this volume is devoted to verification, specification, and testing of protocols.

Heidelberg, January 1987   Gaithersburg, January 1987

G. Müller   R. Blanc

Contents

REQUIREMENTS AND INTERFACES FOR APPLICATION GENERATION

COMMUNICATION REQUIREMENTS FOR OPERATING SYSTEMS

SPECIFICATION, IMPLEMENTATION AND CONFORMANCE OF PROTOCOLS

ANNEX

The work of CCITT in the field of Telematic Services

Rolf Rüggeberg
Fernmeldetechnisches Zentralamt

Abstract

The paper is opened by a classification of the different
"non-voice-services" where the characteristics of each service
are described and the differences between the services are shown
in their typical facilities.

The organization of CCITT in general and the study program of
Study Group I in particular are described. The study work on
existing services, new services already in operation and new
services for the future will be presented as well as the whole
study program of the introduction of non-voice-services on the
ISDN.

A world wide questionnaire on the introduction of the ISDN was
distributed by CCITT. The responses of 34 Administrations/
Recognized Private Operating Agencies (RPOAs) and a summary of
these responses are included in the paper.

A general consideration of the standardization within CCITT will
close the presentation.

1 Classification of the "Non-voice-Services"

The area of the textcommunication and data services are separated
into two sections:
1. Telematic Services
2. Bearer Services

The first section, telematic services, can be separated into two
subsections:
- 1a. terminal oriented telematic services
- 1b. network oriented telematic services

There is a characteristical difference between the services of
the first and those of the second section. For the services of
the first section basic facilities of all seven layers
(OSI-Reference-Model) have to be defined while for the services
of the second section the basic facilities only of the first
three layers (network layers) are to be defined.

It is an essential characteristic of a telematic service that
facilities up to the application layer (man/machine interface)
are standardized.

With the standardization of the network functions within the
bearer services a clear interface to the network is defined. How
to use the facilities of the higher layers is application
dependent and has to be decided by the user.

The following table 1 shows the characteristics of the telematic
and the bearer services. The essential differences between these
sections and subsections are described by the following five
typical criteria:

Compatibility

The most important item within the terminal oriented Telematic
services is the manufacturer independent compatibility between
all terminals participating in a service. Communication between
all users is guaranteed.

The typical characteristic within the network oriented Telematic
services is that communication is always done via a network
storage. Direct end-to-end communication between the terminals
does not exist. Thus compatibility is only necessary between the
terminals and the system in the network. Different interfaces
from the system to the network are possible.

Within the bearer services compatibility is only required
between the terminal and the data processing unit (host) in the
network. The compatibility is application dependent and is only
realized within the "island solution". The traffic structure is
in principle centralized from terminal to the host (star configuration). Direct communication between terminals does not exist.

Addressee

Within the terminal oriented Telematic services the addressee is
always the number of the called terminal which has in general a
fixed location in the organization. In the network oriented Telematic services the addressee is always a person to whom a mailbox
is allocated. Mobile terminals can be used within this service
group.

Dialog

There are two different modes of operation:

1. Dialog man/machine (e.g. access to a data base)
2. Dialog between persons.

In the terminal oriented Telematic services a dialog between
persons is only possible if both partners are at their terminals.
Special facilities (e.g. bell, direct communication between the
screens etc.) have to be provided. Within the bearer services the
man/machine dialog mode of operation is a typical application
while an interactive mode between people is excluded.

Communication

In the terminal oriented Telematic services the network provides
a transparent connection between the terminals and allows a
direct end-to-end communication. The traffic structure is very
similar to a batch transmission which is typical also for the
bearer services.

Within the network oriented Telematic services there is only an
indirect transmission between the terminals. It is an essential
feature of these services that the message is always stored in
the network. The terminals are not necessarily fixed located. A
mobile use of the terminals is possible and in some cases highly
required.

Service group / criteria	Telematic - Services		Bearer - Services
	terminal - oriented	network - oriented	
compatibility	between all terminals within the service	between terminals and system (host), between systems	only to network interface (e.g. X.21, X.25, V.24...)
adressee	terminal	person	system (terminal)
dialogue	person to person	person to system (person to person possible within MHSS)	person to system
communication	direct end to end without storage in the network	indirect end to end with storage in the network	direct end to end from terminal to system
reception	in the terminal	in the network (storage), access from the terminal	in the system
Examples	Telex, Teletex, Telefax	Videotex, MHSS	datatransmission services

Table 1 Classification of the "Non - Voice - Services"

Reception

The differences between the service groups are shown on the receiving end:

In the terminal oriented Telematic services the received message is always in the terminal while in the network oriented Telematic services the message is stored in the network and the user has to access the store in order to be informed that there is a message waiting for him and to get the message out of the store.

2 Organization of CCITT

The CCITT (Comité Consultative International Télégraphique et Téléphonique) is organized in 15 study groups, 7 joint committees (CCITT/CCIR) and 5 special autonomous groups. In all these groups the whole area of international telecommunication is studied and standardized:

Com I	Definition, operation and quality of service aspects of telegraph, data transmission and telematic services (Facsimile, Teletex, Videotex, etc.)
Com II	Operation of telephone network and ISDN
Com III	General tariff principles including accounting
Com IV	Transmission maintenance of international lines, circuits and chaines of circuits; maintenance of automatic and semi automatic networks
Com V	Protection against dangers and disturbances of electromagnetic origin
Com VI	Outside plant
Com VII	Data communication networks
Com VIII	Terminal equipment for telematic services (Facsimile, Teletex, Videotex, etc.)
Com IX	Telegraph networks and terminal equipment
Com X	Languages and methods for telecommunications applications
Com XI	ISDN and telephone network switching and signalling
Com XII	Transmission performances of telephone networks and terminals
Com XV	Transmission systems
Com XVII	Data transmission over the telephone network
Com XVIII	Digital networks including ISDN

Joint CCIR/CCITT Committees administered by CCIR

CMTT   Television and sound transmission

CMV   Definition and symbols

Joint CCITT/CCIR Committees administered by CCITT

WORLD PLAN COMMITTEE	General plan for the development of the World Telecommunication Network
PLAN COMMITTEE FOR AFRICA	General plan for the development of the Regional Telecommunication Network in Africa
PLAN COMMITTEE FOR LATIN AMERICA	General Plan for the development of the Regional Telecommunication Network in Latin America
PLAN COMMITTEE FOR ASIA AND OCEANIA	General Plan for the development of the Regional Telecommunication Network in Asia and Oceania
PLAN COMMITTEE FOR EUROPE AND THE MEDITERRANEAN BASIN	General Plan for the development of the Regional Telecommunication Network in Europe and the Mediterranean Basin

Special Autonomous Groups

GAS 3	Economic and technical aspects of the choice of transmission systems
GAS 7	Rural telecommunications
GAS 9	Economic and technical aspects of transmission from an analogue to a digital telecommunication network
GAS 10	Planning data and forecasting methods
GAS 11	Strategy for public data networks

The study groups I and II deal with the definition of services:

Study group I:	Definition, operation and quality of service aspects of telegraph, data transmission and telematic services (Facsimile, Teletex, Videotex, etc.)
Study Group II:	Operation of telephone network and ISDN.

In this paper the whole area of "Telematic services" is dealt with. The area of "Voice services" which is in general the Telephone service is not considered.

3 Study program of Study Group I

In this section the whole study program of Study Group I is presented in three subsections: Existing services, new services already in operation and new services under Study.

3.1 Existing Services

Telegram service

The telegram service is the oldest telecommunication service at all. In the last century and during the first half of this century the telegram service was of considerable importance as far the transmission speed and security (guaranteed correct transmission and maximum time of delivery) was concerned. Today the telegram service is in competition with new and modern textcommunication services for subscribers. In industrial countries the telegram traffic is decreasing permanently, however it is still necessary for certain applications.

The present study in CCITT is concentrated on simplifying the service and increasing the efficiency. A new telegram type service called "Telemessage" is under study and is considered as an international public service to enable the exchange of messages with electronical input on public networks or by other means for postal delivery in the destination country.

Telex service

The telex service was established first in 1933 as a "private telegram service" between subscribers. This service developed considerably and is today the second biggest telecommunication service after the telephone service.

The CCITT-study is concentrated on improving the quality of service and on interworking aspects with new telematic services (e.g. Teletex, Videotex) and other teleservices like message handling.

3.2 New services already in operation

Telefax service

The facsimile technique is as old as the telegraph service at all. Already in the middle of the last century first facsimile terminals were available.

The international standardization was influenced by the fact that many different facsimile terminals were connected to the network (telephone network) and communication was only possible between terminals of the same manufacturer.

In the study period 1973-1976 three different groups of facsimile terminals were standardized:

Group 1:   6 min transmission per A-4-page, analog, 3.85 lines/mm

Group 2:   3 min transmission per A-4-page, analog, 3.85 lines/mm

Group 3: about 1 min transmission of an A-4-page, digital,
          3.85 or 7.7 lines/mm

A group-4-terminal has been standardized in the period 1981-1984
(240/300 pels/inch, preferable network: data network or ISDN).

The present study deals with the following items:

1. Revision of current operational provisions for Telefax
   services and Telefax 4 service

2. Revision of current operational provisions for the facsimile
   service between public bureaux and subscriber stations and
   vice-versa.

3. Revision of general provisions

4. Service quality

5. Interworking between Telefax 4 and Teletex

6. Telefax 4 on the PSTN.

Bureaufax service

The bureaufax service is a public service between offices for
public acceptance and delivery and with access from and to
Telefax subscribers.

The study covers in particular the following aspects, taking due
account of user's and Administration's needs and of the
development of technical resources:

1. Facilities

2. Quality of service
   - identification
   - error correction system
   - ITU Bureaufax Table

3. More intense standardization of the Bureaufax service
   (e.g. transmittal sheet)

4. Attractiveness of the service for users and Administrations

5. Development of the service between public bureaux and
   store-and-forward systems.

Teletex service

The Teletex service is a new Telematic service characterized by a
terminal (or equipment) which is used for text production as well
as for transmitting and receiving of texts with the whole
character repertoire of an office typewriter.

This service was standardized during the period 1977-1980.

Presently the Teletex service is being implemented in a number of countries based upon the Recommendation F.200 for the service itself; Recommendation F.201 for the interworking between Teletex and Telex and in a number of T Recommendations for technical elements of the service.

The objectives of the present Study (1984 - 1987):

1. Identification and description of the service and operational issues for the enhancement of Recommendation F.200

2. Identification, definition and description of quality of service parameters, both qualitatively and quantitatively.

3. Investigation of and problem solving with regard of Teletex-Telex interworking.

4. Further elaboration of the service requirements for the mixed mode of operation.

5. Investigation and definition-work with regard to interworking with other services, concentrating on MHSS, Videotex and Facsimile.

6. Identification and description of store-and-forward facilities to be provided from bearer resources.

7. Investigation into the problem of description techniques for Recommendations.

8. Investigation of problems arising from national variations in Telex requirements.

9. Teletex in the ISDN

Videotex service

The Videotex service was standardized in a first step during the study period 1977-1980.

The objectives of the present studies are:

1. Development of a basic set of functions/facilities/applications for the international Videotex service

2. Charging and tariff principles

3. Intercommunication with other Telematic services.

3.3 New Services under Study

Teleconference service

The "Teleconference service" is a new study question with the following items:

1. Scope of the teleconference service

2. Operation of the Teleconference service

3. Telconferencing call phases

4. Identification method and format of the teleconferencing call and terminal

5. Quality of service of the teleconference

6. Terminal characteristics

7. Notification of messages

8. Teleconferencing requirements.

Telewriting service

This new study question is close related to the teleconference service. The service aspects will be studied together with the study question above.

Message handling system services

The fundamental ability of this new service is to provide a public interface between originators and recipients to enhance their means of communication. The MHSS will be used when immediate or convenient direct telecommunication services cannot be provided within an existing Telematic service, e.g. in the case of interworking between different services like Teletex and Videotex.

This service may also provide facilities available for the preparation and the presentation of the message.

The following are the objectives in the study during the present period:

1. Develop a new Recommendation on message transfer service

2. Develop a new Recommendation on interpersonal messaging service

3. Determine requirements for any additional Recommendations.

Data transmission service

For this new question a study is required to establish service and operational principles for public data transmission services and to draft Recommendations to include for each type of network (telephone, telex, packet and circuit switched) taking into account existing Recommendations:

- service definitions
- quality of service
- provision of customer support.

It is the aim of Study Group I to define international data transmission services where application oriented facilities are standardized with a guaranteed world wide compatibility.

4   Introduction of ISDN

The "Integrated Services Digital Network" (ISDN) will integrate the existing different networks, in particular the PSTN (Public Switched Telephone Network), CSPDN (Circuit Switched Public Data Network) and the PSPDN (Packet Switched Public Data Network). The most important aspect of the ISDN from the customer's point of view is the he gets access to all different services via one access line and that he has one subscriber number for all defferent applications.

Objectives of the study

The technical aspects of the Integrated Service Digital Network (ISDN) has been standardized during the last study period.

The ISDN will be the common network for voice and non-voice applications. Study Group I is asked to study how existing services may be transferred to ISDN and which new services should be introduced. The following objectives have been agreed:

Relevance of ISDN to existing services

a) the following Telematic services should be studied to be provided in the ISDN:

   1. Teletex service (including mixed mode)
   2. Telefax service (Groups 2/3, 4)
   3. Videotex service
   4. Telex service

b) The following bearer services should be studied to be provided in the ISDN

   1. data transmission services of the PSTN
      (public switched telephone network)

   2. data transmission services of the CSPDN
      (circuit switched public data network)

   3. data transmission services of the PSPDN
      (packet switched public data network).

Introduction strategy and transfer methodology

If existing telematic and bearer services are introduced in the ISDN, the following questions have to be studied:

1. What kind of interworking has to be provided between services in the ISDN and the same service provided in the conventional networks?

2. What will be the international access to the same service in different networks?

3. Three transfer methodologies are possible in principle:

a) complete transfer of a service into the ISDN within a short time

b) Independent parallel development of the service and other networks without interworking

c) Parallel development of the service in ISDN and other networks with service interworking.

It is up to the Administration's decision which strategy will be used.

Which new services should be studied for ISDN?

1. CS-oriented bearer services

2. PS-oriented bearer services

3. New services in a broad sense with following facilities as examples:

a) text facility as e.g. Teletex

b) graphic facilities with e.g.
photographic,
alpha mosaic
vectographic

c) voice (in combination with documents)

d) document structure in accordance with Recommendation T.73 (mixed mode) and document handling in accordance to MHS.

First results of the ISDN-Study

Use of telematic terminals in ISDN

General principles have been agreed for the use of telematic terminals in the ISDN:

1. Interworking of telematic terminals with ISDN is mandatory

2. The migration from non-ISDN terminals to ISDN is a national matter but technical problems should be identified and resolved by Study Group VIII.
If terminals migrate to ISDN they must remain compatible with the terminals on existing networks within the same class of service.

3. Compatibility between ISDN terminals and existing terminals must be retained.

Telex service on the ISDN

At present there are no specific plans to connect telex terminals to the ISDN-network. However it cannot be excluded that there will be a subscriber's demand to connect a telex terminal to the ISDN. In this case the technique and the operational procedure should be standardized. The following principles were agreed:

1. The intercommunication between telex terminals on an existing network with telex terminals on ISDN should use Recommendation F.69 numbering

2. The operational procedures for 1) above should follow Recommendation F.60.

3. The competent technical groups will be requested by Working Party I/1 to specify recommendations necessary to ensure standardized operation.

4. The operational procedure should be standardized in such a way that a call from telex to a telex terminal connected to ISDN is the same as a conventional telex call.

Provision of existing services in the ISDN

Existing services that may be provided in ISDN to intercommunicate with the same services on existing networks including existing conversion facilities.

Existing Services on ISDN	Services on existing networks
1. Teletex	Teletex (PSTN, CSPDN, PSPDN) Telex
2. Telex	Telex Teletex (PSTN, CSPDN, PSPDN)
3. Telefax Gr. 2/3	Telefax Gr. 2/3 (PSTN)
4. Videotex	Videotex (PSTN)
5. MHSS	MHSS (PSTN, CSPDN, PSPDN)
6. Data transmission	Data transmission (PSTN, CSPDN, PSPDN)

Provision of new services in the ISDN

New services that may be provided on ISDN with minimum intercommunication requirements to other services on ISDN and on existing networks.

New Services on ISDN	Services on ISDN and existing networks
1. Textfax (note 1)	Teletex (PSTN, CSPDN, PSPDN, ISDN) Telefax Gr. 2/3 (PSTN, ISDN) (note 2) Telefax Gr. 4 (ISDN, CSPDN) Telex
2. Telefax Gr. 4	Telefax Gr. 2/3 (PSTN) (note 2) Telefax Gr. 4 (CSPDN) (note 3)
3. Videotex 64 (note 4)	Videotex (PSTN, ISDN)

Legend:

ISDN   - Integrated Services Digital Network

PSTN   - Public Switched Telephone Network

CSPDN - Circuit Switched Public Data Network

PSPDN - Packet Switched Public Data Network

MHSS   - Message Handling System Services

Notes:

General note: Additional intercommunication requirements may be
identified subsequently.

Note 1:   Temporary name for the mixed mode operation of
teletex

Note 2:   To be forwarded to Study Group VIII for identification and resolution of possible technical
problems with the interworking with Telefax Gr.2.

Note 3:   The study of this intercommunication should be
delayed until Telefax Gr. 4 is implemented on
existing networks.

Note 4:   Videotex service using the full capacity of a 64
kbit/s B-channel of the ISDN access.

Assignment of priority to the future work

The current assignment of priorities to the future work is:

a) Priority for non-voice Teleservices:

   1) Existing Telematic Services on ISDN
   2) New non-voice Teleservices

b) Priority for bearer services:

   1) New bearer services on ISDN
   2) Use of bearer services of ISDN by existing terminals.

The Terminal Identification within the Telematic Services

1. The problem

Normally the ISDN will be developed from the digitalized
telephone network and will possibly replace the analog telephone
network in some countries at a later date. In the meantime
however both will be provided simultaneously and they will be
linked together by a common numbering plan. For international
(telephone-)traffic they will have the same country-code
according to Rec. E.163/164 (e.g.: 49 for D, 33 for F etc.)

A subscriber may have a multifunctional terminal or separate
terminals for several services (Telephone, Teletex 64, Telefax
group 4 etc.) via one subscriber line (2 x 64 kbit/sec) linked to
the ISDN. There will only be one subscriber number for all
services.

In the case of teletex service terminals 2.4 kbit/sec will be
connected to the CSPDN, to the PSPDN and to the PSTN and
terminals 64 kbit/sec to the ISDN.

An interworking between the 64 kbit-Telematic services and the
services in todays data networks and the PSTN is absolutely
necessary.

Apart from the conversion (2.4/64 kbit/sec) this leads to
problems with the numbering plans and the representation of the
selection information in the TID.

The following examples are to clarify the problem:

Let us assume that France and the Fed. Rep. of Germany have
established ISDN and that they both provide teletex service also
in ISDN. That means, the following types of teletex terminals
will be available:

in France	in Germany
Ttx 2.4 on PSPDN	Ttx 2.4 on CSPDN
Ttx 2.4 on PSTN	Ttx 64 on ISDN
Ttx 64 on ISDN	

Table 2 shows some international connections, the appropriate figures to be dialled and the TID of the called terminal as it is today (according to F.200) and as proposed.

In cases 2, 6, 9, 10 and 12, figures appear in todays TID which are not part of the number to be dialled.

If table 2 mentions two simular cases (in one case access in country A and in the other case access in country B, e.g. cases 1+2, 5+6 etc.), this does not mean an alternative routing, but two possibilities, one of which may technically not be available (e.g. case 2, if country A does not provide an ISDN).

It is acceptable that a subscriber has to dial some special figures in addition to the figures given in the TID, but it is unacceptable from an operational point of view, that the subscriber has to decide whether or not some of the figures given in the TID are to be replaced by other figures.

The figures to be dialled do not only depend on the destination network but also on the network of origin (see cases 10 and 13).

The existing numbering plans (X.121, E.163/164) cannot solve that problem satisfactorily. The revision of these numbering plans is under study in the responsible Study Groups.

2. A possible solution by CCITT

Draft Recommendation F.351

GENERAL PRINCIPLES ON THE
PRESENTATION OF TERMINAL IDENTIFICATION
TO USERS OF THE TELEMATIC SERVICES

The CCITT

Considering

a) that the terminal identification (TID) in the telematic services should provide the called subscriber not only with an unambiguous identification of the calling subscriber, but also with the essential information needed to establish a call to the former;

b) that transfer of network addresses, terminal identifications and other session management functions are covered in Recommendations specific to individual networks and telematic services;

c) that detailed provisions on TIDs may be also found in specific telematic service Recommendations (e.g. F.200);

Table 2

Examples for international teletex connections

(xx = access codes, not yet fixed)

case	country	from network	country	to network	Access to the destination network	figures to be dialled	TID of called terminal today	TID of called terminal proposed
1	F	Ttx 2.4 (PS)	D	Ttx 64	in F	xx49 22859246	49-22859246=BPM	TID-22859246=BPM
2	F	Ttx 2.4 (PS)	D	Ttx 64	in D	02627xx 22859246	49-22859246=BPM	"
3	F	Ttx 2.4 (TN)	D	Ttx 64	in F	xx49 22859246	49-22859246=BPM	"
4	F	Ttx 64	D	Ttx 64	-	0049 22859246	49-22859246=BPM	"
5	F	Ttx 64	D	Ttx 2.4 (CS)	in F	xx2627 228300	2627-228300=BONN	TCD-228300=BONN
6	F	Ttx 64	D	Ttx 2.4 (CS)	in D	0049xx 228300	2627-228300=BONN	"
7	D	Ttx 64	F	Ttx 64	-	0033 1234567	33-1234567=PTT	TIF-1234567=PTT
8	D	Ttx 64	F	Ttx 2.4 (PS)	in D	xx2080 1345678	2080-1345678=PAR	TPF-1345678=PAR
9	D	Ttx 64	F	Ttx 2.4 (PS)	in F	0033xx 1345678	2080-1345678=PAR	"
10	D	Ttx 64	F	Ttx 2.4 (TN)	in F	0033xx 1654321	933-1654321=SIB	TTF-1654321=SIB
11	D	Ttx 2.4 (CS)	F	Ttx 64	in D	xx0033 1234567	33-1234567=PTT	TIF-1234567=PTT
12	D	Ttx 2.4 (CS)	F	Ttx 64	in F	02080xx 1234567	33-1234567=PTT	"
13	D	Ttx 2.4 (CS)	F	Ttx 2.4 (TN)	in D	0933 1654321	933-1654321=SIB	TTF-1654321=SIB

d) that, from the operational viewpoint, it is acceptable for a subscriber to add selection digits to those presented in the TID, but it is unacceptable for a subscriber wishing to make a call to have to decide whether some selection digits in the TID might have to be replaced by others depending on the type of the origin and destination networks;

e) that telex network identification codes (TNIC) of one or two letters have been assigned to assist telex subscribers in accordance with Recommendation F.68 and are listed, along with the corresponding numerical destination codes, in Recommendation F.69,

declares the view that the following general principles on the identification of terminals in the telematic services (and, as appropriate, other terminals) should be applied to all new equipment, and to all existing equipment by 1992.

1. TIDs presented to the user should consist of a Service and Network Identifier (SNI) followed by the national subscriber number and possibly a mnemonic identifier, e.g. TCSF - 109009 = TE0int

2. The SNI shall consist of 3 or 4 capital letters. The first letter shall identify the service (e.g. T for Teletex). The second shall identify the network type (e.g. C for CSPDN). The third (and if necessary the fourth) letter shall be the relevant TNIC (e.g. SF for Finland), allocated to the Administration or RPOA in accordance with Recommendation F.68. If none has been allocated for telex, a 2-letter code shall be allocated in accordance with Recommendation F. 96.

3. The SNI is only intended to provide subscribers with a uni-versally applicable, readily understood and easily interpreted alphabetic code to identify the service and network. Its use need has no effect on technical aspects of setting up calls and/or transfer of identification information. Nevertheless individual Administrations may provide facilities within the network or in some or all terminals to translate the SNI to the appropriate numeric address information. Provision of such facilities is a national matter.

4. Each Administration shall provide its subscribers with a list (e. g. in the relevant service directory) showing the selec-tion information corresponding to each SNI, unless automatic translation of SNIs is universally available in the service concerned.

5. Administrations should encourage subscribers to give every possible publicity to their SNIs, for example by including them with the appropriate national subscriber numbers in letterheads or on business cards (e.g. Teletex: TCSF - 109009) and by prominently displaying the SNI (or SNIs) assigned to the network(s) in the relevant service directory.

6. the significance of the first and second letters of SNIs is as shown in the following table (which is provisional at this stage):

FIRST LETTER (Service)	SECOND LETTER (Network Type)
T Teletex	C Circuit-switched PDN
F Telefax	P Packet-switched PDN
V Videotex	T Public switched Telephone Network
D Data	I Integrated Services Digital Network
M MHS	
.	
.	
.	

Notes:

1. The application of this Recommendation to the Telefax 2 and 3 services requires further study since Recommendation F.180 only allows digits and spaces in the TID at present.

2. If there is any significant technical advantage the order of the three elements in the SNI could be changed, but this matter should be settled in principle at the November 1986 meeting of Study Group I.

Plans of Administrations

General

The questionnaire on the introduction of the ISDN was drafted at the WP I/3 - meeting in Rome (October 1985) and was sent to all Admistrations/RPOAs by Collective Letter No. 78.

The responses received are listed in the annex (four pages). A summary of the reponses is given below.

Summary

34 Administrations/RPOAs responded positively to the questionnaire. Only two countries announced that they do not intend to introduce the ISDN.

1. INTRODUCTION OF THE ISDN:

1987 - 1989:	20	58 %	a)
1990 - 1994:	5	15 %	
1995 - 1999:	6	18 %	
later.	-		
not yet decided:	3	9 %	

    a) in many cases Administrations/RPOAs begin
        with an ISDN pilot project

2. INTRODUCTION OF EXISTING SERVICES ON ISDN

yes:	28	82 %
under study	6	18 %

3. WHICH OF THE EXISTING SERVICES WILL BE INTRODUCED (28 = 100 %):

Telex:	13	46 %
Teletex:	22	78 %
Telefax:	26	93 %
Videotex:	22	78 %
MHSS:	17	61 %
Datatransmission:	27	96 %

4. NAME OF THE EXISTING SERVICE ON ISDN
   (13 not yet decided, 21 = 100 %):

same name:	18		86 %
same name + add.:	2	b)	10 %
different:	2		10 %

    b) for modified services under study by one
        Administration

5. ADAPTATION TO THE S-INTERFACE:

either terminal adapter or
adaptation within the terminal:

    a mixture of both possibilities by all
    Administrations/RPOAs

6. ACCESS FROM ISDN TO THE SAME SERVICE ON EXISTING NETWORKS:

yes:	24	80 %
not yet decided	4	13 %
no:	2	7 %

7. INTERCOMMUNICATION BETWEEN DIFFERENT SERVICES

a) within ISDN (27 = 100 %)

yes:	9	33 %
no	9	33 %
under study	9	33 %

b) from ISDN to existing networks: (29 = 100 %)

yes:	14	48 %
no	5	17 %
under study:	10	35 %

8. NEW SERVICES ON ISDN:
(list of all services announced by Adm./RPOAs)

a) New bearer services:

    circuit switched 64 kbit/s
    circuit virtuèlle X.25
    circuit mode 64 kbit/s (speech)
    circuit mode 3,1 khz audio information transfer
    circuit mode 8 khz structured
    384 kbit/s
    1920 kbit/s
    more than 2 Mbit/s

b) New Teleservices

    Telefax Group 4
    Teletex 64 kbit/s
    Videotex 64 kbit/s
    Mixed-mode (Text-Fax)
    Picture Phone
    Audiovideotex
    Audiographic
    Teleconference
    Electronoc Funds Transfer
    Teleaccion

Country/ Administ./ RPOA	Introduction of ISDN 1987–1989	1990–1994	1995–1999	later	Introduction of existing services on the ISDN — Telex	Teletex	Telefax	Videotex	MHSS	Datatransm.	others	Name of the service on ISDN — same name	same+addition	different	Adaptation to the "S+" interface — term. adapt.	within terminal	Access from ISDN to the same service on existing networks — yes	no	Intercom. within ISDN	from ISDN to ex. networks	New services — bearer services	tele services	Comments
AUSTRIA	x				+1)	x	x	x	x1)	x	–	x		2)	+1)	–	x1)		yes1)	yes,1)	1)		1) under discussion 2) not yet decided
BELGIUM	x				–	x	x	x	x1)	x2)	x2)	x,			Ttx Tfx Vtx Data	Ttx Tfx Data	x		no	yes3)	circ.-switched 64 k circuit virtuélle X.25 channel B	Telefax, Teletex videotex, Mixed-mode Ttx/Tfx	1) access from PSPDN 2) circ.-switched and pack.-switched 3) To PSPDN for interworking Ttx/Tx
CYPRUS	x				–	x	x	x3)	x1)	x	–	1)			1)	1)	1)		1)	1)	1)	1)	1) not yet decided
FINLAND	x				–	x	x	x1)	x2)	x3)	–	x			Ttx Tfx Data3) Pph	Tfx Data Pph	Ttx Data		3)	3)	circuit mode 64 k unrestricted	Telephone Picture Phone	1) switched 56, 48, 19.2, 9.6 kbps 2) Picture Phone 3) not yet decided
FRANCE	x				–	x1)	x	x	x2)	x3)	x	x			x4) Tfx2) Vtx Data3)	Tfx Data Pph	x5)		FS	FS	cicuit mode 64k unrestricted circuit mode (speech) circuit mode for .1 khz audio-information transfer	numerical telephone Audiovideotex (alphanumeric + geometric + photographic) Audiographie at 64 kb/s	1) experimental, 2) not defined, 3) TRANSDYN; TRANSCOM; Analog leased lines, numerical leased lines, TRANSPAC 4) depends on terminal, date and manufacturer 5) Telephone PSTN/ISDN Vtx (alphamosaic) ISDN/ Teletel or Eletr. Directory Service on PSPDN or PSTN, Ttx ISDN/ TTX PSTN or PSPDN
GERMANY F. R. of	x				–	x	x	x	x1)	x	–	x			x	Ttx	x4)		x5)	x6)	circuit mode 64 k unrestricted 8 khz structured CCITT I.221, 2.1.1	Teletex 64 k Telefax Gr4, 64 k Videotex 64 k 'Textfax' (Mixed-Mode)	1) under discussion 2) Group 2/3 3) with terminal adapter V.24, X.25, X.21. 4) Ttx 64–Ttx 2.4, Tfx Gr4–Gr3 (TA), Tfx Gr2/3 (TA)–PSTN, 5) Gr4–Gr3 (TA), 6) Ttx 64–Tx, Ttx 64–VOTx
HUNGARY	+1)																		1)				1) ISDN will be introduced, there are certain activities in the field of non-voice services, no firm decisions have been made yet.
IRELAND	x1)				+2)														2)				1) pilot project 2) under study
ITALY PTT	x				x	x	x	x	x	x2)	x2)	x			x	x	x		yes yes1)			Teleconference	1) Ttx–Tx 2) from PSPDN, CSPDN, PSTN

Country/ Administ./ RPOA	Introduction of ISDN				Introduction of existing services on the ISDN							Name of the service on ISDN			Adaptation to the "S"-interface		Access from ISDN to the same service on existing networks		Intercomm. between different services		New services to be provided on ISDN		Comments
	later	1995-1999	1990-1994	1987-1989	Telex	Teletex	Teletex	Videotex	MHSS	Datatransm.	others	same name	same+addition	different	term.adapt.	within terminal	yes	no	within ISDN	from ISDN to ex. networks	bearer services	tele services	
ITALY SIP				x[1]	-	x	x	x	x	x	-	x			x	x	x		-	yes[1]	Analog connect.,3l khz Unrestricted 64 kbit Packet switching	Slow sl. video 64 k Facsimile Teleconference	1) Ttx - Tx
LUXEMBOURG		x[1]			-[2]														[2]				1) 1988/89 ISDN-Switch (island) 2) under study
NETHERLANDS				x	-	x[1]	x	x	x	x	x	x			Ttx Tfx Data	Ttx Tfx Vtx MHSS Data	x		x[2]	[2]	circuit switched service	for further study	1) provisional 2) for further study
NORWAY				x	1)	x	x	1)	1)	x[2]			x[3]		x[4]	x[4]	x[5]		x[1]		64 kbit/s unrestrict. 64 kbit/s audio 384 kbit/s 1920 kbit/s	Electronic Funds Transfer	1) Further study 2) circuit switched, packet switched, 3) modified services 4) mixture 5) Teletex circuit switched, packet switched 6) will be provided in the public data network
SPAIN				x	-	x	x	x	x	x[1]	x[2]	x			Ttx[3] Tfx Vtx Data[4]	Ttx[5] Tfx MHSS Data[6]	x		yes	yes	64 kbit/s 2 Mbit/s more than 2 Mbit/s	FAX 4 TEX-FAX Videotex 64 TELEACCION	1) CS and PS, 2) semipermanent 3) Gr. 2/3, 4) packet 5) Gr 4 6) circuit
SWEDEN				x	-	x	x	x	x	x	-	[2]			-	-	x[3]		[2]				1) CSPDN, PSPDN 2) under study 3) datatransmission
SWITZERLAND				x	-	x	x	x	x	x	x[1]	x[2]	x[2]	x[2]	Ttx[3] Tfx Vtx Data	Tfx[4]	x		x[5]	x[5]	SWISSNET I (1988)[6] SWISSNET II (1990)[6]		1) X.25, 2) not yet fixed, 3) Group 3, 4) Group 4, 5) not yet decided, 6) provisional name
UNITED KINGDOM				x	x[1]	x	x	x	x[2]	x[3]	x			x[4]	x	x	x		no	yes	bit transport up to 64 kbit/s for speech or data,switched or leased line up to 64 k 64 kbit for packet		1) indirect via PSS, 2) x.21 CCT switched or leased services 3) packet switched services 4) integrated digital access "IDA"

Country/ Administ./ RPOA	Introduction of ISDN				Introduction of existing services on the ISDN.							Name of the service on ISDN			Adaptation to the "S"-interface		Access from ISDN to the same service on existing networks		Intercomm. between different services		New services to be provided on ISDN		Comments
	1987-1989	1990-1994	1995-1999	later	Telex	Teletex	Telefax	Videotex	MHSS	Datatransm.	others	same name	same+addition	different	term.adapt.	within terminal	yes	no	within ISDN	from ISDN to ex. networks	bearer services	tele services	
AUSTRALIA	x[1]				–	x	x	x	x	x[1]	x[1]		x[2]		[3]		x		yes	yes	switched 64 kb/s	photo videotex Tfx Gr4, 7 khz voice, voice mail	1) exact nature not yet determined 2) not yet determined 3) probably a mixture of both
HONGKONG (CAW)	x				x	x[2]	x	x	x[3]	x[1]	x[1]	x			Tx[2] 4)	Tx[3]	x		yes	yes	4)	––––	1) switched and leased line up to 56/64 2) initially 3) eventually 4) FS
JAPAN NTT	x[1]				x[2]	x	x	x	x[2] x[3]	x[1]	x[4]		x[2]		Ttx Tfx MHSS Data	x	x[5]		x[6]		64 kb/s, 128 kb/s, 384 kb/s, 1536 kb/s	64 kbit Telefax Videotex	1) from 1984 (1), 2) under study, 3) circuit switched, packet switched 4) telewriting, 5) PS, Telefax, 6) media protocol conversion
KDD	x				–	x	x	–	x	x[1]	x[2]			x[3]	Ttx Tfx MHSS Data		x		yes	yes	most of essential bearer services	digital telephone subrate speech 32k 64 kbps telephone	1) virtual call, 2) leased circuit 3) under study
NEW ZEALAND	x				x	x	–	–	x	x[1]	x[1]		x[3]		[2]		x		[2]		––––	––––	1) Packet service, access to digital leased data service, 2) not yet decided
THAILAND			x		[1]	x	x											x			––––	––––	1) still in consideration period
PHILIPPINE			x		x	x	x	x	x	x[1]	x[2]	x			Ttx Vtx Data	Tx MHSS		x	no	no	under study	under study	1) Packet Switching
BRAZIL	x				–	x	x	x	x[2]	x[1]	x[2]	x			Ttx Vtx Data[1]	Ttx Tfx[1]	x		x[2]	x[3]	64 kbit unrestricted 64 kbit for 3.1 khz audio, 64 kbit speech, 64 kbit alternated speech	64 kbit Videotex[2] 64 kbit Teletex[2] 64 kbit Telefax[2]	1) Packet mode 2) under study 3) Teletex/Telex, others under study
PAKISTAN			x		x	x	x	x	x	x	x	x			[1]	x[1]		x[2]	no[2]	no[2]	will be considered after successful introduction of ISDN[1]	1)	1) for all services 2) preferably
SURINAME			x		x	x	x	x	x	x	x	x			x[1]	x[1]		x	no	no	1)	1)	1) not yet decided

Country/ Administ./ RPOA	Introduction of ISDN				Introduction of existing services on the ISDN							Name of the service on ISDN			Adaptation to the "S"-interface		Access from ISDN to the same service on existing networks		Intercomm. between different services		New services to be provided on ISDN		Comments
	1987–1989	1990–1994	1995–1999	later	Telex	Teletex	Telefax	Videotex	MHSS	Datatransm.	others	same name	same+addition	different	term.adapt.	within terminal	yes	no	within ISDN	from ISDN to ex. networks	bearer services	tele services	
CANADA (Telecom)	┤1)				–	x 2)	x	x 2)	x 2)	x 1)	x 3)	x	2)				x		2)		64 kb/s restricted 64 kb/s for speech Virtual Call Packet	Wideband Audio Audio Visual Conf. Enhanced Videotex	1) 64 kb/s, 2) not known yet 3) Encrypted Data Trial
USA RCA	x				x	–	x	–	x	x		1)			Ttx Tfx MHSS	–	x		no	yes	1)	––	1) no answer
MCI				┤	1)				┤	x		1)			2)	–	x		no	no	1)	––	1) not yet defined, 2) not relevant for MCI

Country/ Administ./ RPOA	Introduction of ISDN				Introduction of existing services on the ISDN							Name of the service on ISDN			Adaptation to the "S"-interface		Access from ISDN to the same service on existing networks		Intercomm. between different services		New services to be provided on ISDN		Comments
	1987–1989	1990–1994	1995–1999	later	Telex	Teletex	Telefax	Videotex	MHSS	Datatransm.	others	same name	same+addition	different	term.adapt.	within terminal	yes	no	within ISDN	from ISDN to ex. networks	bearer services	tele services	
ALGERIENNE				x	x	x	x	x	–	x	–	x			x		x		no	no			
PAPUA NEW GUINEA				x	x	x	x	x	–	x 1)	x 1)	1)		x	2)	x	x		2)	2)			1) whatever applicable 2) not yet answered
ZIMBABWE				x	x	–	–	–	–	x	–	1)		x	2)		x		no	no	none	none	1) no decisions yet, 2) under study, the cheapest alternative will be used.
KUWAIT		x			x	x	x	x	–	x	–	x			1)x 2)x	2)x	x		yes	yes	Electrical Mail	––	1) for all services 2) for datatransm.

5 Some Aspects on CCITT Standardization

The standardization of services in the CCITT is based on the principle that

- the basic service should be as broad as possible

- the definition of the standards should be as far as necessary.

Under these assuptions the standardization is devided into three levels:

1. level: Definition of <u>basic</u> functions

            The basic functions have to be implemented in each system and must be available within the service world wide.

            Compatibility is guaranteed by the availability of these functions.

2. level: Definition of <u>options</u>

            Optional functions don't belong to the basic service, however they are standardized. It is up to the decision of the user/Administration to use or provide options. If they are used, they have to be used in the standardized manner.

3. level: <u>Private use</u> functions

            The service should allow. the use of non standardized functions. It is up to the decision of the user/manufacturer to use or implement such functions.

With this philosophy in standardization the CCITT does an important work to the benefit not only for the user, for the Administrations and manufacturers too.

The CCITT lives up to the purposes of the ITU as stated in the International Telecommunication Convention:

"... to coordinate efforts with a view to harmonizing the development of telecommunications facilities ..." and "... to promote the development of technical facilities ... and making them, as far as possible, generally available to the public."

Quellen:

CCITT - Empfehlungen der F.- Serie
(für alle Telematik - Dienste)

CCITT - Dokument Nr. 70 der 1. Studienkommission
aus der Studienperiode 1984 - 1988
(Planungen der Verwaltungen)

Die Empfehlungsentwürfe für die Telematik -
Dienste im ISDN werden auf der Vollversammlung
des CCITT im Jahre 1988 verabschiedet.
Derzeitig sind die Informationen lediglich als
Arbeitsdokumente verfügbar und können vom Verfasser
zur Verfügung gestellt werden

Adresse: Fernmeldetechnisches Zentralamt
          Referat T 22
          Am Kavalleriesand 3
          6100 DARMSTADT
             Telefon: 06151 83-5220
             Telex: 417419 tbetr d
             Teletex: 6151917=FTZ

NBS Program in Open Systems Interconnection (OSI)

Robert P. Blanc
Institute for Computer Sciences and Technology
National Bureau of Standards
Gaithersburg, Maryland 20899
USA

1. Introduction

The NBS Program in Open Systems Interconnection (OSI) has as objectives the following:

- to define OSI specifications that meet Government and industry requirements
- champion those specifications as international standards
- assist industry in implementing those standards as commercial products
- assist the Government in making use of advanced OSI technology
- perform research to define the next phase of OSI.

The NBS Program cycle is shown in Figure 1.

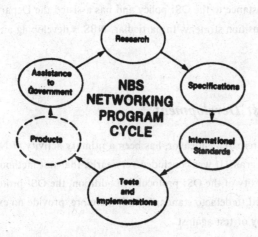

Figure 1. NBS Networking Program Cycle

The networking research effort initially addresses robust protocols for reliable communication, protocols for internetworking and protocol testing techniques. This research led to specifications and specification techniques. The specifications developed were for a robust transport protocol and an internetworking protocol to facilitate the coupling of local nets to each other and to wide-area networks. The specification techniques were developed toward the objective of "testability" of protocol specifications. This research resulted in the NBS Formal Description Technique (FDT) which was the predecessor of the ISO extended state transition language (ESTELLE). Along with this machine readable specification language, NBS developed software for the automatic analysis of the specification, generation of implementations in the C language and generation of test scenarios for testing external implementations.

In the area of transport standards development, NBS was the lead organization in the development of the ISO transport class 4 protocol and the connectionless network service (Internet). In the area of test and implementation, NBS pioneered development of test techniques for complex network protocols working in close cooperation with research laboratories internationally. NBS also organized OSI implementors to make specific protocol implementation decisions through the NBS/OSI Implementors' Workshop. In assisting the Government, NBS has provided technical assistance to the OSI policy and has assisted the Department of Defense (DoD) in its OSI transition strategy. In particular, NBS is developing an application layer gateway for DoD.

2. Conformance Test Development

Conformance testing, as a research endeavor, has been a primary activity of NBS since the start of the networking program. It was concluded very early, that test methods were necessary because of the complexity of the OSI protocols. In addition, the OSI protocols are consensus standards as opposed to defacto standards and, therefore, provide no existing commercial product to copy or test against.

NBS is not a test service and does not intend to develop tests on a production basis. Its objective is to perform research and advance the state-of-the-art in test methodologies. Therefore, we have selected the protocols with very interesting characteristics for test method developments. We have completed the tests for CCITT X.25, ISO transport class 4 and connectionless internet. The latter two were applied successfully and made possible the 1985 Autofact demonstration of the manufacturing automation protocol (MAP) and the technical and office protocol (TOP). The transport test had also been approved earlier to facilitate the 1984 National Computer Conference demonstration of OSI. The current efforts address interoperability tests for the token bus technology (IEEE 802.4). Test development has been completed for the physical layer and is progressing for the media access control layer.

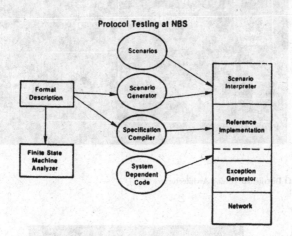

Figure 2.   Test Architecture for Transport Class 4

The test technology developed for the transport protocol is shown in Figure 2. This architecture led to the development of a number of software tools that partially automate the testing as well as the test development process. After the test architecture was defined and implemented, the ISO working group agreed to a different architecture. This architecture is shown in Figure 3.

To demonstrate the feasibility of this new architecture, we applied it in the development of the Internet Protocol Test System. The structure of the test system is shown in Figure 4.

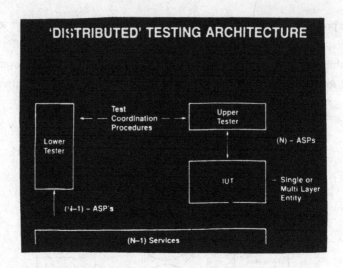

Figure 3. ISO Distributed Testing Architecture

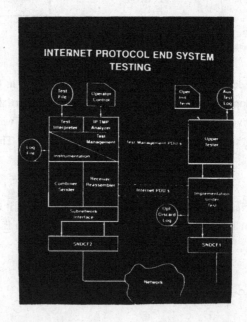

Figure 4. Connectionless Internet Test Architecture

3. Protocol Design and Measurement

The NBS performance measurement effort measures the inherent performance characteristics of OSI protocols in order to increase their applicability to new application requirements, tune their performance to work well with new communication technologies, provide measured guidance to implementors and users and recommend minor changes to international standards for significant increases in performance. The research has five phases, including:

- experimental design
- simulation modeling
- protocol design
- prototype implementation
- live measurement

The performance requirements investigated include those based on applications for file transfer, status reporting, transaction processing and realtime applications. Metrics include user throughput, one-way delay, retransmissions and CPU memory and channel utilization. The last two measures are important in demonstrating if OSI implementation is feasible for certain kinds of devices.

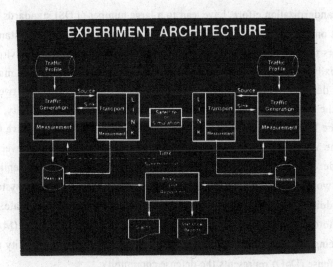

Figure 5. ISO Transport Protocol Measurement Experiment for Satellite Networks

As an example, Figure 5 shows the experimental architecture used for improving performance of the transport over satellite networks. The experiment allows for measurement using a satellite simulator or a real satellite connection. Additional experiments have been conducted for carrier sense multiple access (CSMA) and token bus local networks and X.25 packet switched networks. The real-time performance capabilities of the protocols have also been addressed.

4. OSI Security

NBS has an effort to define an OSI security architecture and to experiment with implementations of parts of that architecture in a laboratory environment. The purpose is to define access procedures and interfaces to security functions. The intention will be to propose those access procedures and interfaces as standards. To assist in this effort, NBS has established a special interest group on OSI security within the NBS Workshop for Implementors of Open Systems Interconnection. The security effort is described and more detailed in an accompanying paper.

5. NBS OSI Implementors' Workshop

International standardization efforts have lead to a large number of OSI standards (too many to implement at one time), multiple classes of services and options within each standard, and many parameters too definite to allow the protocols to work well in different environments. This leads to flexibility which is desirable because it allows the implementors to make the implementation decisions; however, without organization it is unlikely that any two companies will make the same implementation decisions. In recognition of this difficulty, in 1983 NBS initiated the NBS/OSI Implementors' Workshops. The Workshops meet five times a year. In 1983 the Workshop started with thirty companies and today involves over 200 companies worldwide. The participants include computer manufacturers, carriers, semiconductor manufacturers, word processing equipment manufacturers and factory control system manufacturers. In addition, a number of user companies and organizations participate. In particular, General Motors (GM) represents the manufacturing automation protocol (MAP) community, Boeing represents the technical and office protocol (TOP) community and Department of Defense (DoD) represents the defense community.

33

The workshop evolved through three distinct phases:

- Phase I started in February 1983 and concluded in the successful 1984 National Computer Conference demonstration of OSI. Sixteen organizations participated in that demonstration of the file transfer protocol (according to the then current ISO services definition) Transport Class 4, IEEE 802.2, 802.3, and 802.4.

- Phase II began September 1984 and concluded in the successful Autofact 1985 Demonstration of MAP and TOP. Figure 6 shows the topology. Figure 7 shows the protocols that were implemented to make the Autofact '85 demonstration possible. Of particular significance the Internet protocol made it possible to interconnect computer systems connected to three different local networks.

- Phase III began in December 1985 and will lead to the OSINET described in Section VI.

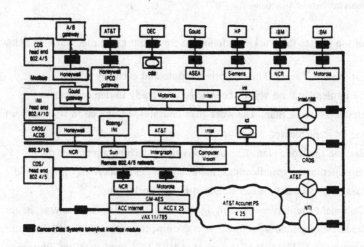

Figure 6. Autofact '85 MAP/TOP Topology

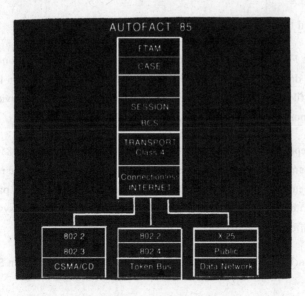

Figure 7. Autofact '85 Protocol Architecture

Within each of the phases the workshop follows a consistent game plan outlined below:

- Step 1 - Establish a Target. The target is the date and event by which
  the implementation will be completed and ready to demonstrate.
- Step 2 - Develop a work plan. The work plan contains the agreed-to standards which
  will be implemented.
- Step 3 - Execute the work plan. Execution involves reaching agreement on
  implementation specifications, including classes of services, options, and
  parameter values for each standard in the work plan.
- Step 4 - Document the implementation agreements. NBS, five times a year, issues
  the latest amended agreements in a document entitled: "OSI Implementors'
  Agreements" (The title of the series is NBS/OSI Implementors' Workshop).
- Step 5 - Implement and demonstrate the completed implementations. This step
  relies on OSI protocol test methods which to date have been provided by NBS,
  but in the future will be provided by organizations like the Corporation for
  Open Systems (COS) and the Industrial Technology Institute (ITI) and other
  organizations internationally.

At this time the NBS/OSI Implementors' Workshop is developing implementation specifications for the file transfer access and management (FTAM) protocol based on the draft international standard and, soon, the international standard when available. It is addressing the message handling services based on CCITT X.400. It will start efforts in the virtual terminal protocol and the directory services protocols. The latest activity is addressed to OSI security. In the near future, the presentation layer protocol and the common application services elements (CASE) will be added.

6. OSINET

In order to further OSI developments and OSI research, NBS proposed the development of OSINET comprised of voluntary, self-funded participants. The OSINET objectives are to facilitate:

- test method verification
- vendor-to-vendor testing
- OSI research

Test method verification is best performed by trying out the test methods on actual implementations. Organizations that develop test methods can connect to the OSINET to verify their methods.

Vendor-to-vendor testing is currently being scheduled. Each participating organization must test its implementations against five other organizations according to established algorithms and procedures. Initial implementations required of all participants include X.25, internet, transport class 4 session basic combined subset and file transfer access and management - FTAM. This is known as the FTAM Connection Project.

Two other projects have been agreed-to by the participants. The first is the MAP/TOP Product Demonstration. This will require the connection to the OSINET of actual products, not prototypes, implementing that portion of MAP and TOP derived from the NBS/OSI Implementors' Workshop. The last project is the DoD/OSI Conversion Project (described in the next session). Figure 8 contains the list of OSINET participants.

OSINET PARTICIPANTS

AMDAHL	NATIONAL BUREAU OF STANDARDS
ATT	NAVY DEPARTMENT
BOEING COMPUTER SERVICES	NCR COMTEN
DEFENSE COMMUNICATIONS AGENCY	OLIVETTI
DEPARTMENT OF AGRICULTURE	OMNICOM
DIGITAL EQUIPMENT CORPORATION	RETIX
GENERAL MOTORS	SYSTEM DEVELOPMENT CORPORATION
HEWLETT PACKARD	TANDEM
HONEYWELL	TASC
IBM	WANG
INDUSTRIAL NETWORKS, INC.	
INTERNATIONAL COMPUTERS, LTD.	

Figure 8. OSINET Participation

7. DoD/OSI Gateway Project

The DoD has developed a transition to OSI strategy meant to preserve interoperability with the Defense Data Network (DDN). The DDN uses protocols which are non-OSI but are functionally similar. DoD has requested that NBS, under contract, develop a DDN/OSI application protocol gateway. The gateway device will contain the full suite of DDN protocols and OSI protocols on the same system with a gateway at the application layer. The gateway will translate between FTAM, X.400, VTP (ISO and CCITT protocols) and FTP, SMTP and Telnet (DDN protocols), respectively. The NBS device will be a prototype leading to commercial implementations. In addition to the prototype, NBS is developing a complete test system for the application gateway. In order to verify the gateway in the test system, participating vendors will provide complete OSI systems connected to the OSINET; DoD will provide DoD protocol systems connected to the DDN, and NBS will provide the prototype gateway connected to both. This architecture is shown in Figure 9.

GATEWAY DEVELOPMENT

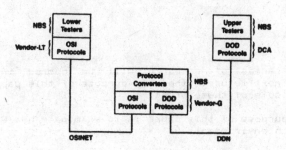

Figure 9.   DDN/OSI Gateway Architecture

8. *Future Work*

Having completed an initial OSI Program cycle (Figure 1.), NBS's resources have been directed to more research. The research falls into six areas:

- network integration
- protocol design and measurement
- network applications
- formal methods
- advanced communication techniques
- security architecture

Future papers from NBS staff will describe initial results in these network research programs.

O S I   AND   RESEARCH

G. Le Moli
C.R.E.I. – Politecnico di Milano
Piazza Leonardo Da Vinci, 32, 20133 Milano, Italia

Summary

Several new areas of research will be opened in the OSI
environment in the next future: the main purpose of this paper is just
to try to indicate some of these areas.

A secondary purpose of this paper is to examinate how OSI can be
used in the research environment.

1. Introduction

The existence and the availability of the OSI Reference Model is
becaming by itself one of the most relevant facts in telematics. Infact
the impact of a world-wide standard on the world of applications is
still to be fully investigated. For example, the amount of new
applications which the sole existence of a world-wide standard will
make possible cannot yet even be imagined.

This paper will mainly discuss which areas will be opened for
research in the next future in the OSI environment. To do this, the
paper starts by examining some aspects of the actual situation,
particulary some of the principles which are at the basis of OSI.

From this analisys, some consequences are derived which evidentiate
some actual problems, and from them it is given the indication of a few
areas which seem more promising for future research.

2. OSI and its basic principles

It is useful to start by briefly recalling some of the principles
which are at the basis of the OSI Reference Model.

The first principle which is here of our interest states
that OSI deals only with the exchange of information: this means
that an OSI system is described only in terms of what is visible
by its remote partners. More precisely, the OSI Reference Model
does not describe how a real system is made, but only how it appears
to its remote partners (or, in other words, the model does not describe
a system, but only its virtual image, as this image is seen by remote
partners which can interact with the system only by exchanging
messages).

This approach is substantially different from the typical CCITT
approach, which rather prescribes the specifications of a physical
system to be connected to a given physical network. One can very well
understand this difference between the two approaches if one considers
that PTTs have the main problem of buying from different manifacturers
systems to be connected to their networks, while ISO has the main
problem to avoid overspecification.

The OSI approach is also different from the classical proprietary approach, where the physical (in the meaning of both hardware and software components) structure of the network is referred under the term "architecture".

A second principle is that the OSI Reference Model allows applications and the communication network to be indipendent from each other. This means that the OSI Reference Model can be used for any application, over any kind of communication network. This result has been obtained by defining a stable Transport Service. A consequence of this principle is the fact that the Reference Model is indipendent from both the application and the communication network.

Other principles could be identified at the basis of OSI: however, the two ones above exposed are enough for deriving the consequences which are here of our interest.

3. OSI and research networks

One first consequence is that the concept of a "dedicated OSI network" is not a technical one, because all "OSI networks" are equal; that is, all OSI networks are made of OSI systems, all of them being technically capable to interact with each other. So, two OSI networks cannot be different for technical reasons: they can be distinguished (e.g. with different names) only for non technical reasons, such as for management ones, or for administrative ones, and so on.

Particularly, there is no need to have an OSI network dedicated to research, or to Universities, or to Research Centres: all what Universities or Research Centres need is to be connected to a pubblic network, by means of which they may interact with anybody else. In the same way, infact, they do not need a specially dedicated telephonic network.

It is interesting to note that in the field of data trasmission, in spite of the evidence of the above statement, somebody who propose to design the "specialized telephonic network" (or more precisely, its equivalent for data trasmission) for Universities (or for Hospitals, or for the Pubblic Administration, and so on) is always available.

So, the question "How OSI can be used for research ?" is a not properly stated one. Infact, if applications useful for research should require some special OSI network or OSI system, technically different from other OSI networks, then OSI would have failed to be general enough to be suitable for every conceivable application (research applications being included).

An example of a well stated question is: "Which applications in the OSI meaning are more suitable for research?". Here are some example of possible answers: X 400, FTAM, VT, JTM, access to remote data bases, etc. Some other specific applications – e.g. libraries data bases, on-going-researches data bases, students data bases, – may also be envisaged. A specially important case – which is less relevant outside the research environment – is when the application process in the OSI meaning is the whole operating system itself of a remote computer: this is necessary to allow a user to access a remote computer and to use all of its capabilities, expecially in terms of scientific software libraries.

4. OSI, management, and proprietary architectures

The problem of network management appears to have not been delt very deeply in the OSI environment. Infact, sometimes people does not well understand why ISO has had that particular approach to the management of OSI networks.

In order to make this matter more clear, let us consider the difference between the approach of the designer of a proprietary architecture and the approach of the designers of the OSI management environment.

A proprietary architecture is designed for being buyed by a single customer, which has usually the need to control and to manage his network as a unique whole (even if often it is also required that the capability to (partially) control must be splitted in several different locations). So the designer of this architecture assumes that the management is a key issue, and that there exists somewhere in the network a manager who wants to be provided with the best possible information for managing his network.

Designers of Open System, on the contrary, must bear in mind that Open System may be used in several different ways.

Firstly, a single Open System, belonging to an Organization, may just be used to interact with other Open Systems, belonging to other Organizations. So, in this case, this Open System does not need to be managed together with other Open Systems: the management of an Open System of this kind is just a local matter.

Or, secondly, the same Organization may have several Open Systems interacting with each other and with other external Open Systems. In this case, there is a need for common management tools. OSI provides room for management problems in the application layer: this means that manifacturers may implement their own management tools as OSI applications.

The third step is that also the management applications be normalized (more precisely, their protocols): when this will be completed, it will be possible, to a relatively large extent, the common management of networks made of really etherogeneous Open Systems.

Several manifacturers have already really good management tools, which very often constitute one of the best characteristics of their proprietary architectures: so it is very natural that some manifacturers are very proud of their management systems, and that several customers are really wondering if there is a danger that moving to OSI may mean to loose such powerful tools.

So, some problems may arize for the third approach. The most likely final result will propably be the following one: let us suppose that a user needs to have, for his own reasons, several Open Systems of n different manifacturers on its network; then he will probably have also n+1 different management systems, one for each manifacturer and one for the network. Fig. 1 shows a frequent example: we have an X.25 network, with its management terminal (all terminals dedicated to management are indicated in black); then we have some hosts and PADs of manifacturer A (indicated as squares), with their management terminal; then hosts, PADs and management terminal of manifacturer B (indicated as triangles); octagones indicate hosts and PADs of manifacturer C (which, for example, may offer two management terminals, one for hosts and one for PADs).

The result could be to have n+1 (or more) management terminals spread all over the network, which may be not considered as the ideal situation by the man who has the responsability to manage day by day that system.

Fig 2 shows a better situation, where all the management terminals of fig. 1 are collected in a single place, the management room. The ideal situation is in Fig. 3, where all the management functions are performed through a unique terminal connected to a unique management system. This topic will however be resumed in Sect. 6.3.

This bring us to another problem. If the customer buys all his systems from the same manifacturer, then it is not necessary that those systems appear to each other as Open Systems: they may simply appear as proprietary systems linked by a proprietary architecture. However, this may bring some problems to the customer.

A first problem is that it is virtually impossible the connection of systems of other manifacturers (of course it is always possible to emulate any proprietary architecture on any system, but this approach has been used too long, and customers start to be annoied with it).

A second problem is that the availability of a common and powerful method for accessing the network (as the Application Service of OSI is) makes much more easy to implement distributed applications and is a tool which users are not yet prepared to use: however, as soon as it will became largely known, it will be required by everybody.

Finally, a third problem is that with this approach it is difficult to implement gateways properly working between the proprietary architecture and the external open world: however, this will probably affect the performance of the proprietary architecture, and it is matter of each manifacturer to deal with this problem.

Also the user knowledge is rapidly increasing: such matters as standards, OSI, programming interface, and so on are becaming widely known.

As a consequence of all the above, users will soon start to require that also in proprietary architectures systems interact between each other as Open Systems; later, they will start to require that also management be standardized. This user attitude may be summarized by the following key question: will systems belonging to a proprietary network offer the OSI application service? will managament be included?

5. OSI and ISDN

The relation between OSI and ISDN is at least as difficult as the relation between OSI and proprietary architectures is (by other hand ISDN is a sort of PTT's proprietary architecture).

ISDN may be viewed from three points of view. Firstly, it is a design phylosophy: the idea of integrating in a unique network the possibility of transmitting all kinds of data - text, voice, fixed and moving images - is very powerfull, and will give a large contribution to the development of telematics.

Secondly, ISDN is a set of protocols for accessing a network: from the OSI point of view, this aspect is dealt in the usual way, that is with the definition of a suitable set of the three lower layers. Moreover, ISDN is a particularly good network: infact, ISDN can offer a large spectrum of services, in such a way that its correspondent Network-Entity inside the host may offer to the Transport-Entity the capability of dealing directly with a larger range of possible values for the Quality-of-Service parameter.

However (this is the third aspect), in the "I" series there is one particular recommandation, namely I 320, which generates a lot of problems. Infact, it prescribes that "connection control" and "data transfer" are treated as separate functions, and that both are delt by means of a seven layer Reference Model (one for each): while the "data transfer" part does conform to the normal ISO OSI Reference Model, on the contrary the "connection control" part of I 320 is such that it seems that the (nodes of the) network are interposed between the end systems.

Fig. 4 shows this proposed architecture: one can see that all the seven layers are splitted into two separate parts, one for "data transfer" (U in fig. 4), and one for "connection control" (C in fig. 4): the second one for the upper four layers is not end-to-end. If it is really so, this means that the second principle of Sect. 2 (separation between the application and the network) has been violated: this may be a severe limitation for ISDN.

Infact, there is no doubt that systems connected to any network (ISDN included) need end-to-end communications and control at level four and above: everything which is interposed between them is, by definition, at the OSI Reference Model level three or below. Infact, T-, S-, P- and A-layers are intrinsecally End-to-End, because the functions which they perform inside the Open Systems are, by definition, End-to-End functions.

So, the whole architecture composed by the seven layers foreseen by I 320 will be considered by OSI systems as the normal three lower levels of a normal communication network. This could probably be a reason for being the use of ISDN networks less efficient that it would be if I 320 was limited to only the three lower layers.

6. OSI and future research

From what has been said up to now, we may now try to extrapolate which can be in the future the more promising areas for researches in OSI environment.

6.1 Applications

The first area concerns applications: infact, the real impact of the sole fact that a world-wide standard exists is still to be fully explored.

It may first of all have a large impact on how to decide which applications should be implemented. Nowadays, infact, some applications are possible, which only a few years ago where not even conceivable: new, possibly unexpected, business opportunities will appear, as a consequence of the sole existence of a world-wide standard.

Secondly, the way in which application processes will be conceived, designed and constructed will probably be largely affected by the concepts and principles which have been adopted in the design of the OSI Reference Model: up to now there is too little experience on how to use the Application Service. It is however very clear that future applications will be designed in a different way than today.

It has been recently presented a good example of an application, where the method of building layers has been adopted inside the application process (which were more precisely a whole class of application processes): in this example, two further internal layers have been defined on the top of X 400, both in terms of protocols and of services provided.

Thirdly, also "where" an application has to be implemented becomes a free parameter. In the Italian Public Administration, for example, there are several large systems located in Rome, each of them with a large network of terminals spread all over Italy. If a new application has to be installed, presently the obvious solution is to add some hardware and some software in Rome. In the future, on the contrary, the new application can be located anywhere, and the impact of this fact is to be fully investigated.

Methodologies and techniques are required and should be investigated for all the above mentioned problems, which can be summarized as follows: to determine which applications, and how and where, are to be implemented.

In this area there is also the very important item of distributed data bases: infact methodologies of their design should be revisited for being introduced in the OSI Reference Model, particularly having in mind to use as far as possible the Application Service.

6.2 Reference Model and protocols for large band applications and network

Another promising investigation area concerns the use (and the applicability) of OSI in the context of applications which require, and networks which offer, very large transmission capabilities, such as interactive TV applications and optical fiber (or satellite) networks (ISDN is just the beginning of this business).

Infact, up to now, telematics has been dealing with typical EDP oriented applications, such as file transfer, access to remote data bases, messages exchange, and so on. With large band networks, new applications will diffuse, such as voice and video signal processing, vocal messages handling, high definition video, and so on. In the following, we will use the name "HDP oriented applications" in order to distinguish them from the classical EDP ones, where information is generated and used by EDP machines, such as computers or terminals: if men are involved, keybords and alfanumerical videos are the interfaces.

The acronimous HDP means "Human Data Processing", and refers to the fact that HDP sources and destinations are Human beings: suitable terminals (microphones, laudspeakers, cameras, videos, etc.) provide translation of information from form suitable to human beings to form suitable for processing.

The world of HDP is still to be investigated: in particular, the suitability to HDP applications of the OSI Reference Model (which has been developed just of EDP applications) is still to be investigated.

A first answer to this question can be given as follows. OSI is simultaneously three things.

Firstly, it is a set of design principles, as the ones described in Sect. 2 (other principles being, e.g., structuring, virtuality - e.g. of terminals, data structures - modularity, layering, etc.). They are surely suitable also for large band environment.

Secondly, it is a structure (seven layers, with some given functions for each layer, with some precise sublayering, and so on). This is very likely to be suitable also for the large band environment, but needs to be confirmed by a deeper investigation.

Thirdly, it is a set of protocols: very likely, they are not suitable for large band environment, and need to be redefined.

A final requirement is that compatibility between EDP and HDP applications must be insured.

Before leaving this subject, some words should be spent on the matter of integration, and on the meaning of this word.

The more natural meaning of the word "integration" refers to the fact that in the user's mind all information is integrated: he can, infact, process sounds, images and text simultaneously.

The next aspect of integration is when the user interacts with an application running on an informatic system: this application may infact, or may not, be capable of dealing simultaneously with several kinds of information: if it does, this is the integration of the application process.

A further aspect of integration concerns the terminal(s) by means of which the user interacts with the application: this interaction may infact require just only one terminal or several. In the first case we have the integration of the terminal.

Finally, if the terminal (or the terminals, if not integrated) are far from the application, then they may be connected by just one network capable of carrying all kinds of information, or by several networks: in the first case we have the integration of the communication network.

The order in which these aspects have been here presented is also the priority order which they have in satisfying user needs, at least in the author's opinion.

It may be now of interest to try to find an answer to the following questions: why the concept of integration is so popular in the communication world? and why it is commonly diffused the idea that "the integration" starts from the communication? The answer may be that communication people has been since the beginning very active in pushing the idea of the integration, and of the fact that integration of communication media is a sort of "sine qua non" condition for all the other kinds of integration. All this may have caused confusion and misunderstanding.

6.3 Management

As already stated, network management will became a key point in OSI future development: the various management situations which were described in Sect. 4. are just one issue; other relevant aspects are here discussed.

6.3.a) Management of several networks interconnected

In real network life, today we have often a large proprietary network with one or more hosts, several lines and terminal concentrators, possibly with more than one level of concentration.

To this purpose, it may be useful the following remark. When planning the development of an informatics system, one must clearly distinguish three states:
- the initial state of the system;
- the target final state;
- one or more intermediate states.

So, it is strategically important to choose correctly the target state, while the importance of intermediate states is only tactical. In the particular case of network standards, OSI is the strategic choise, while a large variety of intermediate states are tactically acceptable. From this point of view, even solutions non architecturally perfect are acceptable, if it is clear that they are adopted only as intermediate states, and if it is sure that they will not later become an obstacle on the way to the target state.

In the target state every system is fully OSI: hosts are Open Systems, the network is, e.g., an X.25 network (no matter wether this is a private or a public one, or both, e.g. because the public network is used as back-up); terminals are replaced by computers, fully OSI equipped, connected to the X.25 network either directly, or through local networks; terminal concentrators and PADs will not exist any more.

In the intermediate states, hosts are not always fully Open Systems, the network is again an X.25 network, but terminals remain just the previous old terminals, connected to the X.25 network through some dedicated local host, or through some PAD, or through some LAN.

So, in the intermediate states, we have several possible solutions which have different degrees of OSI conformity. Some of such solutions are very bad from the point of view of the purity of the architecture: but real life is somewhat different from theory, ad makes acceptable, at least to help transition, some compromises which could not be the best from a theorical point of view.

So here we have two management problems: in the target state the network management becomes the management of an X.25 network and of several LAN (hopefully, but not necessarly, of the same manifacturer). The intermediate states are even worst, because we have also lack of standards. Fig. 5 shows which could be, at present, the actual situation of some (may be several) really existing systems: several hosts (ot manifacturers A, B, C) are connected to a private X.25 network (the public X.25 network is used as a common global back-up), either directly or through some local network (which on their turn may be of several manifacturers say a, b and c). Each system is managed by its own management terminal.

So, research is necessary to deal with all these situations: it will become a key issue for the development of OSI.

6.3.b) Management of OSI upper layers

Also the part of the OSI Reference Model lasting in layers from Transport to Application needs to be managed: to this part applies what has been said in Sect. 4..

6.3.c) Application management

The management of application processes has been till now done as a part of the management of the network. With the OSI Reference Model the application management must be separated by the management of the network and of the upper layers: this will probably bring a new set of problems and of ways for solving them: this is a new style of management and some research is needed.

6.3.d) Integrated Management

So we have obtained several different sub-systems, each of them having its own management system and/or terminal: therefore, a large user should have a management Center made of several different management terminals: one for the main network, one for each local network, one for each private network, one for the upper layers of manifacturer A, one for the upper layers of manifacturer B, and so on for all the manifacturers; then one for (each group of) applications, etc.

It is now clear why some research is also needed for integrating all these systems in a single, acceptable, integrated management system, as Fig. 6 shows.

6.4 FDTs, testing and certification

One may remind that for several years the word "standard" has had no meaning in the world of informatics; suddenly, a few years ago, standards have become a very important thing, and now everybody is interested. One may wonder why this happened.

One of the reason is the need - and the business - to make systems to be capable of communicating with each other. Now, this goal risks to be severly penalized if it is not certain that any two Open Systems will be able to communicate in any circumstance, no matter to their manifacturers, their location, their internal characteristics, and so on.

So, methods for testing, for certification, for debugging, and generally for helping implementors will be another important area for future research.

It is also to be noted that OSI protocols are typically very reach of options: this helps a lot in fitting OSI to a large variety of environments. Moreover, the availability of a large spectrum of options allows to delay some decisions for which some experience is needed. However, just this large spectrum may become a source of problems in implementations, particularly for not skilled implementors (which in the future may become a common case).

So, testing and certification by part of qualified laboratories may become a very useful tool for helping OSI to diffuse correctly.

As a consequence, also techniques for formal description (FDTs) of protocols will play an important role: even if large improvements have been achieved, still a lot remain to be done in this field.

FDTs may also probably have a new hint from their usage in the Application layer. Application protocols and services, infact, have several characteristics which are new for most of the FDT world. Here are some examples of such characteristics: colloquy among more than two partners, services obtained by assembling the services offered by a set of Service Providers (in the lower layers, usually a service is offered by just a single Service Provider), services describing complex objects such as a Virtual Filestore (and its management rules), services implying simultaneous negotiation among several Service users and the Service Provider, etc.

6.5 OSI and local networks

This is also an interesting problem: up to now, several people have believed that local networks are different from normal networks (the fact itself that a different name - LAN - is used is significant).

In the future, LANs will be used as PADs (or terminal concentrators) are used now. As said before, terminals will disappear and will be substituted by full OSI systems: this means that the capability of easy connecting complete OSI systems (or more precisely, the capability of properly fitting in the OSI environment) will become one of the more important requirements for LANs.

Also large band gateways with the large band public network will became a key research issue in this field.

7. Conclusions

The above conducted analysis has put in evidence a large spectrum of research activities which seem to be promising in this field: we have tried to indicate some of them, but of course it may be that other researchers may have different opinions.

What seems to be solid enough, is that surely the amount of work to be done is very large, and that no European Institute will probably be in a position to efford to tackle by itself alone the whole problem.

Here is clearly a case in which international cooperation and coordination become essential, and play a key role in the whole European strategy in this sector.

L1:lechii (860901)

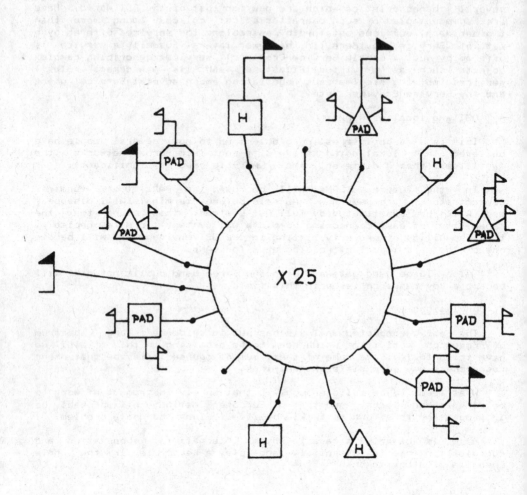

FIG. 1

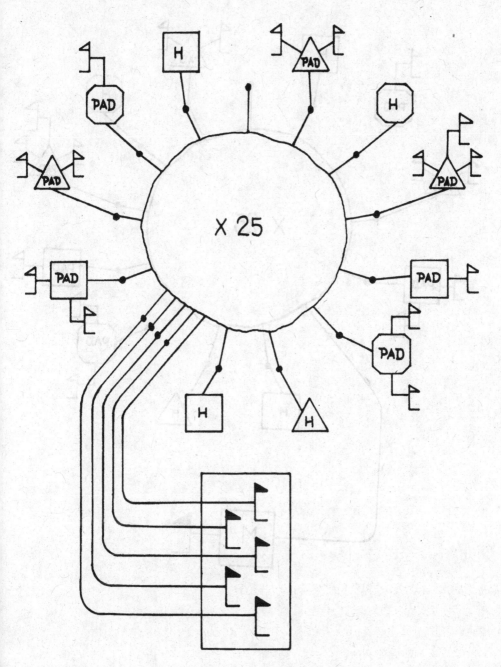

FIG. 2

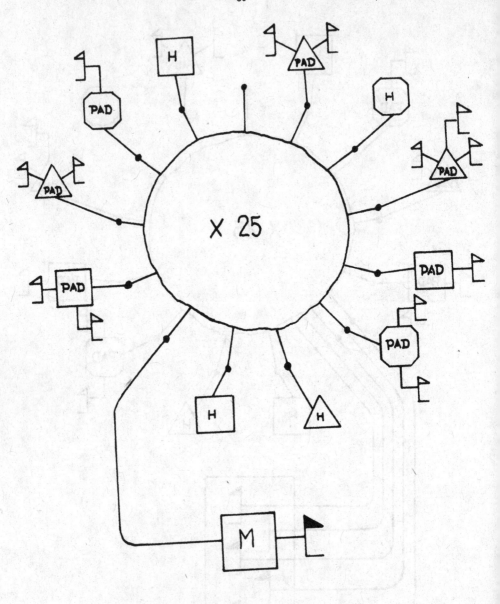

FIG. 3

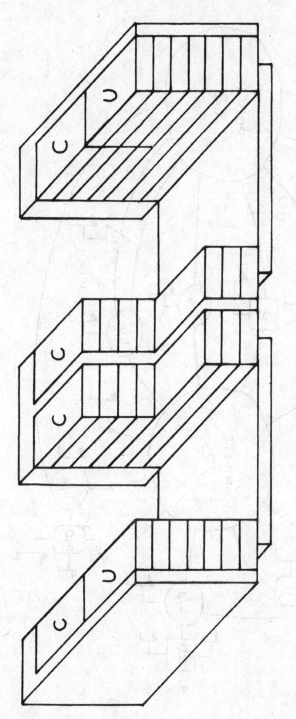

FIG. 4

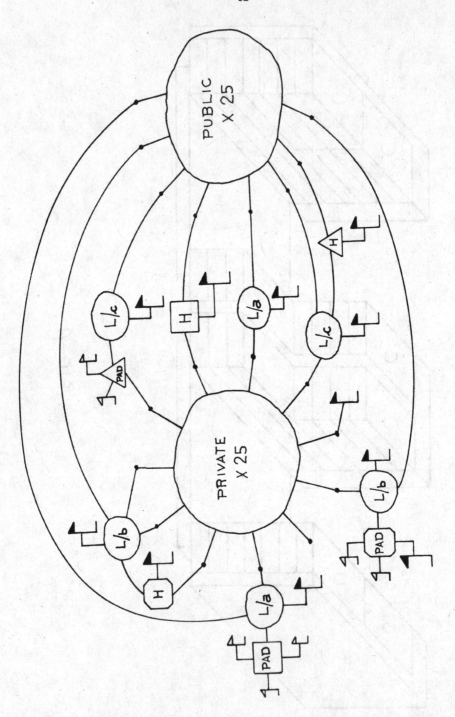

FIG. 5

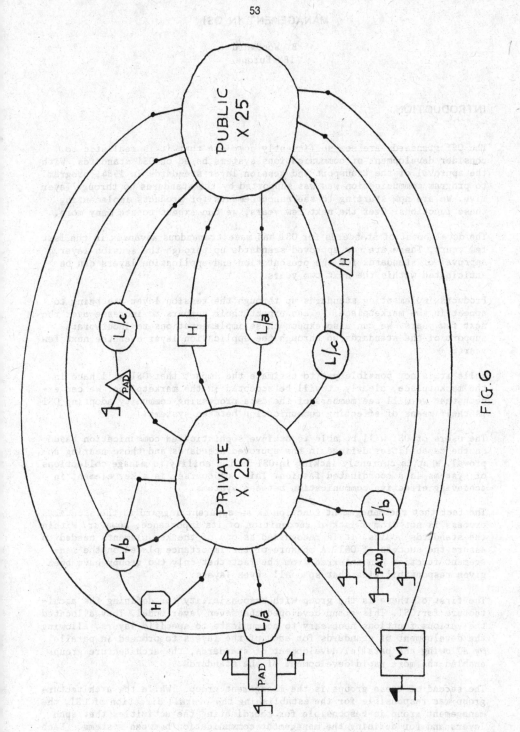

FIG. 6

MANAGEMENT IN OSI

B. Woodward
IBM Europe

INTRODUCTION

The OSI standards are now sufficiently complete that it is realistic to consider development of communications systems based on OSI standards. With the approval of the transport and session layer standards in 1984, program to program communication was was supported by the standards up through layer five. We are now starting to see announcements for products implementing these functions. Over the next few years, we can expect to see many more.

The development of standards for OSI has made tremendous advances in the last few years. There are now approved standards up through the session layer. Approval of standards for the presentation and application layers can be anticipated within the next two years.

Products implementing standards up through the session layer are being to appear in the marketplace. We can expect their numbers to increase over the next few years. We can also expect these implementations to incorporate support of the standards up through the application layer over the next few years.

While it is not possible yet to estimate the impact that OSI will have in the marketplace, clearly it will be accepted in the marketplace. We can expect that we will see members of the data processing community adopting OSI as their means of effecting communication between systems.

The users of OSI will be able to achieve sophisticated communication based on the capabilities defined in the approved standards and those nearing approval. What is currently lacking in OSI is the ability to manage collections of systems in a coordinated fashion. This, of course, is a key element in achieving effective communication between systems.

The fact that the management function is an apparent laggard in the standards process is not from a lack of recognition of its importance, however. Within the standards bodies, it is recognized as one of the key elements needed to assure the success of OSI. A measure of the importance placed on the management function can inferred from the fact that only two groups have been given responsibilities that span all seven layers.

The first of these is the group with responsibility for defining the architecture for OSI. This group developed the seven layer model. They allocated the various functions necessary to communicate to specific layers, allowing the development of standards for each of the layers to proceed in parallel. By allowing the parallel development of standards, the architecture group enabled the more rapid development of the standards.

The second of these groups is the management group. While the architecture group was responsible for the establishing the overall direction of OSI, the management group is responsible for coordinating the activities that span layers and for defining the management communication between systems. Each of these requires that the management function consider the standards produced for each of the layers. Thus, the management activity is, by its nature, one that must lag behind the work of the other groups. Its function is, in part, to tidy up the jagged edges that must result from the parallel development of the standards.

OVERVIEW OF MANAGEMENT

Within the ISO committee structure, the responsibility for the definition of OSI management standards has been placed in Technical Committee 97 Subcommittee 21 Working Group 4 (TC 97 SC 21 WG 4). TC 97 SC 21 has been assigned the responsibility for layers 5 through 7 and topics spanning the seven layers (architecture and management). Responsibility for the lower four layers has been assigned to Subcommittee 6 of TC 97. Also, responsibility for message handling standards has been assigned to Subcommittee 18 of TC 97. Each of these groups has at least some requirement for support and interaction with the group responsible for management.

As the scope of the management function is potentially very wide, WG 4 has explicitly limited its scope to a subset of the overall management function. This scope is defined to be two fold:

* management of the OSI resources of each participating system

* participation in the overall management of the global OSI network

This is to limit the scope of the management activity to the communications environment. The key to understanding exactly what the scope of management is is to have a clear definition of what the OSI resources are. This has proven an elusive definition to obtain so far.

There are two areas that the management group will address with its standards.

The first of these is the definition of service and protocol standards for the exchange of management information between systems. This includes the specification of both the semantics and syntax of the information to be transferred. This is the activity that will result in the standards that effectively define the management activities that can occur between participating systems.

The second of these is the definition of a service standard only that can be used to define information that must be shared between layers, but that may not be used between systems. The need for this activity is a result of the structure of the standards activity. As noted above, there are many groups involved in the creation of OSI standards. These groups are spread over three subcommittees in ISO, each subcommittee having several working groups. Each working group is further subdivided into rapporteurs groups concentrating on a single topic. The standards process has divided the work into discrete layers, with each layer being assigned a carefully defined segment of the overall work. The standards for each layer are restricted to the scope of the layer and are constrained to interacting with the adjacent layers via defined service standards. Unfortunately, not all information that must be shared between layers is available in the defined service standards.

However, in addition to interacting with adjacent layers, each layer is permitted to provide information to and request information from the management function. A service standard (without an accompanying protocol standard) will be defined for the exchange of information between management and the layers. This standard will serve as the mechanism by which the layers will be able to incorporate within their standards the definition of information that must be made available to other layers, but is not defined within the service standard.

Organization of work: To achieve the standards in the areas outlined above, WG 4 has has organized into three rapporteurs groups: architecture, management information systems (MIS), and directory.

Architecture: The architecture group is responsible the definition of an extension to the OSI Basic Reference Model (ISO 7498) that further clarifies the role and scope of management within OSI. To achieve this, it will defines an abstract model of management. This model is used to explain the concepts and terminology, the objectives, the general principles, and the components of OSI management.

MIS: The MIS group is responsible for the definition of the service and protocol standards for all areas of OSI management except for the directory function. This includes:

* Configuration Management

    The definition, collection, monitoring and use of configuration data that describes the inventory of OSI resources, their capabilities and attributes, and their interrelationships constitute configuration management.

* Fault Management

    Detection, reporting, and corrective procedures for abnormal events that occur within the OSI environment are the scope of fault management.

* Accounting Management

    The collection of cost information is the scope of accounting management.

* Performance Management

    Collection and evaluation of statistical data describing the performance of the communication between open systems.

* Security Management

    Access control, password management, and encryption services for the protection of OSI resources are the scope of security management.

* Name and Address Management

    Support for the registration and access to name and address data is the scope of name and address management.

Directory: The directory group is responsible for definition of the information to be contained in the directory and the protocols for the exchange of this information.

STATUS OF WORK

Each of the rapporteur's groups has made substantial progress at recent meetings. Each has addressed and resolved major issues in its area. Finally, each group either has already proposed, or plans to propose at its next international meeting, progression of a document from working draft status to draft proposal status. This is a significant step since a request for draft proposal status is a statement that the work is considered to be technically stable.

Directory: The directory group submitted a working draft for advancement to draft proposal status as an output of the meeting in October, 1985, in Philadelphia. This includes both a service and protocol standard for the directory. The salient features of the directory are that it is tree structured with 'limitless' levels, but only the leaves may be qualified with attributes. The major issue facing the directory proposal is that it is not consistent with the current work going on in CCITT. The current CCITT proposal also is based on a tree structure with 'limitless' levels, but permits each node to be qualified with descriptors. This is an issue that will inevitably have to be resolved, given the almost universal desire to have consistency between the ISO standards and the CCITT recommendations.

Architecture: The architecture group plans to propose that the working draft of the Management Framework be advanced to draft proposal status. The Management Framework is the proposed addendum to the Basic Reference Model describing OSI management. Its salient features are descriptions of the components of management (system management and layer management), definition of their roles, and and explanation of the relationship between them.

There were originally defined three major components of management. These were system management, layer management, and application management. Application management was, after much discussion, dropped as a component of management since it seemed to be focused more on the management of the applications than of the communications service. This placed it outside the scope of management. It was recommended to SC 21 that the topic be placed in some other group.

The definition of system management was fairly straightforward. It includes the system to system exchanges of management information necessary to support the communications environment. This covers information exchange in the areas outlined above. Many of these topics (e.g., fault management) are also relevant to the groups defining layer standards. It was important to provide clear guidelines about role of each group in the development of standards. As there were many different views about the distribution of responsibility, this involved considerable discussion. A simple example of the disagreement is shown in the discussion of connection establishment. There were many who contended that connection establishment was clearly a management function. Others contended that it was clearly the responsibility of the layer groups. Other examples can be imagined in each of the functional areas covered by management.

To resolve this discussion, WG 4 endorsed the current allocation of standards work with a clear statement that all activities necessary within a protocol to allow its correct operation are the responsibility of the layer standard groups, whether or not the activity is management related. Protocol standards may incorporate whatever functions are regarded as necessary by the developers of the protocol standard. The layer groups have an option, under this definition, to specify that the information be carried within the systems management protocol. The system management group is obliged to provide facilities to support these exchanges.

The above decision does not resolve another major issue about the role of the layer functions relative to management. This is the question of whether or not the layer groups may define separate protocols dedicated to support solely of a management function for that layer. Current sentiment on this issue seems to favor allowing these protocols to be developed. The alternative would be to require these functions to be included in the systems management protocol, rather than have a proliferation of special management related protocols. The restriction placed on the layer specific management protocols is that they may not incorporate functions reserved to layers above it.

Although only recently addressed, the relationships that may be established between systems may be reaching consensus. The concept of domains was introduced to deal with the grouping together of systems. Domain definitions, while not yet complete, allow systems to group themselves together to cooperate in some management functions. System management must support the capability of negotiating these domains for each of the functional areas of management.

MIS: The MIS group plans to propose that working drafts for both the services and protocols for Common Management Information Service be advanced to draft proposal status at the next international meeting (September 1986). The salient features of these are that they propose a 'subprotocol' that supports three major ways to exchange management information. These are event notification (generally unsolicited information exchanges), information transfer (exchanges of a request/response nature), and control information (exchanges that are executive in nature). These protocol elements are defined as general management service functions to be used by the protocols defined within each of the functional areas of management previously defined. It provides a syntax for information exchange, while the standards for the functional areas will provide the semantics for the information exchange.

AREAS OF FUTURE WORK

Though each of the groups has work that has advance to the point of technical stability, much fundamental work remains to be done before OSI will have a true management capability.

The single most important area that needs to be addressed is the addition of semantic content to the directory and MIS proposals. Each currently provides an abstract carrier that can be adapted to many uses. Until the use of these is made specific, they will not be able to serve as the basis for an standard that can be implemented into product.

As noted earlier, OSI management is concerned with the management of OSI resources. It must be able to exchange information about the status of these resources and initiate actions to change the status of them. To do this, it is necessary to have a framework for describing the OSI resources. Until recently, it was not clear just what this fundamental framework is. Recent advances in the naming and addressing work, though, has provided the basis for the development of this framework. The naming and addressing work has identified the basic elements that are the architectural basis for describing a configuration. With this in place, the management group is now has a basis on which to elaborate a description of a configuration. Once this is in place, it will be possible to develop the protocols that specify the management activities that can take place on these resources. With this, it will be possible to have implementations of OSI management.

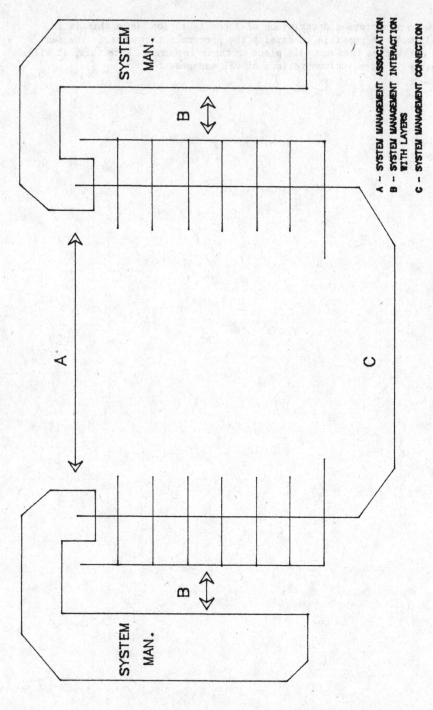

OSI SYSTEM MANAGEMENT

A – SYSTEM MANAGEMENT ASSOCIATION
B – SYSTEM MANAGEMENT INTERACTION
      WITH LAYERS
C – SYSTEM MANAGEMENT CONNECTION

A minimum set of desirable security goals in OSI identified by the author is:

1. Protection of data against unauthorized modification.
2. Protection of data against undetected loss/repetition.
3. Protection of data against unauthorized disclosure.
4. Assurance of the correct identity of the sender of data.
5. Assurance of the correct receiver of the data.

As a memory aid for these five basic security goals, the following five terms starting with the letter "S" have been selected to represent the security achieved by satisfying these goals. They are, respectively:

1. Sealed
2. Sequenced
3. Secret
4. Signed
5. Stamped

Achieving these security goals in the OSI architecture will assure that data being transmitted from one OSI system to another will not have been modified, disclosed, replayed, or lost in the network without the sender and/or the intended receiver being notified and that the participating parties in the communication have been correctly identified.

Other security goals that have been identified [11] as being desirable include: labeling of data according to its sensitivity, source, etc.; not disclosing the identities of the sender and recipient of data, and the quantity of data exchanged, except to each other; providing security audit trails of network communications; assuring the availability of communications under adverse conditions; assuring that data inside an OSI system cannot be transmitted using covert information channels, even of very low bandwidth; proving to an independent third party that a communication did occur and the correct contents were received; obtaining explicit authorization for access to a system before making a connection to the system.

C. Role of the National Bureau of Standards

The National Bureau of Standards (NBS) has fostered the development of the OSI architecture and the implementation of commercial products implementing the standard protocols defined for the architecture. NBS has had a program in computer security since 1973 and has fostered the development of numerous security standards [7, 8, 9, 10] since that time. It has assisted in the development of several security standards in the banking community [4, 5, 6] and the information processing community [1, 2, 3] through the American National Standards Institute. It is now supporting the development of an OSI security architecture [11] via the ISO/ TC97/ SC21/ WG1 and the OSI SIG-SEC.

CONSIDERATIONS FOR SECURITY IN THE OSI ARCHITECTURE

Dennis K. Branstad
Institute for Computer Sciences and Technology
National Bureau of Standards
Gaithersburg, Maryland, 20899, USA

I. Introduction to OSI Security

The Open Systems Interconnection (OSI) computer network
architecture has given computer network designers and
implementors a common vocabulary and structure for building
future networks. It has also given network security designers a
foundation upon which desired security services can be defined
and built. This paper discusses several goals of security in
the OSI architecture as well as where and how the security
services that satisfy them could be implemented.

A. Need for a Security Architecture

A standard security architecture is needed in OSI in order
to begin the task of implementing security services in commercial
products so that not only can one OSI system communicate with
another, but also it can do the communication with the desired
security. The security goals and services discussed in this
paper are predicated on the assumptions that sensitive or
valuable data are being transmitted between systems in the OSI
network, that changes in the network between the systems could be
made by an unauthorized person or persons in order to obtain or
modify the data, and that security services are to be available
in the network to prevent the unauthorized disclosure of
sensitive data and to detect (and report) the unauthorized
modification of data.

For this paper, security is defined to be the protection of
the confidentiality and integrity of data. Privacy, often
combined with security or confused with security, is a social
issue regarding protection of personal information from
undesirable use and is not discussed in this paper. Security is
often defined as including protecting the availability of data
but is not included in the scope of this paper.

B. Requirements for Security

A large number of potentially desirable security goals in
computer networks have been identified in the literature. The OSI
Implementors Workshop Special Interest Group in Security (OSI
SIG-SEC) is establishing a desirable set of security goals for
implementors of OSI and the resulting list of desirable services
to implement. This SIG is sponsored by the U. S. National Bureau
of Standards and is open to anyone interested in OSI security.

II. OSI Network Security Perimeters

A useful notion in the development, implementation and use of security in a computer network is that of a security perimeter. This logical structure in a computer network is the equivalent to a physical structure in a secure facility such as a bank vault. In actuality there are multiple security perimeters around highly secure facilities where a principal of "security in depth" is practiced. Similar analogies can be drawn in computer networks. For simplicity in this discussion, a single security perimeter concept will be used in which each OSI system will have a security perimeter. The overall goal of OSI security is to communicate data from within one security perimeter to another. Loss of security within a perimeter is beyond the scope of this paper.

A. One Security Perimeter around Network

If a security perimeter is drawn around the entire network (Figure 1), either because no sensitive or valuable data are ever communicated in the network, because no threats are believed to exist in the network, or because security it provided through non-OSI methods, then no OSI security services are needed. Many networks are presently being operated in this manner. This is acceptable as long as everyone and everything inside the perimeter is "trusted." Trust implies that no intentional or accidental event will occur which will result in an undesirable disclosure, modification or loss of data. A simplified definition of trust is used in this paper with trust being a binary valued parameter (i.e., multi-level security is not considered). Trust can also be assured within the system through the use of a "Trusted Operating System." This system assures that adequate security is provided within the security perimeter.

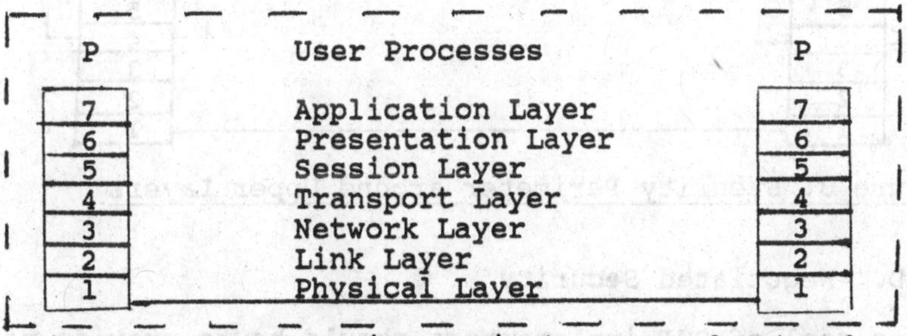

P	User Processes	P
7	Application Layer	7
6	Presentation Layer	6
5	Session Layer	5
4	Transport Layer	4
3	Network Layer	3
2	Link Layer	2
1	Physical Layer	1

Figure 1: One Security Perimeter around Network

B. Security Perimeter around each User Process

A security perimeter could be drawn around each user process which provides high granularity security (Figure 2) since each user process provides its own protection and nothing within the OSI architecture needs to be trusted. However, this requires that all desired security services be implemented in every user

process or program. While possible, this approach is contrary t
the goal of OSI for performing services in the layers of OSI
rather than in each user process.

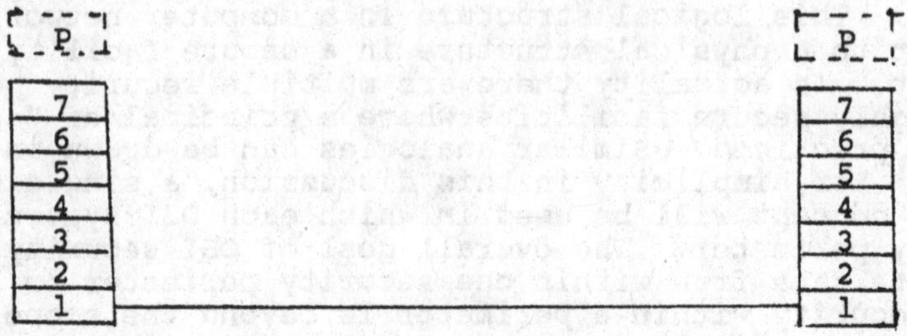

Figure 2: Security Perimeter around each User Process

C. Security Perimeter around Upper Layers

A security perimeter can be drawn between these two extreme
around the upper layers of the OSI architecture. Different
granularities of security result from selecting different
placement of the security perimeter. In actuality, a hierarchy
of security perimeters will be implemented, each providing
security against a different perceived threat. A security
perimeter has been drawn at the transport layer (layer 4) of the
OSI architecture (Figure 3) for subsequent discussion in this
paper

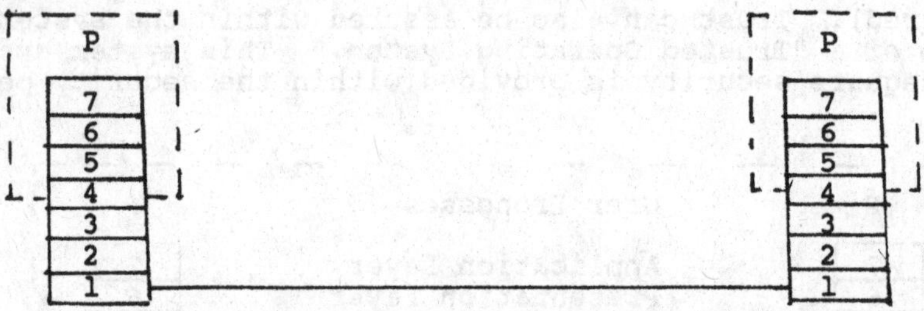

Figure 3: Security Perimeter around Upper Layers

D. Negotiated Security

One goal of OSI implementors should be to provide maximum
flexibility for users of an implementation. An implementation
should provide for negotiation between users in selecting an
optimum set of OSI services, including security services.
However, security may be somewhat unique in this regard in that
some organizations may not desire to negotiate certain security
services, especially if the negotiation could result in security
less than some predetermined minimum. Other organizations may
accept negotiating away all security services if those services
are temporarily causing functionality or throughput to drop belo

a minimum. Some organizations may add to the basic security services provided in standard implementations and not desire other organizations to use or know about the additional services.

An extensible security architecture is desired which will provide for these special services without causing an unacceptable overhead on those not requiring these services.

III. Placement of Security Services in the OSI Architecture

A. Security Addendum to the OSI Architecture

A draft security addendum to the OSI architecture [11] has been developed by Ad Hoc groups of the American National Standards Institute (ANSI) and the International Standards Organization (ISO) TC97/ SC21/ WG1. The draft security addendum presents a glossary of computer security terms, describes a number of security services for OSI, and presents a matrix of where in the seven layer OSI architecture the security services may be located (See Below). It then presents the rationale for why the security services are placed in those layers. Recent work [12] defines an authentication framework for the layer 7 directory service for which User Agents are authenticated before they are granted access to sensitive information in the Directory.

While the draft addendum satisfies the goals of defining a number of security services and discussing where they could be placed, the addendum is not adequate for an implementor desiring to implement security in the OSI architecture. First, it would be too expensive to provide all security services at all possible layers allowed in the addendum. Second, if one implementor chose to implement a service at one layer and another implementor chose to implement the same service at a different layer, the goal of compatability between peer layers of OSI would not be achieved. Finally, standards for implementing the services are not currently specified.

B. OSI Security Categories and Services

The following security categories and services are defined in the draft security addendum to the OSI architecture. The OSI layers in which the services could be implemented are shown in the matrix next to the services. The services need not be implemented in all of the layers that are specified.

OSI LAYER 1 2 3 4 5 6 7	CATEGORY OF SERVICE SERVICE
	1. IDENTIFICATION/AUTHENTICATION
|_|_|3|4|_|_|7|	A. Data Origin (Connectionless)
|_|_|3|4|_|_|7|	B. Peer Entity (Connection)

2. ACCESS CONTROL

|_|_|_|_|_|_|7| A. User Agent Authorization

|_|_|3|4|_|_|7| B. Peer Entity Authorization

3. INTEGRITY

|_|_|3|4|_|_|7| A. Connection (w/wo error recovery)

|_|_|3|4|_|_|7| B. Connectionless (wo error recovery)

|_|_|_|_|_|_|7| C. Selective Field Integrity

4. CONFIDENTIALITY

|1|2|3|4|_|_|7| A. Connection

|_|2|3|4|_|_|7| B. Connectionless

|_|_|_|_|_|_|7| C. Selective Field

|1|_|3|_|_|_|7| D. Traffic Flow

5. NON-REPUDIATION

|_|_|_|_|_|_|7| A. Originator

|_|_|_|_|_|_|7| B. Recipient

C. Factors in Placing Security Services

Many factors must be considered in selecting the layer(s)
for implementing selected security services. First, a basic set
of security services to be implemented must be chosen. Second,
minimum number of layers should be chosen in which to implement
the services to minimize the number of layers affected by
security. Third, use of existing services of a layer may be
utilized by the security service if a proper layer is chosen.
Fourth, the overall cost of providing the selected security
services will be minimized if the layer is properly selected.
Fifth, a set of primitive security functions need to be defined
and implemented (hardware, software, firmware) in such a way tha
they can be performed at one or more layers of the architecture
in providing the desired security service.

D. Primitive Security Functions

OSI security services could be implemented utilizing a set
of primitive functions similar to the ones below. The primitive
functions would be called with a set of parameters enclosed in [
and return the results enclosed in {} following execution.

I. AUTHENTICATE [ID; AUTHENTICATOR] {RESULT; STATUS}

This primitive verifies that the AUTHENTICATOR does correspond with the claimed ID by searching the local Secure Management Information Base and responding with the correct RESULT and STATUS.

II. AUTHORIZE [ID; TYPE; RESOURCE] {RESULT; STATUS}

This primitive verifies the authorization of ID with the indicated TYPE for access to the requested RESOURCE and sets the correct RESULT and STATUS.

III. ENCIPHER [PT; LENGTH; KEYNAME] {CT; LENGTH; STATUS}

This primitive enciphers plaintext beginning at PT for the indicated LENGTH into ciphertext beginning at CT for the indicated LENGTH and sets the resulting STATUS using the KEY associated with KEYNAME.

IV. DECIPHER [CT; LENGTH; KEYNAME] {PT; LENGTH; STATUS}

This primitive deciphers ciphertext beginning at CT for the indicated LENGTH into plaintext beginning at PT for the indicated LENGTH and sets the resulting STATUS using the KEY associated with KEYNAME.

V. COMPUTEMAC [DATA; LENGTH; KEYNAME] {MAC; STATUS}

This primitive computes a Message Authentication Code (MAC) on the DATA of indicated LENGTH using the KEY associated with KEYNAME and sets the resulting STATUS.

VI. VERIFYMAC [DATA; LENGTH; KEYNAME; MAC] {RESULT}

This primitive computes a Test Message Authentication Code (TMAC) on the DATA of indicated LENGTH using the KEY associated with KEYNAME and sets the correct RESULT to indicate if TMAC is identical with the input MAC.

VII. SIGN [DATA; LENGTH; USERID; KEYNAME] {SIGNATURE; STATUS}

This primitive computes a SIGNATURE on the DATA of indicated LENGTH for the user indicated by USERID using the KEY associated with KEYNAME and sets the resulting STATUS.

VIII. VERIFYSIGNATURE [DATA; LENGTH; USERID; KEYNAME; SIGNATURE] {RESULT; STATUS}

This primitive computes a Test Signature (TSIGNATURE) on the DATA of indicated LENGTH for the user indicated by USERID using the KEY associated with KEYNAME, compares it with SIGNATURE, and sets the correct RESULT and STATUS.

E. Initial Recommendations for Placement

Based on the simplifying assumptions stated at the beginning of this paper, the transport layer (4) of the OSI architecture was chosen by NBS for initial implementation of a selected subset of security services. This layer was chosen after several years of participating in the development of standards for security at layers 1/2 [2], layer 4 [13] and layer 6 of the OSI architecture by the accredited ANSI Technical Committee X3T1. The layer 1/2 standard was developed for protecting data in each link of a network. However, it does not provide security from one OSI end-system computer to another through a general network. A layer 4 standard was drafted to provide security for all data in a layer 4, class 4 connection. A layer 6 standard was drafted to provide security for selected fields of data specified by an application in such a way that it need not be unprotected even at the intended destination. Early development of the layer 4 standard was facilitated by an early definition of services at layer 4 and the existence of standard protocols and implementations of layer 4. It was also facilitated by using existing services of layer 4 for security purposes.

IV. Protocols for Transport Layer Security Services

A. Integrity Service

A connection integrity service protocol has been defined for class 4 of the transport layer (4) of the OSI architecture. The integrity service can achieve two security goals, sealing and sequencing, and assures that all data in a connection are transferred from one OSI security perimeter to another without being intentionally or accidentally modified, lost or repeated. Such security is especially important in Electronic Funds Transfer (EFT) transactions. EFT messages are vulnerable to modification; deposit and withdrawal messages are vulnerable to loss or repetition. While present EFT security standards specify security services at layer 7 of the OSI architecture, a wide variety of other applications could utilize similar security services if they are implemented at layer 4.

The integrity service protocol utilizes the sequence number provided by layer 4, class 4 service. This is a 31-bit number defined as 4 octets in the header of each layer 4 Protocol Data Unit (PDU). The sequence number is provided by layer 4 for resequencing the PDUs if they arrive out of order and for flow control on a connection. The integrity service also utilizes the existing layer 4, class 4 mechanisms for recovery from errors (i.e., lost or modified data). Connectionless network layer (3) services can then be used if a class 4 integrity service is provided and used at layer 4.

The PDU integrity protocol specifies how an electronic data integrity seal, called a Message Authentication Code (MAC), is computed for each PDU. The seal covers either the user data only for simple data integrity or both the user data and the header (including sequence numbers) for data stream integrity. The seal is typically a 32-bit number that is computed using cryptographic functions on the parts of the PDU that are to be sealed so that its integrity can be verified when it is received at the corresponding security perimeter (layer 4 peer entity). If any part of the PDU has been accidentally or intentionally modified, including the address and sequence number, the test value computed on the received PDU will not match with the seal computed by the transmitter on the transmitted PDU and transmitted in the security header of the PDU. If the value is not correct, the suspected PDU is discarded and a retransmission is requested. If the value is correct, the PDU is accepted. Sequence numbers are also verified to assure data stream integrity.

B. Confidentiality Service

Data can be protected against unauthorized disclosure in a network with encipherment (encryption). The ISO/OSI security addendum calls this a confidentiality service. Enciphering is a transformation of data into a form that is not usable or readable but preserves the information content. The resulting ciphertext is transmitted. The authorized receiver must perform the correct inverse operation, called deciphering (decryption), in order to obtain the original, usable, readable form of the data. Typically, a cryptographic algorithm, implemented in a computer with either hardware, software or both, and a cryptographic variable called a key are used to perform the two required transformations. A requirement of this service is that something be kept secret or available only to authorized communicating parties. Details of this service are beyond the scope of this paper.

The confidentiality service requires that the user data of a PDU be enciphered before leaving the security perimeter of the transmitter and be deciphered only after entering the security perimeter of the intended receiver. Other portions of the PDU need not be enciphered since they contain no user data. If enciphering is performed only on the user data, the addresses or identities of the communicating parties are not enciphered and hence a monitor in the network can determine who is communicating and how much data in being communicated, even though the contents of the data cannot be determined. The OSI security architecture specifies a traffic flow confidentiality service at layer 1 to protect against traffic analysis if this protection is desired. Encipherment at this layer would protect all data on a communication link, including the addresses of the communicating entities. However, it would be unprotected in all intervening gateways.

C. Peer Authentication Service

The two communicating transport layers are called peer entities and must perform equivalent services in order to communicate. Simplistically, what one does the other must check and/or undo. The security protocols that have been defined to date at layer 4 will assure that the peer layers are mutually identified and that a connection between them is a current connection and not a replay of a previous connection. This

protocol relies on cryptographic procedures during the
establishment of a connection. Once a connection is established
data intended for the peer layer 4 can only be used by that peer
entity. It can be accidentally or intentionally destroyed,
delayed or misrouted, but it cannot be used by the unauthorized
receiver if encrypted.

Peer authentication is performed by a connection procedure
often called a three-way handshake. Using proper cryptographic
procedures, a challenge-response-verification is performed by
both peer entities of a connection. Random numbers are used in
standard procedure to assure that both peer entities have the
correct key and that a replay of a previous connection is not
being attempted. The user data is not signed with this
technique. The personal identities of the users of a connection
or the applications using a connection are not involved in this
service. It merely assures that an entire stream of data is not
replayed to an unsuspecting recipient.

V. NBS Laboratory Implementation

    A. Local Area Network Environment

The National Bureau of Standards initiated an experiment in
implementing these security protocols in the transport layer of
several computer systems in a local area network environment.
The experiment was to determine the adequacy of a proposed ANSI
standard for the security protocols, the ease of implementation
and impact on the operation of the network.

The network was based on one of the IEEE 802 standards often
called Ethernet. Six IBM Personal Computers were used with the
Disk Operating System. Ethernet circuit boards were added to the
Personal Computers and connected together using "thin" coaxial
cable for four computers and "thick" coaxial cable for two
computers. Software supplied with each Ethernet board was used
to provide layer 1, 2 and 3 functionality. A transport layer
protocol that was implemented (but not tested) on a time-shared
mini-computer was used as the basis of the experiment. Null
layers 5 and 6 were used. A simple layer 7 application was used
to demonstrate connections and data transfers among the
computers.

The National Bureau of Standards Data Encryption Standard
(DES) was used for the cryptographic functions. Six circuit
boards each containing DES devices were obtained from two
companies and plugged into the six personal computers. These
boards were used by the layer 4 security services. Cryptographic
keys for each of the six computers were manually installed in the
computers for demonstrations. No automated key management was
performed during the experiment.

    B. Lessons Learned

The difficulty of converting a protocol designed for a
time-shared, interrupt driven mini-computer to a single-user,

event driven personal computer was not anticipated. Even though
the programming language was the same on both systems, it was
found to be very difficult to convert the program from one system
to another. A completely new system interface had to be
developed in order to use the services of the transport protocol.

It was found to be easy to integrate the security services
into the transport protocol once the protocol was working. The
confidentiality service was the easiest to implement. The
integrity service was the most difficult as it required more
modifications of existing layer 4 functions. The peer
authentication service was trivial after implementing the
integrity service. Since the system was designed only for
demonstration, there was no attempt to verify the correctness and
trust of the implementing code itself which would be necessary
for operational systems.

It was difficult to effectively demonstrate security of the
network. Good security implementations should have minimal
effects on the user and the network. It was often impossible to
tell if the security services were being performed since they
caused negligible overhead on the network. A network monitor was
finally designed to observe the data on the network so that
security services, or lack thereof, could be observed.

It was acceptable to have special applications to
demonstrate the security services and the transport services but
it was apparent that original equipment and software implementors
and vendors have to support the enhanced security functions as a
basic feature of their product in future products in order to
gain the desired security and user support. The interface to
security enhancements has to be standard and integrated into the
product or security will often be bypassed.

VI. Summary and Conclusions

A security architecture is needed as a fundamental part of
the OSI architecture. Standard security services must be
defined, standard security protocols must be developed and
standard security interfaces for applications programs must be
specified. Optional security services must be defined and
standard implementations must be available to be used on an
optional basis. All security services need to be negotiated but
with provisions for default services and enhanced, user defined
services. The user should not be aware of the operation of
security services other than the need for providing initial
information for the service (e.g., the set of services required,
specific parameters for the service if default parameters are not
acceptable).

While only a small subset of the possible desirable security
services were selected for discussion in this paper, there is a
need for research in providing additional services and for
standards activities for specifying implementations of them. The
National Bureau of Standards is seeking interest and assistance
in providing these necessary activities.

VII. References

[1] ANSI X3.92, American National Standard for Information Systems - Data Encryption Algorithm, American National Standards Institute, New York, NY, 1981.

[2] ANSI X3.105, American National Standard for Information Systems - Data Link Encryption, American National Standards Institute, New York, NY, 1983.

[3] ANSI X3.106, American National Standard for Information Systems - Data Encryption Algorithm Modes of Operation, American National Standards Institute, New York, NY, 1983.

[4] ANSI X9.8, American National Standard for PIN Management and Security, American National Standards Institute, New York, NY, 1982.

[5] ANSI X9.9, American National Standard for Financial Institution Message Authentication - Wholesale, American Nationa Standards Institute, New York, NY, 1986.

[6] ANSI X9.17, American National Standard for Financial Institution Key Management - Wholesale, American National Standards Institute, New York, NY, 1985.

[7] Federal Information Processing Standard 46: Data Encryption Standard (DES), National Bureau of Standards, Gaithersburg, MD, 1977.

[8] Federal Information Processing Standard 74: Guidelines for Implementing and Using the Data Encryption Standard, Nationa Bureau of Standards, Gaithersburg, MD, 1980.

[9] Federal Information Processing Standard 81: DES Modes of Operation, National Bureau of Standards, Gaithersburg, MD, 1980.

[10] Federal Information Processing Standard 113: Computer Data Authentication, National Bureau of Standards, Gaithersburg, MD, 1985.

[11] ISO 7498: Proposed Draft Addendum Number 2 - Security Architecture, ISO/ TC97/ SC21/ WG1, 1986.

[12] The Directory - Authentication Framework, ISO/CCITT Directory Convergence Document #3, ISO/ TC97/ SC21/ WG4, 1986.

[13] Transport Layer Protocol Definition for Providing Connection Oriented End-to-End Cryptographic Data Protection Using a 64-Bit Block Cipher, X3T1 Draft Document forwarded to IS TC97/ SC20/ WG3, 1986.

ISDN-TECHNOLOGY, NETWORKING CONCEPTS AND APPLICATIONS

P.J. Kuehn
Institute of Communications Switching and Data Technics,
University of Stuttgart, FRG

Abstract

ISDN - the Integrated Services Digital Network - evolves from the all-digital Telephone Network. Its advantages are the integration of voice, text, facsimile and data services within one network, the so-called narrow-band ISDN based on 64 kbps circuit switched channels. The usage of the existing subscriber lines which will be operated at 192 kbps, allows for 2 B (64 kbps) and 1 D (16 kbps) channels for each subscriber (Basic access). The B-channels can be used for any voice, data, text or facsimile connections simultaneously, whereas the D-channel allows for signalling at any time. Within the ISDN, interoffice signalling is also separated from the user information paths; thus, an end-to-end packet switched signalling network controls all connection management. Therefore, the narrow-band ISDN can easily be extended to a broad-band ISDN by extending the switched network by broadband switching and transmission facilities.

Most countries start their ISDN pilot and regular services during the next years. It is anticipated that ISDN will quickly grow and take over a number of non-voice services which are today operated on different networks. Particularly, it will be of highest interest, how packet switched services can be integrated into the ISDN. There are several options in the subscriber access area (D-channel, B-channels) and within the network itself (packet switched subnetwork, signalling network or switched B-channels).

This paper addresses the following aspects: Transition from analog to digital, service concepts, network concepts, technology, interworking and research.

1. Transition from Analog to Digital

The ISDN evolves from the all-digital telephone network. Digitalization is currently pursued by the introduction of digital transmission facilities within the trunk network and computer controlled switching exchanges in the local and toll network. Both, transmission and switching use digital time division multiplexing for full duplex 64 kbps channels which are interconnected in the circuit switched mode. The subscriber and subscriber loops are still analog and digitalization takes place within the peripheral units of the local exchanges, see Fig. 1.

The digitalization will now be extended to the terminals on the subscriber premises allowing for an end-to-end all-digital connectivity. The prerequisite of all-digital connectivity forms the basis for the integration of many services within one network which have been operated so far on individual networks. The future ISDN will therefore be characterized by a number of technical and economic aspects as:

- use of existing subscriber loop cuircuits
- universal interface between subscriber terminals and network for all ISDN-services
- multiple terminal configuration with interchangeable terminals
- unique subscriber number, independent of the service being used
- integration of voice, text, facsimile and data services
- simultaneous usage of several services
- change of services through a connection
- terminal adaptation for non-ISDN compatible equipment
- use of separate signalling channels for the subscriber access (D-channel) as well as for interoffice signalling (signalling network No.7)
- circuit switched channels with different transmission rates (64, 384, 1920, ... kbps) channels (B, H, ...) respectively
- packet switched connections for data and teleaction applications
- provisioning of higher services with storage and processing capabilities within the network
- provisioning of various supplementary attributes
- interworking between the ISDN and other existing networks.

In 1984, an initial set of new ISDN-recommendations has been issued by CCITT. National recommendations have been released based on the international ones upon which now the ISDN-equipment and networks are developed in many countries. ISDN services and operations will first be experienced in field trials before the final introduction will take place. A number of countries have announced that their regular ISDN network will be put in operation by 1988.

2. Service Concept

The notion of service is used in different meanings dependent on the context. Both meanings are important for ISDN.

2.1 Service Concept in the Sense of Layered Protocol Architecture

The communication between different application processes (users, terminals, end systems) is logically structured into functions with a layered architecture. The principles of this architecture are defined by the well-known Basic Reference Model for Open Systems Interconnection [1]. The layering concept partitions the system functions into subsets of manageable pieces which are ordered in a hierarchical manner as shown in Fig. 2 for the general case of a level N.

The (N+1)-Entity is the service user of the functionality (service) of the (N)-Entity. The (N)-Entity is a service provider for the (N+1)-Entity. The services of the (N)-level are controlled by a quadruple of (N)-Service Primitives REQUEST, INDICATION, RESPONSE and CONFIRMATION. The adjacent layer communication is executed by the exchange of (N)-Service Data Units, (N)-SDU.

The communication between peer (N)-Entities is governed by a set of well-defined procedures, the (N)-Protocol. The protocol uses a number of particular protocol elements which are encoded by (N)-Protocol Data Units, (N)-PDU which consist of (N)-User Data and the (N)-Protocol Control Information, (N)-PCI. The exchange of (N)-PDU's can be greatly enhanced by the establishment of one or more (N)-Connections between the (N)-Entities; connections are identifiable by (N)-Connection End Points and their corresponding addresses (identifiers). Connections may support the communication through functions as sequence control, flow control or error recovery.

Within each of the architectural layers particular problems have to be solved as addressing, connection management (establishment, multiplexing, splitting) data unit management (segmenting, blocking, concatenation) and control of the protocol function. The Basic Reference Model defines seven fundamental layers which can be further subdivided for particular services or network types:

```
7   A    Application
6   P    Presentation
5   S    Sessions
4   T    Transport
3   N    Network
2   DL   Data Link
1   Ph   Physical.
```

Layers 1-4 refer to the communication functions, whereas layers 5-7 belong to the processing and storing function of the end systems. The lower layer functions and protocols are basic for use of many application services; their standardization has reached a quite stable state. The higher layer functions (HFL) are much application-dependent and their standardization is still subject of international standardization bodies.

2.2   Service Concept in the Sense of Network Operation

A "Service" comprises all technical, operational and legal aspects for a particular type of communication between users or between users and the provider of a public network. Within the ISDN various types of services are distinguished:

a) Bearer Services

Bearer Services provide the circuit or packet switched transport of information between two terminal-network interfaces irrespective of the compatibility of the terminals. A typical example of these services are switched or non-switched 64 kbps (B-) channels for text, data and graphic applications.

b) Standard Services

Standard Services provide the transport of information between two terminals with assurance of compatibility. The functionality comprises all 7 layers. Typical representatives are:

- ISDN Telephone
- ISDN Teletex
- ISDN Telefax
- ISDN Textfax.

Again, these services are based on switched B-channels with 64 kbps.

c) Higher Services

Higher Services generally use centralized storage and processing capabilities of the ISDN. Typical examples of such services are

- ISDN Videotex
- ISDN Voice, Text and Fax Mail
- Protocol conversion.

d) Services on the D-channel

The D-channel basically carries signalling information (s-data). These data require only a small part of the available 16 kbps capacity so that low rate user packet data (p-data) may be transferred additionally over virtual connections or telemetric data (t-data) over permanent virtual connnections.

e) Supplementary Service Attributes

Additionally to the basic service attributes of the standard services the users may optionally subscribe for further service attributes as, e.g.,

- abbreviated dialling
- automatic repetition of calls in case of blocking
- inward dialling into PBXes
- automatic call back
- call redirection
- conferencing
- reverse charging, etc.

3. Network Concepts

The network concept bases on the existing network structure in the local networks, in particular on the use of subscriber loops and existing digital trunks. The computerized (stored program control, SPC) operation of the control and the digital transmission and switching facilities allow, however, for a much higher functionality of the ISDN.

3.1 Physical Structure and ISDN-Capabilities

Fig. 3 shows the basic structure of the public ISDN and its major capabilities. The main aspects are:

a) Multiple terminal configuration on the customer premises comprizing

- Terminal Equipment (TE 1) with ISDN compatibility for voice, text, etc.

- Terminal Equipment (TE 2) not compatible to ISDN with Terminal Adaptation (TA) for protocol conversion

- Connection of Local Area Networks (LAN) through Gateways (GY)

- Free connectivity through a local bus system and an ISDN-socket

b) Unified Network Access (Basic Access) through

- Usage of the existing subscriber loop

- Network Termination (NT) as the end point of the public network

- Exchange Termination (ET) as the interface between the subscriber access and the switching exchange

- Provisioning of 2 B-channels (information channels) with 64 kbps FDX each for circuit switched connections or (optionally) packet access

- Provisioning of one D-channel with 16 kbps FDX for signalling (s), user packet (p) and teleaction (t) data; s-, p-, and t-connection management is transparent to the NT and subject to the D-channel protocol between TE(TA,GY) and ET (levels 2 and 3).

- Extended Network Access through the introduction of multiplexors (Basic Access Multiplexor) and concentrators in the subscriber area; both are connected with the switching exchange through PCM 30/32 transmission facilities (not shown in Fig. 3).

c) Multiple ISDN Capabilities within the ISDN through

- Circuit Switched (CS) B-channels
- CS H-channels of higher bandwidths
  (H_0: 384 kbps, H_{11}: 1536 kbps, H_{12}: 1920 kbps,...)
- Non-Switched B and H-channels
- a uniform signalling network based on the common signalling channel concept of CCITT No.7
- Packet Switched (PS) facilities either by providing access to a separate PS-network or by integration of PS-services into the ISDN.

d) Provisioning of facilities for information storage, information processing and interworking with other networks through

- Data Bases for information (voice, data, text, ...) storage and retrieval
- Hosts for Information Processing (server functions as, e.g., protocol conversion)
- Gateways (GY) for interconnection of ISDN and non-ISDN networks.

Fig. 4 summarizes the major ISDN capabilities once more. Fig. 5 shows a more detailed structure of the subscriber access, including the various reference points as it is planned for the German ISDN.   Fig. 6 illustrates the subdivision of the ISDN network with respect to the CS-information network and the PS signalling network with associated, quasi-associated and non-associated signalling links, signalling points (SP) and signalling transfer points (STP).

3.2 Logical Structure

The logical structure of the ISDN is reflected by three major aspects,

- the information flow and communication contexts
- the protocol architecture
- the international standardization.

3.2.1 Information Flows and Communication Contexts

Networks supporting a large number of services are characterized by various flows of user and control information. For the modelling of the information flows specific Reference Points are considered as, e.g., Terminal Endpoints (TE), Network Terminations (NT), Exchange Terminations (ET) or Network Interworking Units (Gateways). Fig. 7 shows the general configuration as recommended by CCITT for the ISDN.

The user-network signalling context is fully covered by the D-channel protocols, whereas the network-internal signalling context is covered by the
signalling system No.7. A rather new aspect is added by the user-user signalling context which uses both the D-channel and the No.7 signalling func-
tionalities.

3.2.2 Protocol Architecture

The exchange of information between users, application programs or any particular level entities is controlled by a set of rules subjected to a protocol
definition. Based on the layered protocol architecture of the ISO/CCITT OSI
Basic Reference Model [1], CCITT has developed a generalized model for ISDN
protocols. Whereas for packetized communication control and user information
are combined in the respective PDU's, circuit switched communication with
separate signalling channels and networks needs a multidimensional approach
where different Protocol Planes are distinguished for User, Control and
System Management information.

The multiple plane protocol architecture is illustrated in Fig.8 for a
CS-connection through an ISDN with D-channel signalling for the network
access and No.7 common channel signalling network for network-internal
signalling.

The relevant standards of CCITT are defined in the I-series for ISDN, particularly I 431, I 441 and I 451 for levels 1, 2 and 3 of the D-channel protocol, and in the No.7 Signalling System. Both the D-channel and No.7 signalling protocols have been developed for their particular purposes, the
control of the network access and network-internal control. Therefore, they
differ considerably and an interworking is necessary at the origination and
destination exchanges. For the higher layers 4-7 this concatenation of the
signalling system is not visible.

3.2.3 International Standardization

The extension of network capabilities and services implies a detailed structuring of network functions and application processes. This is reflected by
the intensive activities of international standardization bodies ISO, CCITT,
ECMA, IEEE and IEC. Through these activities a set of services and protocols
have been standardized (or are in process of standardization) as a necessary
prerequisite for the development of networks and terminal equipment.

Standards may be classified into two classes: lower layers for communication-
oriented functions and higher layers for processing, and storage-oriented
functions. The lower layers 1-3 differ substantially due to the different
network and access properties; their corresponding standards are referred to
in Fig.9. Level 4 (Transport)is generally considered as a network-indepen-
dent layer where connections between end systems are provided irrespective
from the underlying special network type. For references to these protocols
see [5-11].

The higher layer protocols differ mainly because of the different application services. Fig. 10 gives a survey of their corresponding standards [12-18].

In the CCITT-Recommendations on ISDN the functional capabilities are subdivided according to Low Layer Capabilities which refer to the Communication Path covering levels 1-3 (Bearer Services) and Higher Layer Capabilities which refer to Information Processing and Storage covering levels 4-7 (Teleservices).

4. Technological Concepts

The introduction of the ISDN requires a completely new technology for the subscriber access and terminal equipment. Within the switching exchanges which are already digital and computer controlled, mainly peripheral control equipment has to be added. Two examples will be addressed as representatives of the technological concepts, Terminal Equipment and Switching Exchange.

4.1 Terminal Equipment

Fig. 11 shows the particular chip set developed for the German ISDN. The various components are:

CODEC   Analog/Digital Converter

ILC   ISDN Link Controller for the D-channel protocol, level 2

SIC   S-Interface Circuit for the level 1 of the S_0-interface

UIC   U-Interface Circuit for the level 1 of the U-interface.

The ILC, SIC and UIC circuits have been developed particularly for the ISDN in a full custom design comprising roughly 20 000, 12 000 and 110 000 transistor functions.

Fig. 12 shows the basic structure of an ISDN-Telephone set where the ILC and SIC modules are incorporated. Other Terminal Equipment, as for Teletex or Telefax services, differ in the residual part according to their particular service functions.

4.2 Switching Exchange

The modular design of todays switching exchanges allow an upgrading of the existing equipment with ISDN-modules. Fig. 13 shows an example of ITT's System 12, where new modules have been developed for ISDN-Terminals, Basic Access Multiplexers and No.7 Interoffice Signalling.

Similar concepts have been developed for PS-access. In case of PS-access through the D-channel, the p-data of the various D-channels are extracted and multiplexed on one common p-data channel which is connected with an ISDN-internal PS-module. For higher transmission rates switched B-channels can be used for the access to the PS-network, as well.

4.3 Broad Band ISDN

A second phase of ISDN will be started in the 90ies when broad band switching facilities are added. The current plans are to add optical fiber transmission and switching facilities in the subscriber loops, trunk circuits and switching exchanges, whereas control can be performed completely by using the existing narrow-band ISDN. At this time, no common agreement has been reached for the transmission rates for videophone service (140, 70, 34 or 2 Mbps). This depends largely on the success of information processing equipment for redundancy reduced encoding. Current activities direct even to solutions which could possibly be introduced within the narrow band ISDN using one or six B-channels corresponding to 64 kbps or 384 kbps for videophone services. The latter option requires multiple B-channel switching.

As an alternative to CS broad band switching fast packet switching (FPS) techniques will become interesting in the future. The switching equipment for such ultra-high capacities has not yet reached a satisfactory stage and will be subject to further research.

Broad band services will have another application field for fast data exchange in connection with document transfer. The interactive exchange of textfax documents requires transmission capacities beyond 1 Mbps.

5. Interworking of Networks

The success of ISDN during the introductory phase will much depend on the subscriber numbers for the various types of services. Thus, interworking of new ISDN- and existing non-ISDN-terminals has to be provided. One solution is the Terminal Adaptation at the subscriber side of the ISDN allowing for operation of non-ISDN terminal equipment on the ISDN. Another solution will be protocol conversion as a public service to interconnect ISDN-terminals within the ISDN with non-ISDN-terminals operating on existing CS or PS networks.

Since the new terminal equipment is more expensive, commercial users will dominate in the beginning, i.e. users who are connected to ISDN-PBXes. On the other hand, data-oriented equipment is already widespread and operated on Local Area Networks (LANs). The particular advantages of LANs for PS computer communication and ISDN-PBXes for CS point-to-point communication allows the conclusion that both types of inhouse networks will coexist for a long time. Fig. 14 shows the basic structure of an ISDN-PBX. The full use of both network-specific advantages requires an interworking between LANs and ISDN-PBXes through Gateways (GY).

The main Gateway functions are as follows:

- transformation of connection management within levels 2 and 3 (LLC1, LLC2 or LLC3 within the packet switched LAN, D-channel call set up within the circuit switched PBX)
- transformation of user data from packet mode to circuit mode
- intermediate buffering for speed matching and mode conversion

Fig.15 shows an architectural approach how LAN according to IEEE 802 can be operated on an ISDN with D-channel signalling and B-channel information transfer according to CCITT's I-series. The dual plane (user, control) of the ISDN is schematically sketched by parallel blocks of the protocol architecture.

6. Research Topics

The introduction phase of ISDN allows only very few features out of all possible features a mature ISDN will offer. There is a great number of open questions left for research, as for example:

a) PS-services within ISDN
   - D-channel or B-channel access
   - Separate PS network or integration of PS within ISDN
   - Fast packet switching

b) Broad band ISDN
   - Definition of standard bandwith for videophone
   - Video-conferencing with pre-reservations
   - CS or FPS solution for broad band services

c) Planning of multi-service networks
   - Traffic measurements and subscriber behavior
   - Traffic analysis methods
   - Development of planning tools
   - Cost optimization.

83

REFERENCES

|1| ISO 7498/CCITT Recommendation X.200: Reference Model of Open Systems Interconnection.

|2| Potter, R.M.: ISDN Protocol and Architecture Models.
Proc. 11th Int. Teletraffic Congress (ITC), Kyoto 1985.
Elsevier Science Publ. B.V. (North Holland), paper 1.3-4.

|3| Duc, N.Q. and Chew, E.K.: ISDN Protocol Architecture.
IEEE Comm. Magazine, Vol. 23 (1985) 3, pp. 15-22.

|4| ECMA TR/XX: OSI Sub-Network Interconnection Scenarios Permitted within the Framework of the ISO-OSI Reference Model. First Draft (1983).

|5| CCITT Recommendation X.21: Interface between Data Terminal Equipment (DTE) and Data Circuit Terminating Equipment (DCE) for Synchronous Operation on Public Data Networks. Yellow Book, Fasc. VIII.2, Geneva (1980).

|6| CCITT Recommendation X.25: Interface between DTE and DCE for Terminals Operating in the Packet Mode on Public Data Networks.
Yellow Book, Fasc. VIII.2, Geneva (1980).

|7| CCITT Recommendations on ISDN (I-Series).
I.100 Series: General ISDN Concept
I.200 Series: Service Aspects
I.300 Series: Network Aspects
I.400 Series: User-Network Interface Aspects
I.500 Series: Internetwork Interfaces
I.600 Series: Maintenance Principles.
Red Book, Geneva (1985).

|8| CCITT Recommendations on Signalling System No. 7.
Q.701 - Q.710: Message Transfer Part (MTP) and PABX Applications
Q.711 - Q.714: Signalling Connection Control Part (SCCP)
Q.721 - Q.725: Telephone User Part (TUP)
Q.761 - Q.766: ISDN User Part (ISDNUP)
Q.795: Operations and Maintenance Application Part (OMAP).
Red Book, Geneva (1985).

|9| IEEE Project 802. Local Area Network Standards, Rev. D (1982).
IEEE Standard 802.2: Logical Link Control
IEEE Standard 802.3: CSMA/CD Access Method and Physical Layer Specifications
IEEE Standard 802.4: Token-Passing Bus Access Method and Physical Layer Specifications
IEEE Standard 802.5: Token Ring Access Method and Physical Layer Specifications (1985).

|10| ISO DIS 8348: Network Service Definition.
ISO DP 8648: Internal Organisation of the Network Layer.

|11| ISO 8072: Transport Service Definition (1983).
ISO 8073: Connection Oriented Transport Protocol Specification (1983).
CCITT T.70: Network Independent Basic Transport Service for the Telematic Services (1984).

|12| CCITT T.62: Control Procedures for Teletex and Group 4 Facsimile Services (1984).

|13| CCITT T.73: Document Interchange Protocol for the Telematic Services (1984).

84

|14| CCITT Recommendations
    T.60: TTX Terminals
    T. 5: Facsimile Terminals
    T.72: Mixed Mode Terminals
    T.61: TTX Character Set
    T. 6: Facsimile Coding.

|15| ISO DIS 8326 and CCITT X.215: Connection-Mode Session Service Definition.
    ISO DIS 8327 and CCITT X.225: Connection-Mode Session Protocol
    Specification.

|16| ISO DP 8822: Connection-Oriented Presentation Service Definition.
    ISO DP 8823: Connection-Oriented Presentation Protocol Specification.

|17| ISO DP XXXX: Virtual Terminal Service and Protocol (VT).
    ISO DP 8571: File Transfer, Access and Management (FTAM).
    ISO DP 8831: Job Transfer and Manipulation Concepts and Services (JTM).

|18| CCITT X.400-Series Message Handling Systems (1984).
    X.411: Message Transfer Layer.
    X.420: Interpersonal Messaging User Agent Layer.

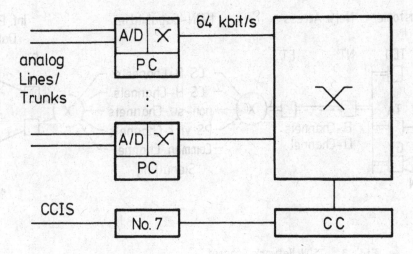

Fig. 1. Digital Switching Exchange (Non-ISDN)

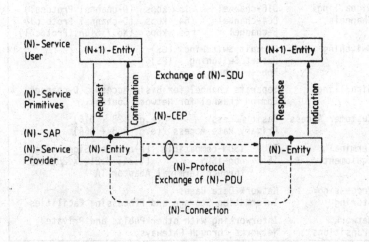

Fig. 2. Protocol Service Concept

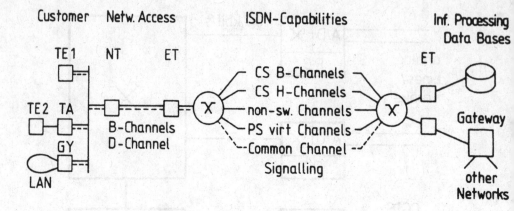

Fig. 3.   ISDN Network Concept

```
┌─────────────────────────────────────────────────────────────────────────┐
│ Information   B-channel      64  kbps  (CS or non-switched)               │
│ Channels      HO-channel    384  kbps  (CS or non-switched)               │
│               H11-channel  1536  kbps  (CS or non-switched)               │
│               H12-channel  1920  kbps  (CS or non-switched)               │
│               ----                                                        │
│                                                                           │
│ Signalling    D16-channel    16  kbps  (D-channel Protocol)               │
│ Channels      D64-channel    64  kbps  (D-channel Protocol)               │
│               E-channel      64  kbps  (No.7 Sign. Protocol)              │
│                                                                           │
│ Switching     Circuit Switching  (CS)                                     │
│               Packet Switching   (PS)                                     │
│               Non-switched                                                │
│                                                                           │
│ Signalling    Separate channel for Basic Access  (D-channel)             │
│               Common channel for Network Control                          │
│                                                                           │
│ Customer Access  Basic Access         (e.g.  2B + D16)                    │
│                  Primary Rate Access  (e.g. 30B + D64)                    │
│                                                                           │
│ Terminal      TE 1  ISDN-compatible  (D-channel Signalling)               │
│ Equipment     TE 2  non-ISDN         (V.24, X.21, X.25,...)               │
│                     through Terminal Adaptor TA                           │
│                                                                           │
│ Processing/   Network Data Bases                                          │
│ Storing       Information Storage and Processing Facilities               │
│                                                                           │
│ Network       Interworking with other Public and Private                  │
│ Transitions   Networks through Gateways                                    │
└─────────────────────────────────────────────────────────────────────────┘
```

Fig. 4.   ISDN Capabilities

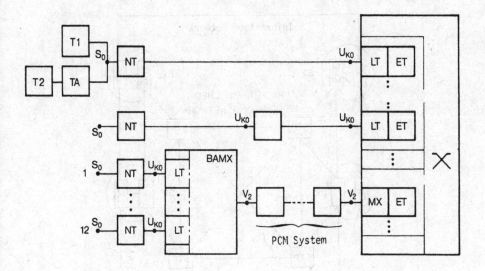

Fig. 5. Subscriber Access in the ISDN

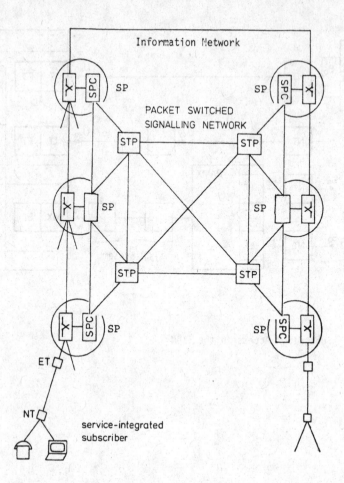

Fig. 6. ISDN Information Network and Signalling Network

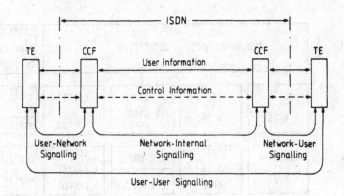

Fig. 7. Information Flows and Communication Contexts
      TE   Terminal Equipment
      CCF   Connection Control Functions

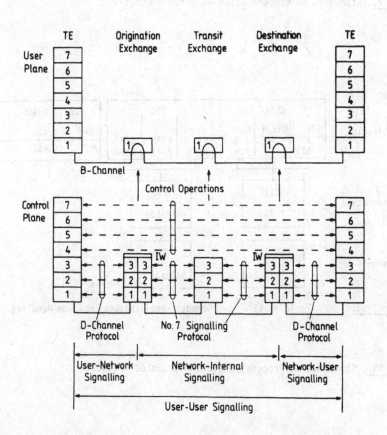

Fig. 8. ISDN Protocol Architecture for a CS Connection

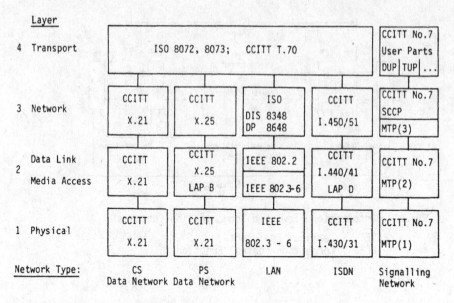

Fig. 9. Standards on Communication-Oriented Protocols

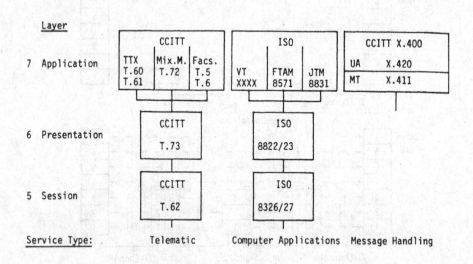

Fig. 10. Standards on Processing/Storing-Oriented Protocols

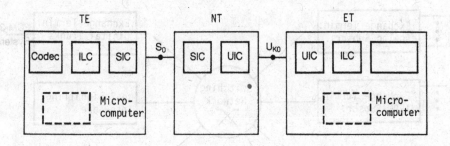

Fig. 11. Chip Set for ISDN Basic Access

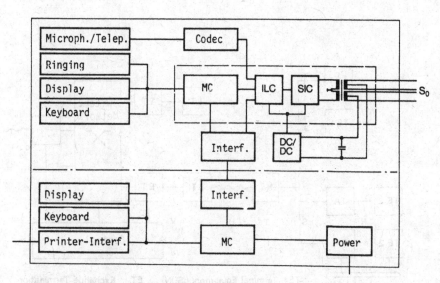

Fig. 12. ISDN Telephone Terminal

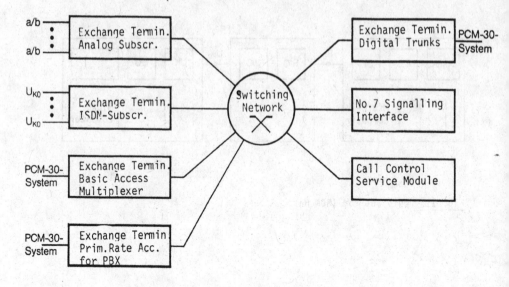

Fig. 13.   ISDN Switching Exchange (System 12)

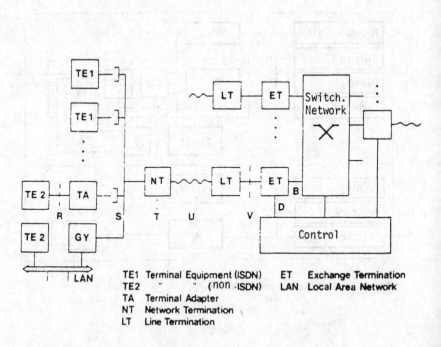

TE1   Terminal Equipment (ISDN)   ET   Exchange Termination
TE2   "   "   (non .ISDN)   LAN   Local Area Network
TA   Terminal Adapter
NT   Network Termination
LT   Line Termination

Fig. 14.   ISDN-PBX with Central Control

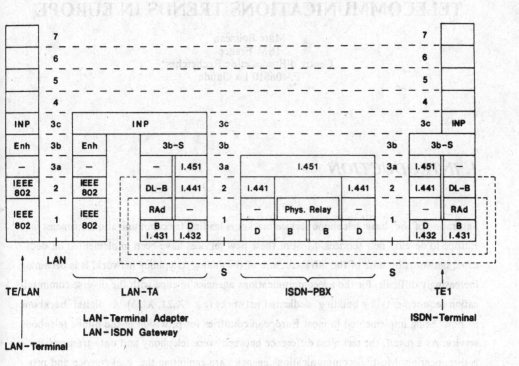

TELECOMMUNICATIONS TRENDS IN EUROPE

Fig. 15. Architecture for LAN-ISDN Interconnection

TELECOMMUNICATIONS TRENDS IN EUROPE

Marc Boisseau
IBM France
Centre d'Etudes et de Recherches
F - 06610 La Gaude

1. INTRODUCTION

Saturation of the basic telephone service market is leading telecommunications agencies in Europe to develop new services. Most of these new services have been implemented on dedicated networks because of the infrastructure of the analog telephone network. It is becoming increasingly difficult for the telecommunications agencies to cope with the diverse communication requirements by building dedicated networks (e.g. X.21, X.25). A digital backbone is now being implemented in most European countries for provision of the public telephone service. As a result, the technical difference between voice telephony and data transmission is disappearing. Most telecommunications agencies are exploiting the convergence and promoting the concept of Integrated Services Digital Networks -ISDN.

ISDN is generally viewed by the telecommunications agencies as an evolutionary development resulting from the increased use of digital technology within existing telephone networks, with the local loop expected to be the last link to be converted. As networks evolve to the stage where digital transmission and switching are employed internally, they are referred to as Integrated Digital Networks.

Such networks will then evolve into ISDN providing Transmission service for all types of information, including digitized voice, data, text and image, and allowing the development of a wide range of new services. Thus, ISDN is not a new and separate network but, rather, the result of increasing use of digital technology in the public telephone networks.

This paper will examine the evolution towards ISDN in two parts:

- Network Digitization: migration from analog to digital technology for the transport and routing functions.

- Service Integration: data processing and storage capacities within the network.

2. NETWORK DIGITIZATION

Digital technology has created an environment of dynamic change in telecommunications networks. The possibility of digitizing voice, which is and will remain the predominant share of the communication needs, offers the possibility for all types of information to be transported in a single universal network. Compared to analog techniques digital techniques represent less costly and more efficient methods of transmission and switching. Digitization will affect the three major elements of the telecommunications networks:

- Transmission

- Switching

- Network control

This first section will examine the progress of digital technology in these elements with special focus on two new transmission technologies: fiber optic and satellite.

Transmission

Early public voice channels could handle very limited data rates and were limited to 2400 baud without special conditioning. Today's digital networks are capable of handling voice and data over shared transmission facilities at rates over 1 million bits per second. Optical fiber transmission systems can carry information at even higher rates.

Digitized voice was introduced in the 1960s and first appeared in Time Division Multiplexing equipment that digitally multiplexed several voice signals on a single wire by sampling the voice analog signal at a rate of 8000 samples per second. Each sample is converted into an 8 bit binary number giving an information rate of 64 k bits per second (8 bits times 8000 = 64000 bits per second). Then several digitized voice channels are time multiplexed onto a single physical link. The process is reversed at the receiving end to convert the binary back to voice analog form.

The level of digitization is already high in most European transmission networks. However, there is still a need to digitize the link between the customer premises and the first exchange within the telecommunications network. ISDN will capitalize on the conventional pair of copper wires and provide direct digital transmission using Very Large Scale Integration (VLSI). Instead of a large number of interfaces (telephone network, telex, specialized data networks, leased lines,...), ISDN will provide a single interface to the network. The access capacity has been fixed and recommended for standardization by the CCITT at 2x64 kbit/s (called B channels) and 1x16 kbit/s (called D channel). The two B channels can be used transparently for voice and non-voice services. The D channel will be operated on a packet-oriented basis for either signalling or data transport.

With a user capacity totaling 144 kbit/s, the backbone network should not only become fully digital but capacity should also be improved. Two new transmission technologies will contribute to this evolution towards end-to-end digital transmission:

- Fiber optic

- Radio communications including satellite

The following sections will examine the main characteristics of these new technologies.

Fiber Optic

Fiber optic is transmission technology which has a wider field of application than just telecommunications. This section will be limited to its introduction in telecommunications networks. Fiber optic will provide very reliable and very high speed point to point links at reasonable costs. Optical fiber transmission is possible in all parts of a telecommunication network. However, the constraints for successful application differ widely in the different network planes. Four major network planes can be identified:

- The subscriber network (also called the local loop)

- The intra-city trunk network (also called the junction network)

- The inter-city trunk network (also called the trunk network)

- The inter-continental trunk network (e.g. submarine cables)

The widespread use of fiber optic for telecommunications subscriber network as well as for TV distribution, awaits further falls in the cost of launch and optical devices. Wideband subscriber access systems, providing such services as television, picturephone, and interactive data bases, have great commercial potential. However component cost must drop considerably below today's levels to make widespread use of these broadband services economically feasible. Current costs are well above $1000 per subscriber, preventing any rapid development at the moment. An investment of $300 per subscriber is considered the maximum which can be justified.

The use of fiber-optic cable in all the other network planes is primarily justified on maintenance cost savings together with increased capacity. Because the applications for fiber optics span a wide range of line lengths and channel capacities it is not likely that one type of light source or fiber will emerge to dominate the industry:

- typically 10 to 20 km and 480 voice channels per link for the intra-city network (34 Mbit/s in Europe)

- typically 10 to 300 km and 1920 voice channels per link for the inter-city network (140 Mbit/s in Europe)

- over 1000 km and 7680 voice channels per link for the intercontinantal network (560 Mbit/s)

A mix of short and long wavelength light emitting diodes (LED) and laser diodes, and multimode and monomode fibers will be used. Most of the current systems consist of single-channel point-to-point links, with multiplexing and switching functions performed by conventional electronic equipment. There will be a progressive migration towards the optical provision of these functions (integrated optics).

The development of monomode optical systems capable of exploiting bandwidths with the sophistication currently applied to electrical and radio signals will ensure an abundance of bandwidth and bit capacity at all levels of the network. Integrated optical devices will further provide for cost reduction. If bandwidth and bit capacity become substantially cheaper then the incentives for complex techniques to reduce the amount of information to be conveyed is removed (e.g. compression techniques).

If abundance materializes, the laws of supply and demand will expose the revenue of common carriers. This will have two major consequences on common carrier operations:

1. added incentive to obtain a larger proportion of their revenue from other telecommunications areas such as equipment supply and Value Added Networks (VAN).

2. obsolescence of the statutory monopoly, where existing, for the provision of transmission services.

Radio communications

Another major trend in transmission is the increasing part played by radio. Traditionally radio had been used for broadcasting (Radio and Television). The development of higher frequency techniques means that radio can be viewed as an alternative for the provision of short distance point to point link. It is slowly emerging as a competitor to wired techniques. For obvious reasons, it is first exploited in the growing market for personal and mobile communications (wireless telephone / cellular radio). Satellites is an other growing sector of radio communications. Satellites have two extremely attractive characteristics:

- They may cover from a geostationary orbit 100% of a given territory of up to several thousands of kilometers and any two points -within that territory - can communicate with each other.

- Information can be broadcast simultaneously from one source to a number of end users, or collected from several points to one receiving node.

Satellites will support four types of communications services:

- Trunk telephony

- TV programme exchanges (CATV Network feed and/or studio to studio)

- Specialized business services (high speed data, video-conference,...)

- Direct broadcasting of TV programme

The trunk telephony services require large earth stations with antenna diameters above 10 meters. These are expensive and to be cost effective must handle high traffic flows. For this reason they are each associated with international gateway exchanges and are relatively few in number (one or two per country).

The TV program exchange services (also called "TV relay") require smaller earth stations with antenna diameters below 10 meters. These are still expensive and to be cost effective must handle sufficient traffic flows. They will be used mostly to feed Cable Television networks with TV programs and will grow with the growth of these networks (20-40 per country).

The specialized business services require smaller earth stations with antenna diameters around 3 meters. These are less expensive but the traffic flows are totally unpredictable. They will be shared among a rather small number of business users. They are not planned to be installed at customer premises but rather used to feed existing local distribution networks. (under 100 per country). Specialized business services include several services such as leased line for telephony, high speed data transmission and video conference services.

Direct broadcasting services require only small dishes with diameters below 1 meter. These are not expensive and will be installed at customer premises. Such services will directly compete with Cable Television networks fed by the above mentioned TV program exchange services.

While trunk telephony traffic is easy to predict, demand for TV program and specialized business services is uncertain:

- For TV programs, there will be competition between Cable TV Networks fed by low power satellites (about 20 watts per transponder requiring large antennae), and direct TV broadcasting from high power satellites (over 200 watts per transponder requiring small antennae).

- For specialized business services, the nature, volume and user location are very uncertain. The most unpredictable portion of the traffic is in video-communication (mostly videoconferences). These services will consume bandwidth if they begin to sell in volume.

By the end of the decade, it is likely that European skies will be saturated with broadcast satellites resulting in an excess of transponders: at least 70 transponders (33 already committed to TV broadcasting among the 77 planned to be used). However, the incentive to establish regional communications is much lower in Europe than in North America, for both legislative and technical reasons.

Legal Constraints

The growth of national satellite communications in North America has been favored by the independent private carriers. As in any country, the operation of the long distance terrestrial network used to be more profitable than the local distribution and the profit from long lines was compensating for the loss from local plant. Business users were better served than "homes". Satellite communications are now able to compete on the profitable long distance services. As a result, the tariffs of terrestrial services have been restructured (more cost-oriented). Satellite communications in North America has grown as the by-pass technology for the terrestrial long distance network.

In most European countries, satellite communications will be operated by the same adminis-tration which operates the terrestrial network and in most cases, holds the telecommuni-cations monopoly. Competition is limited to the equipment providing the service and is not allowed for the provision of the service itself. Therefore, satellite communications will com-pete with terrestrial media e.g. coaxial and/or fiber optic cables, on a strict cost basis. They will be integrated in the national network where cost justified and not compete with it. They will be used to speed up the availability of ISDN as they can provide end-to-end digital con-nections irrespective of the end point location.

Technical Constraints

The growth of national satellite communications in North America has also been favored by the topology:

- Satellites bridge long distances and unaccessible areas at a cost independent from distance and geology.

- Where rapid development is required, satellites achieve a quicker area coverage.

Neither of these two justifications is usually found in Europe. Terrestrial networks are smaller in average length and include well developed advanced services (Teletex, Videotex, Telefax, ...).

The average break-even distance is presently over 500 km. The terrestrial link cost is de-creasing at 3% to 4% per annum. The likely decrease of satellite links could be up to 5% per annum. This would result in a reduced break-even distance -about 500 km within 10 years- still in excess of average national links in a typical European country .

Wired or wireless communications ?

Wireless networks are rather unexpensive. However, the increase use of radio communications generates two basic problems:

- a significant growth of interference problems. Within the radio domain, the problem is partly solved by plans for radio frequency allocations revisited on a timely basis, but problems of interference between radio and other devices are increasing.

- Security is a problem intrinsic with radio also encryption could be used.

Wired networks are costly to installed. Fiber Optics are free from electro-magnetic interference, and subject to suitable care impervious to water. A variety of infrastructures can be used for fiber optic networks such as canal beds, electricity pylons and railways.

Switching

The switching system is the basis of connectivity for all information communications. Technological convergence has created a situation where the major difference between a computer and a switching system is the architecture of how the electronics are applied to the function. Each has a job of moving bytes of information from place to place at a high speed while performing some data processing on the information as moved.

Early switching systems were controlled directly from dial pulses created by the telephone set or by wired logic relay control systems. Stored Program Control (SPC) systems were introduced into the network in the late 60's. Due to stringent reliability requirements (downtime not exceeding one hour in 20 years) early electronic switching systems based on SPC were specially designed: -dual configuration in either active stand-by or micro-synchronism mode. With the availability of reliable and powerful microprocessors, the design of distributed switching systems became viable. Computer controlled digital voice matrix switches were introduced into the network in the late 70's and virtually all new switching systems today are time-division digital switches fully compatible with digital transmission systems.

Switching systems are now designed with a computer-like architecture with sophisticated software carrying out control operations and network management.

The transportation of packet data in either B or D channels via the user line in ISDN is of particular importance for switching system design. Packet data transmitted in the D channel, like packet data on B channels, should be connected to a packet switching module of the switching system. At the other end, received packets should be distributed among appropriate channels (B or D). A new generation of switching systems combining circuit and packet switching will progressively emerge in both public and private domain.

Network Control

Microprocessors are changing not the way the way switching and transmission equipment is designed, but also network control. Network control is the glue that holds a telecommunication network. It will be the most important factor in the development of the service integration concept.

Traditionally, decadic and multifrequency compelled signalling were the only means to communicate between switching systems. The development of SPC systems initiated the development of advanced signalling systems known as Common Channel Signalling (CCS) systems. The last CCITT version -CCS 7- includes not only telephone signalling, but also functions required for ISDN signalling and non-circuit related communications such as operation and maintenance functions. It is mostly a packet network for signalling and where implemented, is already larger than dedicated packet data networks. Compared with the current signalling systems, CCS-7 offers a fast and virtually unlimited signalling capability including the possibility of signalling while the communication is established:

- call set-up in less than 2 seconds compared with an average of 10 seconds in to-day networks

- up to 1000 messages per second in each direction compared to 1 message per second with most current systems

It is now possible to decouple numbering and routing: in each node, the dialled digits used to determine the outgoing route. Now through the employment of a centralised computer data base together with the fast signalling system based on CCS-7, the relation between numbering and routing could become dynamic providing a more efficient use of network resources. It will also be possible to provide new services such as:

- transfer the caller's identity to the called subscriber

- centralized data base inquiry for call charging on customer's credit card

CCS-7 was initially designed for network control, but is extending to network management. Although distributed processing was applied successfully to switching, processing and interface functions, it was apparent that the operation and maintenance functions were best performed centrally. For these centralised functions, a general purpose computer is needed performing data storage and retrieval, file handling, simultaneous multiple access, and data link communications. CCS-7 messages will be used to connect network equipment and operation and maintenance functions implemented in centralised general purpose processors.

3. SERVICE INTEGRATION

Advanced technologies such as fiber optic and microprocessor erode the established carriers' advantage: alternative to their infrastructures can be developed if the regulatory situation permits. Common carriers should move from a hardware based strength to a knowledge based one. Common carriers have over the years developed operational and managerial skills in the planning, installation and operation of communications networks. Sub-contracting private communications to common carriers may become an alternative for companies with large private networks.

The cost of data processing and storage continues to fall: the number of devices on a VLSI chip doubles every two years. Data processing and storage can now be located anywhere:

- at customer premises

- within communications networks

The decisions as to where it will be located will have a significant effect on the success of the companies competing for the provision of data processing and storage, in other words the 'added-value'. If it is imposed for regulatory reasons to place a particular function at a given point in the system, then the company which controlled that segment of the market under granted regime will be at an initial commercial advantage.

The establishment of ISDN will offer a varied range of 'value-added' possibilities. For better understanding the notion of service integration, it is possible to structure it into:

- Transmission services

- Compatibility-oriented services

- Processing and storage services

Transmission services

Compared with the reference model for Open System Interconnection (OSI) set up by the the ISO (International Standards Organization) the telephone network only offers its users a network connection: layer 1 to 3 of the reference model. The access to ISDN provides a standardized connectivity tool common to several types of information (voice, text, data, ...) This group of services includes the circuit- and packet-switched transmission services operating at a bearer rate of 64 kbit/s.

In the case of circuit switched services, the standards cover the signalling protocols for the user-to-network signalling (e.g. the D channel protocols) and the layer 1 functions for the user-to-user information transfer (e.g. physical interface, bearer rate of the B channel).

In the case of packet switched services, the standards cover in addition to the previous case, the layer 2 and 3 protocols of the user information packet transfer.

Compatibility-oriented services

Some services for user-to-user communication based on standardized transmission services will include customer premises equipment, mostly the terminal devices such as telephone, text and facsimile services. With the advent of the "information economy" and the growth of information services, the compatibility between communications services appears extremely important. For controlling the communications processes, the higher layers of the OSI reference model (layer 4 to 7) have to be standardized. These services provide compatibility between all terminals of a specific service, e.g. character sets, document format, etc. Such services could be called compatibility-oriented services and are similar to telecommunications services offered today where transport and sometimes communications services are bundled with customer premises equipment.

For text-oriented services (e.g Teletex and Telefax), the bearer rate of 64 kbit/s will reduce the transmission time of an A4 page by a factor of 10, while the availability of two B channels at the user-to-network interface will provide a simultaneous voice and text capability. For image-oriented services (e.g Videotex), the bearer rate of 64 kbit/s will guarantee a faster image build-up with more complex modes and a faster input of videotex frames into the videotex centers by the information suppliers.

Processing and storage services

In the longer term, with its stored program control exchanges and its associated high signalling capability, an ISDN will form the basis for integrating processing and storage capability. The network providers will be in a position of offering data processing and storage services potentially bundled with transport services.

Voice mail, text mail, fax mail and multi-media mail are storage services for voice, text, facsimile and data with centralized shared storage and mailbox functions to permit individual information retrieval by the end user. Conversion and adaptation functions for interface, transmission and protocol conversion may be incorporated with these retrieval and mailbox services.

So as to increase substantially their revenues the European PTTs are trying to massively enter the services' market (transmission services, compatibility oriented services and data processing services) by putting as much data processing and storage as possible within the public networks.

Service Integration concerns and issues

In most European countries a legal monopoly/franchise has been granted for the provision of transmission services. Competition is prohibited in the provision of these services. Integrated services of ISDN cover a wider range of services and in the current legal environment, could be offered with tariffs which are not cost-based. This would create an unfair competition with data processing services outside the telecommunications networks. This is the case today for some compatibility-oriented services where the terminal device is bundled with the service, is provided at a non cost-based price.

A clear definition of the boundary between telecommunications networks and user provided customer premises equipment is necessary in order to establish the point where network service interface specifications are to be specified.

Transmission services are services where the information is delivered in real time to the adressee identical in form and content to what was received from the sender. These are also called basic services. Basic services transport (transmit) and route (switch) information from the sender to the addressee.

The services which only transmit and switch information can be provided by monopoly or regulated telecommunications service providers. Services which do more than just move information should be provided on a competitive basis by information processing or data base service providers and in some countries by "value added" service providers.

Requirements for Competitive safeguards

The economic employment and social impact of the progressive dissemination of value added network services are considerable. It opens up new economic activities and employment opportunities. The growth of this new industry segment requires competitive safeguards.

The basic-enhanced dichotomy permits those services considered to have natural or legal monopoly characteristics, scale economy and equality of access, to be provided under franchise with government control or regulation. Other services should be provided through the competitive marketplace.

Telecommunications agencies wishing to provide enhanced services on a competitive basis should be permitted to do so, but only if there are safeguards to protect users of their basic transmission services from cross-subsidizing such competitive ventures.

A clear line of demarcation of the monopoly services permits all equipment located on customer's premises to belong to the user and to be competitively provided, while information content transparency maximizes the capability of a user to select the transmission services and CPEs which best meet his needs.

Some references

Fiber optic networks or communication satellites - Alternatives?
M. H. Ross, Arthur D. Little, edited by Kathleen Landis Lancaster.

ISDN and Value-Added Services in Public and Private Networks
L.A. Gimpelson - Computer Networks and ISDN System 10-1985, pp. 147 to 156

ISDN Protocol and Architecture Models
R.M. Potter - Computer Networks and ISDN System 10-1985, pp. 157 to 166

Traffic Characteristics of Control Signals in an ISDN Switching System
H. Murata - Computer Networks and ISDN System 10-1985, pp. 203 to 210

SNA: RECENT ADVANCES AND ADDITIONAL REQUIREMENTS

R. J. Sundstrom
IBM Corporation
Research Triangle Park, North Carolina 27709

Introduction

SNA was introduced in September 1974. The initial implementation was limited to single-host, tree-structured networks and little higher-level function was provided. However, the architecture was designed with a layered structure to allow for orderly enhancement and evolution over time. This has proven to be a key factor that helped ensure the success of SNA, because all the layers have had significant extensions and some, for example the routing layer (path control), have had their initial formats and protocols completely replaced.

Since its introduction, SNA has been under continuous development. Enhancements have been driven by the need to satisfy customer requirements, to meet the communication needs of departmental processors and personal computers, and to use the capabilities of new communication interfaces and services such as X.25, satellites, and local area networks (LANs). These forces will continue to shape SNA developments.

Since 1974, SNA networks have grown in size and function and they have become critical to the operation of many enterprises. In fact, many businesses consider the design of their computer-communication networks to be proprietary information, since it provides them with a competitive advantage. As more capability is delivered, customer expectations rise. A good example of customer reliance and expectations is in the area of high availability, where we are seeing increasingly stringent requirements.

As we look toward the future and try to map the directions SNA will take in response to current requirements, it is useful to also look at the past. By reviewing the past we can appreciate the multi-dimensional aspects of SNA's evolution and develop a feeling for the pace of that evolution. Some of the aspects of SNA's evolution are configuration flexibility, support for larger networks, network management, incorporation of national and international standards, link-level connectivity options, the inclusion of peer-oriented protocols, high availability, and office functions. In the following sections, I will review the SNA activity in each of these dimensions and discuss current requirements and directions.

Further discussion on SNA directions and an extensive set of references can be found in the work of Sundstrom et al. (1).

Configuration Flexibility

SNA configuration flexibility has grown tremendously from 1974 to the present. SNA initially supported only a simple configuration consisting of a host computer with channel-attached communication controllers and non-switched links to the outboard distributed processors and terminals. Support for remote communication controllers and switched links for attaching the SNA peripheral nodes soon followed. In 1977 these tree-structured networks could be combined into multiple-host networks, and host-to-host communication was possible.

The next major advance came in 1980 when the capability for meshed networks with alternate routes became available. This release also introduced traffic priority (three levels for user traffic and a fourth for network-control traffic) and global flow control. The global flow control scheme is completely dynamic and self-regulating. The rate that traffic is allowed to enter the network from each source is determined by the congestion conditions along the route that traffic will take. If there is no congestion, the rate will automatically be increased to accommodate generated traffic. Various levels of congestion will cause the rate to decrease slowly, decrease quickly, or be stopped altogether until some of the traffic that has already entered the network reaches its destination and queues for intermediate links become shorter.

Most recently, in 1984 SNA Network Interconnection (SNI) was introduced. SNI allows for multiple, self-contained SNA networks to be connected for the exchange of user traffic. Many businesses have found it invaluable during corporate mergers. It is also useful for connections between companies, between divisions, and even between data centers within a company. It allows the management and configuration of the network to mirror the organization of the network owner or owners. Controls are built into the SNI protocols so that connected networks can be protected from each other. For example, flow control is provided so that one network cannot disrupt

another by sending it more data than it is prepared to accept. Other controls help
prevent unauthorized tampering with the internal network-control functions such as
link activation and deactivation or trace facilities. Since the address space used
by each network is independent, SNI has also been used to alleviate addressing
constraints, discussed later.

Today, SNA provides a high level of configuration flexibility: large,
meshed-connected networks can be created, and these networks can be connected
together through SNI. Nonetheless, there are further requirements in this area.
One requirement is to provide for ease of installation and reconfiguration; another
is to add new routing techniques to SNA that are optimized towards small systems,
allowing for flexible, easy-to-install, and easy-to-change networks of departmental
processors and personal computers.

When an SNA network is installed today, one of the first steps is network design:
how many hosts and communication controllers are needed, where they will be placed,
and how they will be connected with links. Then, the best routes through this
network are determined and a coordinated set of routing tables is developed.
Finally, this information is loaded into each of the routing nodes and the network
is brought up. This provides for very efficient routing, since there is little
computational and communication overhead during the operation of the network.
However, it makes reconfiguration of the network more difficult than we would like,
since the network has to be taken down to load a new set of coordinated routing
tables. The computation and storage of these routing tables also becomes more
difficult as networks grow in size. Elimination of this offline, definitional
process would help to move us to our goal of providing for continuous operations.

One technical solution we are exploring involves the dynamic exchange of current
topology and load information throughout the network. When a user at one location
asks for a session connection to another location, the network node serving that
user would have the information available to compute the best available route for
the class of service requested.

To meet the communication needs of users with departmental processors and personal
computers, we are investigating architectural solutions appropriate for networks of
small systems. One approach under investigation builds on the node type 2.1
protocols introduced in 1983 and uses the same type of topology-data exchange and
route-selection algorithm described earlier. The node type 2.1 protocols support
direct peer-to-peer communication between SNA nodes without going through a host or

a communication controller. Dynamic, distributed-directory techniques are also being studied. The combination of an automated dynamic directory and automated route-selection techniques could allow for networks of small systems to be installed or reconfigured without the need for coordinated system definition between the network nodes.

Of course, there is also a need for networks of small systems to connect to high-capacity transport networks provided by traditional SNA networks with hosts and communication controllers.

Larger Networks

The trend in SNA towards support for larger and larger networks parallels much of the configuration-flexibility history we have just reviewed. During the first decade of its history, SNA used 16-bit addresses. The address space was split into two portions: subarea and element. The subarea portion was used to route between the subarea nodes (hosts and communication controllers) of the network. The element address was used by the destination subarea node for local routing. The boundary between the subarea and element addresses was determined by the owner of the network and allowed for 1-8 bits of subarea address and 8-15 bits of element address. As networks grew in size during SNA's first 10 years, it became clear that 16 bits, no matter how they were split, were not enough. SNI provided some relief, since the individual networks connected by SNI could each use independent 16-bit address spaces. The constraint was attacked directly with extended network addressing (ENA), which was announced in 1984. ENA increased the size of the address space to 23 bits -- 8 bits for subarea addresses and 15 bits for element addresses.

While ENA is a recent enhancement, we view it as an interim step and recognize a need to further extend the SNA addressing space, particularly to provide for more than eight bits of subarea address. Since 1980, SNA headers have had room for 48 bits of addressing, so the foundation for this further extension is already in place. However, simply expanding the size of SNA addresses does not completely provide for support of very large networks. As mentioned earlier, route calculation and storage become increasingly difficult as networks grow. The automatic route-selection algorithms discussed earlier could help to solve this problem.

Network Management

The first areas of network management that SNA focused on were problem determination and operator control for multi-host environments. This early support included microprocessor-based modems that could diagnose link problems, provisions for sending the results of the link tests on SNA sessions to a host focal point and an application program (Network Problem Determination Application) that analyzed the data and helped interpret it for the network operator. Later support included another application program, the Network Logical Data Manager, that collected and analyzed information about logical resources such as sessions and routes. Both of these facilities are representative of SNA's approach to network management. They sample the relevant data in the network where it is most naturally obtained, use the facilities of SNA to move those individual pieces of information to a focal point that can assemble a critical mass of information, and analyze this information with the help of application programs.

This process has been refined recently so that an SNA node can perform its own local error analysis. If a problem exists that cannot be handled locally, the node can then send an alert message to the host focal point identifying the failing component and providing a description of the probable cause of the error. Another example of this distributed, yet hierarchical, philosophy is the network management support for the IBM Token-Ring Network. Today, it is provided by an IBM Personal Computer that collects and analyzes network management data for an attached ring. We see a future need to have the results of that local analysis forwarded to the host focal point.

Problem management is but one facet of network management; other disciplines are performance and accounting management, change management, and configuration management. The goal for SNA is to provide full end-to-end management of a network across all the network management disciplines, including telecommunication, non-SNA, and multiple-vendor equipment.

Link-Level Connectivity

Early link-level connectivity options in SNA included switched and non-switched SDLC links and System/370 channels. Over time, many other options have been added. Messages between SNA nodes can now be exchanged over X.25 virtual circuits, X.21 connections, high-speed satellite links using SDLC modulo 128, and token-ring LANs. Because of its layered structure, adding these new forms of communication has been natural and straightforward. Another new technology, ISDN, is on the horizon and is being analyzed for inclusion into SNA.

SNA link-level connectivity has been strongly driven by the rapid development of standards and the introduction of new common-carrier offerings. One trend that is evident is the emergence of high-speed "switched" services such as provided by X.25, ISDN, and LANs. These services will increase the use of direct peer-to-peer protocols such as node type 2.1 and logical unit 6.2 (discussed later). They will also have long-term effects on route selection techniques and foster the need for dynamic distributed directories so that workstations and other SNA nodes can, for example, plug into a local area network through a receptacle in an office wall and quickly join an SNA network without network operator intervention.

Standards

IBM has been an active participant in the development of international and national standards, and many of these have been embraced by SNA. SDLC conforms to the HDLC standard, the IBM Token-Ring Network is based on the IEEE standard for token-ring LANs, attachment to X.21 and X.25 services is supported, and SNA uses the DES national standard for its encryption algorithm. Recent, new SNA X.25 support allows SNA networks to be providers, as well as users, of X.25 communication services. A number of other activities have taken place that underline our support for the goals of OSI:

- IBM Europe announced OTSS (Open Systems Transport and Session Support), which implements OSI layers 4 and 5.

- We established the European Networking Center in Heidelberg, Germany, to perform research into the higher layers of OSI.

- We joined OSI Net, sponsored by the U.S. National Bureau of Standards, and we joined the Corporation for Open Systems (COS), a group of over 40 manufacturers who are interested in fostering full connectivity between systems from different vendors.

IBM recognizes the widespread interest on the part of users to interconnect networks and systems from different manufacturers. IBM favors such interconnections. To this end, we have published and will continue to publish extensive information about

SNA. We also actively support OSI. We see OSI and SNA as complementary, not competitive.

Peer Protocols

In the first release of SNA, the protocols were almost totally hierarchical. This was appropriate for a single-host network where the communication was device-to-host oriented. Over time, however, SNA protocols have become more peer oriented. This was a natural response to the changing nature of the network, with multiple peer hosts and increasing amounts of processor power and storage outboard of the System/370 hosts. Communication technology is playing a role, too, with facilities such as local area networks providing a natural environment for direct peer communications.

SNA today cannot be easily categorized as peer or hierarchical; it has aspects of both to provide the best form of function distribution for the task at hand. As discussed earlier, network management is hierarchical and naturally so. However, even here the trend is to distribute the function in the network and invoke the hierarchy only for reporting purposes and in cases where a problem cannot be contained locally.

The global flow-control algorithm introduced in 1980 is fully distributed between the routing nodes of the network. However, these routing nodes are in a two-level hierarchy with the peripheral nodes of the network. This hierarchy is important because it allows larger networks with less overhead to be built. If all the nodes in the network had to participate as peers in the route-selection process, the size of the topology data base and the change activity against that database would severely limit the size of networks. Directory services are likewise both distributed and hierarchical. Peripheral nodes get their directory services from host nodes, thus creating a service-requester/service-provider hierarchy. The hosts, on the other hand, cooperate as peers and collectively provide the directory service.

The SNA cryptography-key-distribution mechanism is another good example of this phenomenon. Having each node know the encryption key of numerous other communication partners creates security exposures and key-change coordination problems. To solve these problems, SNA provides a key-distribution facility. Each key is known in only two places in the network: the node owning the key and its key server. However, rather than having a single centralized key server, which could cause security problems, especially in an SNI environment, multiple key servers cooperate as peers to provide this function.

SNA also contains purely peer protocols. Node type 2.1, providing for direct peer communication between two adjacent SNA nodes is one example. Another is logical unit 6.2 (LU 6.2), also called Advanced Program-to-Program Communication (APPC), which provides facilities for symmetric peer communication between programs. SNA Distribution Services (SNADS), which provides the SNA non-interactive, or mail-like, transport mechanism, is also fully distributed.

High Availability

Reliability and availability have always been high-priority considerations in the development of SNA. In 1974, SDLC was selected for SNA's primary data link control for telecommunication links because of its superior error-detection and correction capabilities. (It also possessed superior cost and performance characteristics.) In addition, a pause-and-retry algorithm was devised that could provide continuous link-level connectivity across periods of transient errors. In 1980, further link-level availability could be obtained by bundling multiple links into a single logical link called a transmission group. A transmission group fails only when the last remaining operational link in the group fails. Alternate routes were also introduced in 1980, allowing a session to be restarted over a backup route in the event of a failure along the original route.

The need in some networks for backup applications is being addressed through the extended recovery facility (XRF), which will be available for Information Management System (IMS) applications this year. XRF uses a database that can be shared between two hosts and a "heartbeat" protocol that allows the backup IMS system to monitor the operational status of the primary system.

Another requirement that some of our customers have is non-disruptive switching of routes. SNA today provides for backup routes, but should a problem develop along a route, the sessions using that route are deactivated prior to reestablishing them over an alternative route.

To achieve our goal of continuous operation, we have to continue to incorporate fault-tolerant protocols such as those discussed above. These protocols reduce or prevent outages resulting from the failure of network components. In addition, we also have to reduce the scheduled down time required for maintenance of the network. An example of this is the need today to stop the operation of a network to load a new set of routing tables. One of our long-term objectives is to reduce and, wherever possible, completely eliminate system definition. Some of the protocols currently under investigation, such as for automatic route selection, address this requirement. As extensions to SNA are developed, system-definition reduction is one of the key design considerations. Node type 2.1, for example, introduced new node-to-adjacent-node parameter exchange and negotiation protocols. LU 6.2 also included new end-to-end parameter exchange and negotiation protocols with its session activation command and response.

Office Architectures

To meet the needs of the automated office, IBM has developed Document Interchange Architecture and Document Content Architecture. These architectures are at the highest layer, the transaction services layer, of SNA. As such, they reside on and can take advantage of the SNA services and capabilities described earlier. We expect SNADS, LU 6.2, node type 2.1, and token-ring networks to be particularly important in the office environment.

Document Interchange Architecture (DIA) provides a set of protocols that define how SNA nodes cooperate to provide common office services. These include filing, searching, and retrieving of documents as part of DIA's document library services. Application-processing services are defined for formatting and processing documents. DIA also provides document-distribution services for the sending and receiving of mail, either through SNADS or directly to another user through an underlying LU 6.2 session.

Document Content Architecture provides the formats for describing the form and meaning of office documents. Two forms are currently defined: final-form text and revisable-form text. Final-form text allows a sender to control the presentation format of a document at the destination. If further editing or revision is desired by the receiver, revisable-form text can be used. With this form, text-processing indicators are included in the transmission and may themselves be revised.

While these two forms provide a good beginning, other forms are needed to define documents that contain mixtures of text, graphics, voice annotation, and image data.

Summary

SNA has evolved continuously since its introduction in 1974. It now provides rich levels of function in all of its layers. Numerous SNA networks are installed. From these networks we gain valuable information on the actual operation of SNA networks in a wide variety of environments and receive a continuous stream of additional requirements.

SNA has proven to be a flexible architecture by its capability to change and incorporate new solutions to meet the growing needs of our customers. It has also been a versatile vehicle for responding to requirements dictated by the emergence of desk-top computing, the development of new communication technologies, and the unfolding of national and international standards.

SNA has been an open architecture; detailed specifications have been published regularly, beginning as early as 1976. SNA implementations are available from numerous vendors. Many of these products support SNA's newest enhancements such as LU 6.2, SNADS, DIA, and Document Content Architecture, which are all described in available publications as part of IBM's support of attachment of non-IBM products to IBM SNA networks.

SNA is, and will remain, the foundation of IBM's networking direction: the basis for responding to the forces that shape the advanced computer-communication solutions we continue to deliver to our customers.

Reference

1. Sundstrom, R. J., J. B. Staton, G. D. Schultz, M. L. Hess, G. A. Deaton, L. J. Cole, and R. M. Amy, "SNA Directions -- a 1985 perspective." AFIPS Conference Proceedings, Vol. 54, 1985 National Computer Conference, Chicago, pp. 591-603.

Integration of Group Communication into

CCITT X.400 Message Handling Systems

Wilhelm F. Racke, Thomas E. Schütt
IBM Germany
European Networking Center
Tiergartenstrasse 15
D-6900 Heidelberg, Germany
Tel. (+49) 6221-404-218

Abstract

This paper discusses the basic concepts of the X.400 specification for an electronic Message Handling System with respect to the integration of group communication processes. Three group communication scenarios are distinguished: distribution-oriented, collection-oriented and coordination-oriented communication. Possible solutions for the integration of the distribution-oriented features into a X.400 based MHS are described and some light is shead onto the problems involved with the more complex collection-oriented and coordination-oriented communication.

1 Introduction

The current version of the CCITT X.400 recommendations for Message Handling Systems (MHS) /1/ deals mainly with electronic mailing of individuals, i.e. an individual user sends a message to one or more other individual users by explicitly specifying all the recipients of the message. Other features being provided by already existing message systems have not been standardized so far.

One aspect of particularly great importance, as proven by the experience with existing electronic mail and conferencing systems /2/, is not covered by the X.400 MHS recommendations: facilities for the support of the communication needs within groups of people.

In this paper we will give a short survey of the functional model and the protocols of a X.400 MHS. We then describe the requirements for group communication and introduce three classes of features for the support of electronic cooperation within a group of people: distribution-oriented, collection-oriented and coordination-oriented features. We will concentrate on the distribution-oriented features which are under discussion now in standardization bodies. We show how different group constellations can be modelled with the support of a directory service and discuss three possible solutions for the integration of distribution-oriented features into X.400 Message Handling Systems. The paper will close with an outlook on research activities to support the more advanced collection- and coordination-oriented features of group communication.

2 Survey of the X.400 Recommendations

A X.400 MHS comprises different functional components that work together to provide the message handling services. To describe the services, the components and their interworking, a MHS Model was introduced in the X.400 recommendations. A functional view of the MHS Model is shown in Figure 1.

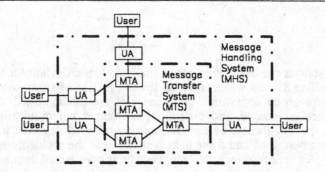

Figure 1: Functional View of the MHS Model

The functional components of a X.400 MHS are User Agents (UAs) and Message Transfer Agents (MTAs), with the collection of MTAs forming the Message Transfer System (MTS). Users of the MHS (people or application programs) request the provided services through a User Agent which interacts with the Message Transfer System. The MTS provides the general, application-independent, store-and-forward message transfer service. Several MTAs can cooperate to transfer a message from the MTA serving the UA of the message originator to the MTA(s) serving the UA(s) of the message recipient(s). To transfer a message from the originating UA to the recipient UA(s) several MTA-UA-interactions and MTA-MTA-interactions are necessary:

- During the submission interaction the originating UA transfers the content of the message plus a submission envelope to the MTA. The submission envelope contains the information the MTS requires to provide the requested services.

- During the delivery interaction the MTA transfers to a recipient UA the content of the message plus a delivery envelope. The delivery envelope contains information related to the transfer and the delivery of the message.

- During the relaying interaction the contents of the message plus a relaying envelope are transferred from one MTA to another. The relaying envelope contains information related to the operation of the MTS.

In the submission and delivery interactions, responsiblity for the message is passed between the MTS and the UA. MTAs transfer messages containing any type of binary coded information, the three types of envelopes reflect the history of the message transfer.

UAs are grouped into classes based on the type of content they can handle. The Interpersonal Messaging System comprises the MTS and a specific class of cooperating UAs, the Interpersonal Messaging UAs. The services provided by these UAs are described in the X.400 MHS recommendations. The users of these UAs are typically people.

According to the ISO Reference Model for Open Systems Interconnection all entities and protocols of the X.400 MHS are located in the Application Layer. The Message Handling functions can be considered to be divided into two sublayers of the Application Layer:

- The User Agent Layer (UAL) contains the UA functionality associated with the contents of messages.

- The Message Transfer Layer (MTL) contains the MTA functionality and provides the general, application-independent message transfer.

The layered description of the MHS is illustrated in Figure 2.

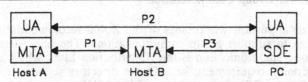

Figure 2: Layered Model and Protocols of the MHS
(without TTX,TTXAU and P5)

The protocol used between MTAs for relaying messages is called P1, and is independent of the protocol used between UAs. The protocols used by the UAs are application-dependent. At this time the only UA protocol defined is the Interpersonal Messaging protocol P2. Other protocols could be defined for new applications, such as electronic funds transfer or library services. Figure 2 shows also the possible physical realizations of the MHS Model. There can be systems with UA- and MTA-functionality (Host A) and systems with MTA functionality alone (Host B). On a PC there can be a standalone UA. The message transfer services for these UAs are provided by Submission and Delivery Entities (SDEs) in the MTL which communicate with the responsible MTA using the remote operations protocol P3.

For the organizational structure of the X.400 MHS it is possible to combine one or more MTAs with a number of UAs to Administration Management Domains (ADMDs) or Private Management Domains (PRMDs). ADMDs are provided as public services by administrations like PTTs while PRMDs are operated by a private company or institution.

A name representing a user of the MHS is assigned to the UA of that user. This name is called Originator/Recipient name (O/R name). An O/R name has to contain values for all attributes of a standard attribute set to determine the domain of a user. Two forms of O/R names are supported in the current version of the X.400 recommendations, the architectural and the terminal-oriented form.

For the architectural form the following standard list of attributes is used to form an O/R name:
Country Name
ADMD Name
< PRMD Name >
< Organization Name >
< Organizational Units >
< Personal Name >
< Domain-Defined Attributes >
At least one of the optional attributes must be selected.

The terminal-oriented form of the O/R name is used to enable the exchange of messages between a X.400 MHS and Teletex terminals. The O/R name consists of the X.121-Address and a Telematic Terminal Identifier.

3 Requirements for Group Communication

One of the most important weakpoints of the X.400 recommendations is the absence of services to support group communication. The only element currently referring to group communication is the "Distribution List", which is only formulated as a functional requirement for a future directory service. But at this time all aspects of the use of directory services within MHS are for further study.

Practical experience with existing computer based message systems (CBMSs) clearly indicate the significance of group communication for users of a CBMS /2/. Group communication is supported in various forms in different existing CBMSs, but mainly as a means of distributing messages to more than one recipient. Other forms of cooperation within a group are not supported. Common names for these facilities are "Computer Conferencing", "Bulletin Boards", "Mailing Lists", "Distribution Lists".

One of the forerunners to integrate group communication into an X.400 MHS is the German Research Network (Deutsches Forschungsnetz, DFN) who has developed a complete model for its message system. This model incorporates the X.400 recommendations, group communication facilities and supporting directory services /3/. Essentially its concept for group communication is based on services already provided by existing local CBMSs.

With **Group Communication** we mean a much wider set of services to support more aspects of electronic cooperation within a group of people. This includes not only **distribution-oriented** features like simple distribution lists, open or restricted discussion groups on a specific topic, closed conferences, electronic journals etc., but also other new forms of electronic cooperation.

Our experience with existing CBMSs indicates that working groups also make heavy use of **collection-oriented** facilities. Typical examples include the travel agencies that do currently collect customer bookings decentrally and transmit them to a central reservations computer via videotex. This central computer will eventually produce some output messages like occupation lists for the hotels, invoices for the travel agencies etc. Another application is computer integrated manufactoring where some orders from different car manufacturers could be collected and processed by a producer of tires.

Group communication services should support even more complex office and manufactoring communication processes like coordinating production and sales facilities, managing appointments, voting etc. We call these features **coordination-oriented**. Examples include again the area of computer integrated manufactoring, office automation etc. These more advanced applications of message systems, however, are currently not actively considered within CCITT.

Figure 3 shows the potential communication constellations.

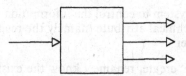

Figure 3a: Distribution-oriented Communication

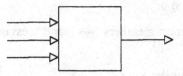

Figure 3b: Collection-oriented Communication

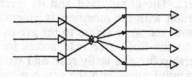

Figure 3c: Coordination-oriented Communication

Figure 3: Communication in Groups

4 Modelling of Group Constellations

Focussing our attention on the distribution-oriented communication, several group attributes must be considered:

Firstly a group may have an internal structure. There may be subgroups, e.g. a standing committee within a working party or even loops, e.g. the ISO being a member of the CCITT being a member of the ISO... Technically this can be modelled by assigning an attribute individual/group to each member of the group.

Secondly a group may or may not be visible. I.e. the existence of a group and/or its membership list can be kept secret or made publicly available. The IRA for example would almost certainly prefer the first alternative with its existence but not their members being publicly known whereas the memberlist of the British JANET should be published to enable scientific cooperation.

Thirdly a group can be open or closed i.e. allow or disallow for input from / output to non members. Typical examples of closed groups include a company corresponding with its salesforce or the attendees of the shareholders annual general assembly. Typical open groups are the readers of a certain newspaper or international standardization bodies who are willing to accept comments or error reports on their standards from non members.

To enable an electronic system to control the information flow within a group there is essentially only one technical attribute (namely the read / write access right) that can be attributed to either

- the group name (-- > create, rename / know the existence of a specific group) or

- the group membership list (-- > accept new members / know the names of the members of the group) or

- communicate with group members (-- > send messages to / read messages circulating within the group).

Let us consider two examples:

JANET: All members are individuals (universities or government sponsored research institutes). The JANET Board of Directors formally set up the group and may accept new members. The group itself and its membership list are publicly accessable. Communication with the members is generally possible. This constellation can be modelled with the write access to the group name and the membership list being restricted to the BoD, free read access for everybody to the group name and group membership list and lastly read and write access granted to everybody for individual communication with the group members.

Salesforce: If a company wants to have a one way communication path to its salesforce to ease the distribution of price lists, competitor information etc, this could be modelled by the group name and membership list access right (read and write) be restricted to the manager of the salesforce, the right to send messages to the members of the group only to be granted to some members of the company and only the read access for messages to the group granted to the salesforce itself. Furthermore the group itself may be substructured into subgroups according to divisional or geographical considerations.

Similarly one could model the IRA, the ISO and the other examples introduced above.

Once we established a suitable directory service to administer these groups and their four attributes, we can go ahead and investigate ways to integrate the group communication processes into an X.400 based message handling system.

Some details of the current CCITT plans and the research network approaches for directory services are given by B. Butscher /4/. L. Wosnitza specifies group management and modelling of group constellations within the framework of an available directory service for the DFN /5/.

5 Distribution-oriented Features in a X.400 MHS

The only element currently referring to group communication is the "Distribution List", which is only formulated as a functional requirement for a future directory service (Rec. X.400, Chap.3.5). But at this time all aspects of the use of directory services within the MHS are for further study. A distribution list is a means for sending messages to a group of people simply by using the O/R name of the group. Some entity will expand the distribution list to the set of recipients of the message.

Three main approaches to integrate the distribution-oriented features into a X.400 MHS are under discussion and shall be introduced in the following paragraphs. All three rely on the availability of a suitable directory service. In the first solution the originating UA performs the expansion of the distribution list whereas in the two other solutions the expansion is done within the MTS on the P1 level or by a special UA, the Group Distribution Agent, on the P2 level respectively.

5.1 Expansion by the Originating UA

This is the most obvious way to integrate distribution lists in a X.400 MHS: the UA of the originator accesses the directory service to check if the sender is allowed to send messages to the group and to get the O/R names of the group members which are allowed to receive group messages. The UA then fully expands the distribution list and sends the message to all recipients, either by using the multi-destination service of the MTS or by sending a separate copy of the message to each recipient.

There are several advantages with this approach. It uses the already existing message handling services in a very straightforward manner. None of the X.400 protocols has to be modified. It is possible to request delivery or receipt reports from all the recipients of the message directly by the originating UA. The originator of the message has complete control about the recipients and the costs for the transfer. In the case of nested groups it is easy to detect and handle loops within groups.

But there are also some disadvantages with this solution. The main disadvantage is that the expansion by the originating UA might generate a lot of overhead in the case of long lists or in the case of nested lists where it is more appropriate to expand the distribution list step by step. The possibility of blind distribution lists where the originator is not allowed to have access to the member list of the group cannot be implemented. Another disadvantage is that the name of the sending group can be lost and the recipients have no chance to see the connection between the message and the group if this name is not explicitly included in the message, e.g. as a comment.

5.2 Expansion by the Message Transfer System

The expansion of the member list and the distribution of the messages to the receiving group members is fully integrated into the MTS and is therefore done totally on the P1 level. All the functions are performed by one or several MTAs in cooperation with the directory service.

One advantage of this approach is that it can be particularly efficient, because the MTS is able to optimize the number of duplicates and the routes of the messages. For example, if a message is sent by a European sender to a nested list with an American sublist, it is easy to send only one copy of the message to America, with the O/R name of the American sublist as recipient. Another minor advantage is that only the P1 UMPDU envelope will be affected by the expansion process and all group members will receive identical copies of the message.

However there are serious technical, ethical, legal and financial objections against this approach:

The main technical problems of this solution are to detect message looping in case of nested distribution lists and to keep relevant transfer information for the recipients within the message, e.g. the name of the group the message comes from, history of the expansion process in the case of nested lists if the expansion is done step by step in the MTS. Both protocols of the MTL, P1 and P3, currently provide no means to solve these problems. Protocol modifications are required to allow for the transfer of the relevant information within the messages.

Another area of importance which is not solved yet is the handling of confirmations. An originator of a message to a group will get a confirmation that the distribution list was expanded by the MTS but will not get reports from the receiving group members. Research work has still to be done to handle the problems with confirmations.

The main objection against this approach however is of an ethical and legal nature: If the MTS is to support the distribution process by doing the expansion of distribution lists itself, it has to have access to the supporting directory systems and must be entitled to obtain the respective lists. It would be in a position to construct personal communication pathes, membership in professional societies, churches, parties, interest groups etc. As the lust for data of government controlled agencies is well known and wide spread throughout Europe, the MTS (i.e. government controlled PTT) approach would be a severe set back for personal privacy and should be discouraged in time.

Another comparatively minor but still serious objection is the unability of the originator of a message to control the distribution process and the resulting costs.

5.3 Expansion by a Group Distribution Agent

The third approach that combines the advantages of solutions 5.1 and 5.2 without the unacceptable disadvantages of the allmighty government is the introduction of a special UA, the Group Distribution Agent (GDA). Such a group distribution agent accepts incoming messages that are bound for a specific group, does the required (most probably local) directory lookups and forwards the accepted message on the P2 level to the members of the respective group which are allowed to receive messages.

A suitable solution is the GDA forwarding the message to the group members as "forwarded IP-message" with the original message contained in the body of the forwarded message. The expansion of messages for nested groups then results in nested forwarded IP-messages. The GDA is therefore able to detect loops by checking all the nested IPM-headers. This solution also keeps relevant transfer information for the recipients within the message. Any modifications to the IPM protocol P2 are not required.

Using the IPM autoforwarding facility, however, is not suitable for the forwarding of group messages because a message is not allowed to be autoforwarded more than once (Rec.X.420, Chap.4.2.2.4). Therefore it would not be possible to have nested groups which is generally not acceptable.

As in the transatlantic example of the MTS approach the distribution can be efficiently implemented with one GDA being located in the US to accept the message on behalf of the American sublist. Only one short copy of the message has to cross the Atlantic instead of the tremendous overhead that is incurred if the expansion were done locally by the originator and a big multi-destination message be created and sent across.

Also membership privacy and costs are completely under control of the respective group distribution user agent.

The handling of confirmations causes problems as in the MTS expansion and is not yet solved. For example, the GDA cannot prevent that a sender who is not allowed to send messages to the group gets a delivery notification from the MTS after the delivery of the message to the GDA. The sender then knows that the O/R name exists despite the fact that he has no access to the group. This may be a security violation which is a general problem and does not only apply to GDAs.

The three possible approaches to integrate distribution-oriented features of group communication into a X.400 based MHS are shown in figure 4. It is assumed that all required directory services are provided by a special directory service agent (DSA) as specified in the CCITT draft recommendations /6/.

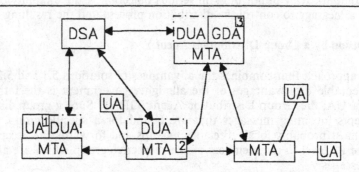

Figure 4: Expansion of Distribution Lists

6 Summary and Future Work

The three possible solutions show that the integration of the distribution-oriented features for group communication into a X.400 MHS is feasible. However, it is dependent on the provision of a suitable directory service. The current CCITT draft recommendations for Directory Systems /6/ do not fully cover the above mentioned requirements (see chapter 4).

The expansion by a GDA can readily be implemented because no modifications of the protocols P1 or P2 are required. The group definition can be contained in a directory system local to the GDA, so that no special directory system protocols are needed.

An expansion by the originating UA is not realizable by standardized protocols so far, because the necessary access to remotely stored public distribution lists is not specified yet.

For the MTS expansion the situation is even worse because modifications and enhancements to the protocol P1 are needed in addition to the required remote directory access.

The more advanced collection- and cooperation-oriented communication systems are still in an early research state. There exist only some prototypes for many promising applications in the office and production environment.

For instance could external newspaper correspondents transmit their articles to a central printing computer via X.400 instead of telex. This could easily be realized as a collection oriented closed user group of a MHS.

Directory System
Requirements and the CCITT-Model

BERTHOLD BUTSCHER, MICHAEL TSCHICHHOLZ

HAHN-MEITNER-INSTITUT BERLIN
BEREICH D/M

Based on experiences with Directory Systems in distributed systems, the paper starts with a summary of requirements to a Distributed Directory Service. The second part describes the actual work of CCITT on Directory Service and gives an outlook to the CCITT/ISO convergence of Directory Systems. Finally an estimation of the CCITT concept and some suggestions for additional features will be given.

1 Introduction

Directory Services and Name Servers have been discussed and implemented for a number of distributed systems or networks. An example of such a national network is the 'Deutsches Forschungsnetz - DFN / German Research Network'. It is the aim of the DFN to create a communication infrastructure for all research institutions. The services offered and planned within this field will contain basic services like dialogue (remote login), file transfer, remote job entry as well as distributed services like message handling or distributed document processing. These services are based on present or oriented on future standards of CCITT or ISO.

There is a common consent today that a satisfying use of any communication service highly depends on the availability of information about the communication partners. It seems useful to provide these informations within the frame of a DFN-wide Directory System in a *unified* and *transapplicational* fashion.

Such an application independent Directory System has to be able to integrate different directories into one. This integration refers to different dimensions:

- integration of the information (directory) on communication entities (e.g. for exchange) on providers (for administration) and users (for retrieval)

- integration of information (directory) of different organizational domains

- integration of information (directory) of different communication services

This integration has to be carried out in a way that the different services or domain administrations can execute their affairs independently. A Directory System in the DFN therefore has to be able to handle different competences. These competences must be modeled in an according organization of the distributed Directory and supported by differentiated access control.

The above mentioned CIM applications could make use of a P1 based MTS but would probably use specialized user agent layer protocols defined by ECMA or ISO instead of the CCITT defined P2 protocol for Inter Personal Messaging.

Restricted to office procedures some work has been done in the COST 11 ter sponsored AMIGO project /7/ that analyzed the environment and describes a functional model for group communication. More general research into these advanced group communication applications has recently started at IBM's European Networking Center.

References

/1/ CCITT Recommendations on Message Handling Systems
     X.400, X.401, X.408, X.409, X.410, X.411, X.420, X.430
     Malaga - Torremolinos, 1984

/2/ Jacob Palme
     Distribution Agents (Mailing Lists) in Message Handling Systems
     IFIP 2nd International Symposium on Computer Message Systems,
     Washington D.C., September 5-7,1985, 109-123

/3/ DFN - Verein, Central Project Coordination (ed.)
     The DFN Message Handling System - Specifications for Realization
     Berlin, June 1985

/4/ Berthold Butscher
     Directory Services - CCITT Plans and Research Network Approaches
     IBM Europe Institute, Oberlech, Austria, August 1986

/5/ Lothar Wosnitza
     Group Communication in the MHS Context
     IFIP 2nd International Symposium on Computer Message Systems,
     Washington D.C., September 5-7,1985, 139-146

/6/ CCITT Draft Recommendations on Directory Systems
     X.ds1, X.ds2, X.ds3, X.ds4, X.ds5, X.ds6, X.ds7
     Version 3, Reston, January 1986

/7/ Jacob Palme, Rolf Speth
     Group Communication in Message Systems - The Amigo Project
     paper to be presented at ICCC '86, Munich, September 15-18,1986

The specific need of a 'Distributed Directory System' (DDS) for the DFN-Message Handling System on one hand and the support and management of further DFN-Services on the other hand led to a DFN-project on Distributed Directory Systems (VERDI). This project has been carried out in cooperation of 'Gesellschaft für Mathematik und Datenverarbeitung' (GMD-Bonn) and Hahn-Meitner-Institut (HMI-Berlin). The results are documented in [DS-D-86] and influenced this paper.

In the second chapter, the requirements to a DDS from users' and applications' viewport are summarized. Chapter 3 gives an overview on standardization activities for Directory Systems, where the CCITT approach is summarized in chapter 4. Finally, the future work items of CCITT and ISO are stated and some remarks and additions to the models are given in chapter 5 and 6.

2   Requirements to a Directory System

Using network applications and services, the *user* must be provided with a lot of information on:

- responsible administration (to become subscriber, required authorization, etc.)

- connection establishment (names, addresses,...) and protocol management

- application specific information (File name structure, job command language)

- addresses of potential communication partners (MHS) or relevant end systems (host, server)

Normally, the user collects the necessary information from different sources in the network environment. It should be the task of a Directory System, to provide all or most of this information to the users of a communication service.

Beside the problems of application usage there is a number of problems that occur with the *operation and management of applications*. Providers of applications are faced to high expenditure of coordination and administration, to guarantee that the applications can be used properly. Following activities have to be carried out:

- setting-up, publishing and deleting subscribers, installations, etc.

- arrangements with subscribers (e.g. for authorization)

- arrangements between different providers / domains

- coordination of protocols used

- accounting of provided services.

The applications are embedded into organizational domains with independent responsibilities. By the definition of a Distributed Directory System with support of domains the provider of applications can be assisted in the administration of the applications and services. A unique definition of directory objects and their categorized representation in entries is required to support such as application management.

User requirements

The requirements from user's view can be summarized as follows:

- information provision about persons, end systems and applications
- functions for reading and manipulation of Directory entries
- comfortable retrieval of objects and classes of objects (white page service, yellow page service)
- user-friendly naming (user-/application-oriented)

    - global, unified, commonly known naming scheme
    - use of public and private alternative names (aliases)

- easy and uniform access from any application
- comfortable handling (user-friendly adaptable interface) of all Directory functions and extensive help facilities
- access-protection for person-related data through differentiated, person-oriented access control mechanisms
- support for maintenance of actuality, completeness and integrity
- directory costs must be covered by aplications
- high quality of service.

Application requirements

The necessity of a unified, global naming scheme for persons and end systems can be viewed as one of the important requirements derived from applications analysis. The naming scheme should be oriented widely at the concepts of the MHS naming, [MHS-C-X4xx]. Additional attributes for naming end systems must be added to support all applications:

- end system specific information (e.g. address, type, OS-version, responsible persons,...)
- information about responsible persons
- information about costs/accounting
- access rights for users.

Apllication instances need following informations in order to guarantee operation without special know-how on the user's side:

- name-address mappings
- description of network path (LAN/WAN access)
- protocol type and version selection

Required Directory entries

Information to be stored in the Directory are summarized to following objects:

- organizational and residential persons
- organizational groups
- organizational roles (manager, personell position)
- end systems
- application entities (MHS-UA, MTA, DSA, ...)
- applications description (DFN-MHS, DFN-FT, RJE, FTAM, ...).

Each object represented as a Directory entry should be structured in the following subgroups of its relevant attributes:

- organizational context (parent CE-classes)
- attributes for naming use
- object specific attributes (properties)
- relations to other Directory objects (entries)
- attributes for Directory internal use.

The examples in Table 1 and 2 (see annex) give an impression of the structure of objects of type 'Organizational Person' and 'End System'.

In order to provide that existing and expected applications can be modeled in the Directory, the Directory System has to be realized *application independent*. The autonomy of both, user and provider, the representation of distributed responsibilities and competences, the operating and quality of service requirements demand for a *Distributed Directory System*.

3 Standardization Activities on Directory Systems

Many research projects worked in the area of name servers or directory systems. Some examples are the Grapevine project at XEROX-PARC, the V-System, the Spice-project. References are found in [LANT-86]. Some research networks provide or plan directory services. A new name service for ARPA Internet - the ARPA Domain Name Service - has been specified and is being implemented.

The 'Deutsches Forschungsnetz' (DFN) has specified a Directory Service in conjunction with the introduction of their X.400-Message Handling Service. It is a loosely coupled distributed Directory System supporting MHS-entities as UAs, MTAs and groups [MHS-D-85b].

The European Computer manufacturing Association (ECMA) has been working on a standard on Directory System during the last years. Due to the unsolved problems, ECMA provides only a Technical Report on Directory Service [DS-E-32]. The ISO/TC97/SC21/WG4 is also working on a Directory Standard and provided two extensive working documents [DS-I-N97], [DS-I-N98] in 1985. Beside this work, the ISO SC21-Working Group 4 is working in the related field of OSI-Management Information Services.

CCITT Study Group VII (Question 35) started work on Directory System for the Study Period 1985-1988. Discussing the first stable proposals, the group changed the data model of the Directory. This led to a considerable delay in the preparation of Draft Recommendations, which are now expected in 1987.

The Concepts of the different standardization bodies are quite similar. They use the same functional model with a hierarchical directory structure. The directory protocols are based on 'Remote Operations'. The Directory Service is mainly seen as an information service for users. The applicability of a Directory Service to distributed systems or applications is - based on the necessity of the Message Handling Systems - recognized by the standardization working groups.

In the following chapters, the concepts of the CCITT Directory System are described.

4   The CCITT-Directory System

The CCITT-Study Group VII is concerned in their study period 1985 - 1988 with the provision of Draft Recommendations, describing Directory Systems. After their meeting in Reston (USA), January 1986, they issued Version 3 of the Draft Recommendations [DS-C-Xdsn] which are entitled as follows:

X.ds1   Directory Systems: Model and Service Elements
X.ds2   Directory Systems: Information Framework
X.ds3   Directory Systems: Protocol
X.ds4   Directory Systems: Standard Attribute Types
X.ds6   Directory Systems: Suggested Naming Practices
X.ds7   Directory Systems: Authentication Framework

The concepts described in this chapter are based on the Version 3 of the Draft Recommendations.

4.1   Key Elements of a CCITT Directory System

The Draft Recommendation is concerned with the Directory System, a widely-distributed, heterogeneous collection of equipment which holds a database of international proportions pertinent to communications, called the Directory. Typically the information helds is such as to facilitate communications between or with *communication entities* (CEs), such as persons, terminals and distribution lists.

The Directory System is not intended to be a general-purpose database system, although it may be built on such systems. It is assumed that there is a considerably higher frequency of 'queries' that of 'updates'. There is also need for instantaneous global commitment of updates: transient conditions where both, old and new versions of the same information in the Directory are available, are quite acceptable.

It is a characteristic of the Directory System that, except as a consequence of differing access rights, the results of directory queries will not be dependent on the identity or location of the enquirer.

Communication Entities and their Names

A primary motivation of CCITT for initiating studies on Directory Systems was the need to provide information about users and providers of CCITT services in a user-friendly way. In the consideration of CCITT services, users and providers are examples of "things" that it is of interest to have a list of, and to maintain some information about. In the terminology of Directory Systems, they are examples of what are referred to as *communication entities* or CEs.

CEs will be identified by *names* consisting of *attributes*. Different types of attributes provide a reasonable way of naming different types of CEs. The ability for a particular CE to be named in many ways, using different attributes, is an important feature of a Directory Service.

A Communication Entity can have multiple names as its attributes. This gives the Directory System the capability such as being able to return a set of names (distribution list) in response to attributes identifying a single Communication Entity.

Part of the Directory System specification includes *standardized attributes* as well as *Recommended name forms* for types of CEs which are expected to be commonly used in the Directory System. The standardized attributes include, among others, personal name, telephone number, and telex number. The Recommended name forms include those for organizational person names and organizational role names. All of these are likely to be widely applicable to users of CCITT services (e.g. message handling service, telematic services, telephone service). In the future, as the capabilities of the Directory Systems are used and applied in various ways, it is likely that other attributes and recommended name forms may also be needed and therefore developed. This may particularly be the case in applying the Directory Systems to providing information on computer resources, or to Open Systems Interconnection, where the CEs of interest are likely to be things such as application-entities (see chapter 2).

Entries and Attributes

Each attribute in a CE's entry represents some particular type of information which is associated with the CE. For users of telematic services for example, information, or attributes, of interest might be the types of terminal equipment the user has, the terminal address(es), optional or variable terminal capabilities, and the other types of services the user subscribes to. The Directory System provides the means of associating attributes with Groups as well as with individual CEs.

User Friendly Capabilities

The Directory Systems has been designed to be user-friendly. This is in addition to any possible user-friendly functionality of a user's local environment. The user-friendly features of the Directory Systems are:

a) it is interactive;

b) it allows the user to interrogate the Directory System by using keywords (attributes) to find any entry;

c) it allows "wildcard" possibilities;

d) in the event of an ambiguous result of an interrogation, it has the capability to either return the number of Communication Entities found or a list of the Communication Entities found;

e) in the case that one part of the Directory System contacted cannot satisfy a query, then it can either automatically route the query to another part of the Directory System or return *hints* to the enquirer suggesting which other part(s) of the Directory System should be tried.

Security Capabilities

It is extremely important to identify a user by a system before allowing him further access to that system. This identifying process is called "authentication" and works in conjunction with access control mechanisms. The Directory System will have the capability of offering various degrees of authentication mechanisms varying from simple user name/user password types to complex types involving encryption and passing of encryption keys.

The Directory System offers access control at several levels. Security (e.g. by the use of passwords) is offered by the Directory System at the organization, Communication Entity and individual "attribute" level. This allows the data held on behalf of the users to be protected at various levels as the user requires.

4.2 The Directory System Model

The Directory System Model comprises several different functional components that work together. The model can be applied to a number of different physical and organizational configurations and it implies the use of the OSI Reference Model to define formally the layered communication structure used between the components.The functional view of the Directory System Model is shown in figure 1.

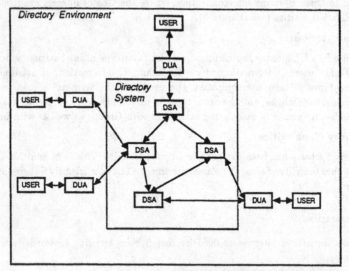

Figure 1 : Directory System Model

Definitions

The *Directory Environment* comprises the Directory System Agents, the Directory User Agents and the Users.

A *User* is generally understood as a person or application process using the Directory System. Each user within the Directory Environment interacts with the Directory System through a *Directory User Agent* (DUA). Each DUA serves a single user so that the DS may control access to directory information on the basis of DUA names. A *Directory System Agent* (DSA) provides directory information to a DUA or to another DSA. A DSA may use information stored in its local database, or interact with other DSAs to provide this information.

The *Directory System* (DS) comprises a number of DSAs which interact with each other in order to provide directory information to DUAs. The complete information contained within and processed by the Directory System is called the *Directory*.

Directory Management Domains

A *Directory Management Domain* (DMD) is composed of at least one DSA and zero or more DUAs. If a DMD is managed by an Administration, it is called an *Administration Directory Management Domain* (ADDMD);if it is managed by another organization, it is called a *Private Directory Management Domain* (PRDMD).

A DUA has to be registered with one DMD. This DMD is called the *home-DMD* for the DUA.

Figure 2 shows a possible configuration of DMDs.

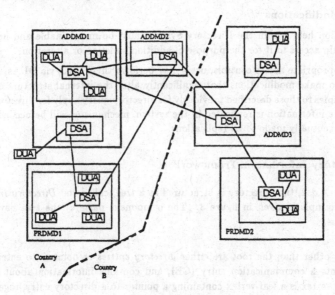

Figure 2 : Directory Management Domains

4.3   Operation of the Model

The DUA obtains information from the Directory by communicating with one or more DSAs. A DUA need not be bound to any particular DSA. It may interact directly with various DSAs to obtain the information it requires. It is also possible that the DUA can access the Directory System through a single DSA. For this purpose, DSAs will need to interact with each other.

Directory Query

There are two types of Directory query interactions: query-requests initiated by a DUA made to a DSA to request for information from directory entry(s); query-responses are the corresponding replies. A DSA may be able to respond based only upon information held in its own database. In this case the response indicates success and the DSA returns the requested information to the *inquirer*.

Query Decomposition involves splitting a query into a number of different queries, and sending them to the appropriate DSAs. It is needed when the information requested is distributed among several DSAs.

Chaining is so-called because the result of one or more DSAs simply passing the query-request on is the formation of a chain of DSAs, along which the query-response is passed in the opposite direction.

Multi-casting involves sending several identical queries to more than one DSA, and is needed when it is not clear which of a number of DSA is most likely to have the answer.

Directory Modifications

The information held within the Directory System must be maintainable and hence it must be possible to gain access to it for the purpose of addition, deletion or alteration.

Subject to appropriate access controls, users may perform these tasks via DUAs. Most users will not be able to make modifications that significantly alter the internal structure of the database. Similar principles to those described previously for directory query apply for directory modification. In cases where information is replicated in the system, mechanisms will be needed to ensure that changes are eventually made to all replications.

4.4   Directory Information Framework

As defined in X.ds2, the Directory is structured as a tree, called the *Directory Information Tree* (DIT). An example is given in figure 3. The component parts of the tree have the following interpretations:

- vertices other than the root are either *directory entries* or pointers to entries. Each entry represents a communication entry (CE), and contains information about that CE. Each pointer vertex is a leaf vertex containing a pointer to a directory entry, together with some information about how that entry may be named.
- arcs define the relationship between vertices, but contain no information. An arc from vertex A to vertex B identifies the CE represented by A as being the *naming authority* which assigns names to B.
- the root represents the existence of the highest level of naming authority for the directory.

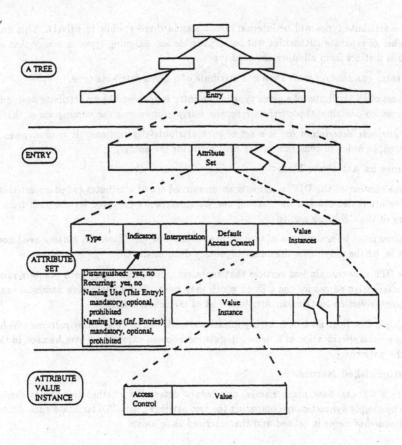

Figure 3 : A Directory Information Tree

Directory Entries

Each directory entry contains a set of *attributes* and a set of *attribute descriptions*. Each attribute consists of the following:

- *attribute type*

  identifies the class of information given by this attribute (e.g. a telephone number).

- *attribute value*

  is an instance of such a class.

- *access controls*

  define permissions and constraints on accesses to the attribute.

Some attribute types will be internationally standardized (public or privat). This implies that a number of separate authorities will be responsible for assigning types in a way that ensures that each is distinct from all other assigned types.

An entry can contain more than one attribute of a given attribute type.

The set of all attributes of a given type in one entry is governed by an attribute description for that type which contains: the *attribute type*, the *interpretation* and the *naming use* of that attribute.

An *attribute description list* is a set of such attribute descriptions. It is used when an entry is created, in order to constrain future definitions for the entry.

Names as Attribute Lists

In the context of the DIT, a name is an unordered set of attributes called an *attribute list* (AL). Those attributes marked by 'naming use' parameters are collected on the path from root to the entry of the CE being named.

A name must be unambiguous, that is, denote just one CE. However, a name need not be unique, that is, be the only name that unambiguously denotes the CE.

The DIT may contain leaf vertices that represent pointers to CEs. Such pointers provide a basis for alternative names for the CEs to which they refer. These alternative names are called *aliases*. A pointer vertice contains an Attribute List of the referred CE.

Multiple CEs (e.g. all CEs in an organizational unit) can be referred by *patterns* which are defined by a partial specification of a name. Regular or boolean expressions may be used in the definition of the patterns.

Distinguished Names

Since a CE can have many names, one cannot determine whether two names denote the same CE by simply syntactically comparing the two attribute lists. To facilitate such determination, a *distinguished name* is defined and characterized as follows:

- every CE has one, and only one, distinguished name;
- a distinguished name for a CE is globally unique;
- a distinguished name should be assigned so as to be long-lived.

In order to ensure that a distinguished name is unique, it is necessary that it is constructed hierarchically from the DIT.

Distinguished names have particular utility in denoting authorized CEs in access control lists. The use of distinguished names in these lists will ensure that the access rights are indeed granted only to the intended CEs.

Standard Attribute Types

The Draft Recommendation X.ds4 defines a number of standard types, which can be used in forming names, and in providing information regarding communication entities and other entries in the Directory System.

The current list of standard attribute types corresponds to those needed for the classes of CEs for which recommended name forms are given in Draft Recommendation X.ds6. The following table lists the standard attribute types.

CE Class	Organizational Unit
Common Name	Owner
Content Type	Postal Address
Country	Postal Code
Description	Protocol
Encoded Information Parameters	PSAP Address
Encoded Information Type	Role Occupant
Facsimile Telephone Number	Serial Number
Function	Telephone Number
ISDN Address	Title
Locality	Teletex Terminal Identifier
Maintenance Provider	Telex Number
Member	User Password
O/R Address	X.121 Address
Organization	

The Name Verification Process

As part of a service request, an attribute list may be submitted that purports to be a name. The formal definition is determined by a path through the DIT. The process by which this attribute list is verified to be a name is the *name verification process*. This process accepts a purported name, compares its attributes to the DIT and determines wether it describes such a path.

When the attribute list identifies a unique path through the DIT, the attribute list is a valid name.

Within a purported name, wildcard characters may appear in the value of an attribute. This facilitates pattern matching where this value is uncertain. Other search techniques, such as 'regular expressions' may also be used.

Authentication and Access Controls

Within the context of telecommunication services, it is important to provide privacy for stored information, by providing protection from unauthorized use of such services. Two types of mechanism are necessary to fulfil this requirement: an access control mechanism, and an authentication service.

The authentication framework provided by the Directory System is intended for use by other services to meet their needs for authentication and other secure services. The Directory System itself uses the authentication framework to meet its own needs for authentication, access control and other secure services. The authentication framework supports both simple and strong authentication.

The Directory System, after having authenticated the credentials of a user, must determine if this user has the right to perform the specific request for service on the referenced information. The access control to directory information thus provided is accomplished by using access control lists associated with portions of the Directory. Each element in an *access control list* (ACL) is a set of distinguished names or patterns, paired with a set of access rights. The Directory System may provide access control at the directory entry, attribute type and value instance levels.

Access categories define those ways in which access to specific attributes may be limited. If an access category has been granted, it allows the described operations to be performed. Access categories are:

know-existence, compare, read, add, update, delete, self-adminster, modify-tree, control.

These access categories may be specified in any combination.

Figure 4 gives an example of an 'Access Control Structure' with two ACL-Elements.

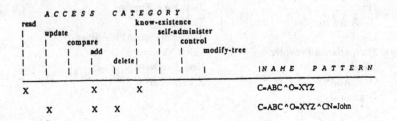

4.5 Service Elements

The services provided by the Directory System to the user are defined in X.ds1, which contains a list of service elements of the directory system and their definitions. The service elements are defined in a general manner so as not to impact the flexibility for detailed formal notations which are contained in X.ds3.

List of Service Elements (SEs)

SEs relating to Entries	SEs relating to Attributes
Add Entry	Add Attribute
Delete Entry	Check Member
Read Entry	Delete Attribute
	Modify Attribute
	Read Attributes

SEs relating to Access Control	SEs relating to Attribute Descriptions
Add Access Control	Add Attribute Descriptions
Delete Access Control	Delete Attribute Description
Read Access Control	Modify Attribute Description
Replace Access Control	

SEs relating to Aliases	SEs relating to Inferior/Superior Entries
Add Alias	Read Inferior Entries
Delete Alias	Read Superior Entries

SEs relating to Authentication	SEs relating to Shadowing
Check Simple Credentials	Initiate Shadowing
	Initiate Shadowing
	Terminate Shadowing

Service Controls

A number of controls can be applied to the service elements. The following controls are applicable:

- control of the return of lists by service elements. A list count limit is set
- setting of the limit on the size of a list
- setting of a time limit for a search operation
- the scope of a search to a portion of the DIT, e.g. local scope, some intermediate scope or global scope
- the specification of the priority of requests
- the applicability of the 'hints' style of DUA-DSA interaction to the operation.

4.6   Information Distribution in Directory Systems

For each directory entry, there shall be one *master copy* to which any updates to an entry are directed and applied. Such a master copy is held by a single DMD. Any number of other DMDs may, in principle, hold(non-master) copies of an entry.

Any copy of an entry which contains sufficient information may be used in responding to requests. A DMD must hold information to allow routing of those requests which cannot be satisfied using only the copies of entries which it holds.

The combination of the copies of entries (master or otherwise) and routing information which a DMD holds, can be considered as a tree itself, with identical structure to a subtree of the DIT. This tree is known as the *knowledge tree* for that DMD.

Shadowing

A DMD may elect to *shadow* a particular directory entry (or set of entries) for which it does not hold the master copy. This requires an arrangement between the DMD and that holding the master copy (the *master DMD*), and may be subject to access control. Shadowing an entry implies that when an update occurs to the entry, the master DMD informs the DMD of the update so that the latter's copy can be kept in step.

Caching

A DMD may elect to enhance its knowledge tree with information received as a result of service requests (queries) of other DMDs. This is known as *caching*. The caching can be overwritten by subsequent requests, but it will not be automatically refreshed as by shadowing. Caching as a generic technique which may of course also be used by the DUA.

4.7 OSI-Architecture of the Directory System Model

The Directory System entities and protocols are located in the *Application Layer* of the OSI Reference Model. This is done to permit the Directory System to use the underlying layers to accomplish the following:

- Establish connections between individual real systems using a variety of network types (for example, CSPDN, PSPDN, PSTN, LAN).
- Establish connections to permit the reliable transfer of directory protocol data units between systems.
- Signal the use of the presentation transfer syntax as defined in Recommendation X.409.

Two *functional entities* appear in the model:

- the Directory User Agent Entity (DUAE) provides those aspects of DUA functionality which relate to communicating with DSAs in order to provide appropriate responses to requests from the User;
- the Directory System Agent Entity (DSAE) provides aspects of DSA functionality which relate to communicating with DUAs and other DSAs.

Two *peer protocols* as defined in X.ds3 appear in the model:

- the Directory Agent Protocol (DAP) defines the exchange of requests and reponses between a DUAE and a DSAE. This protocol is specified in Draft Recommendation X.ds3;
- the Directory System Protocol (DSP) defines the exchange of requests and responses between two DSAEs.

5 Future Developments: CCTII/ISO Convergence on Directory System

Since the basic concepts of the Directory System are quite similar in the ISO-proposals [DS-I-N97], [DS-I-N98] and in the CCITT Draft Recommendations [DS-C-Xdsn] this two standardization bodies agreed to develop a common document, the **CCITT/ISO Directory Convergence Document**.

Main differences in the current drafts are:

- differences in terminology and document structure
- distinguished names, ordering of attributes
- conformance subsets for simple applications
- compability and power of service elements

The scope of the CCITT work is broader than the ISO drafts. It is intended to bring those topics into a commonly agreed standard. For this reason, CCITT and ISO agreed on an action plan for collaborative meetings. Target date for approved final text of an ISO-Standard and CCITT Recommendations will be first quarter of 1988.

The first draft of a CCITT/ISO convergence document [DS-CI-Xds/DP] is now available after a collaborative meeting (Melboune, April 1986) and contains the following parts:

1) The Directory - Overview of Concepts, Models and Services

2) The Directory - Information Framework

3) The Directory - Authentication Framework

4) The Directory - Abstract Service Definition

5) The Directory - Procedures for Distributed Operation

6) The Directory - Access and Distribution Protocols

7) The Directory - Standard Attribute Types and Interpretations

8) The Directory - Suggested Naming and Administrative Practices

6 Shortcomings of the CCITT/ISO concept and suggestion

The above described CCITT-Directory System Recommendations will not be sufficient for supporting the follwing tasks:

- support of the management of applications
- support of the maintenance of Directory-entries
- usage of Directory System by various applications
- assistence of users at their Directory System interworking

The essential shortcomings of the current concept are:

- CCITT defines the Directory System as an information retrieval system for Message Handling and additional Telematic Service. It is not intended to use it as a supporting tool for the management of additional services or applications.

- The definition of new Directory entries for other applications can only be realized outside of the scope of the standard. This led to changes in both, the standard and the existing Directory implementations.

- The proposals have no concepts for supporting the creation and maintenance of Directory entries (e.g. categories, scheme) which would provide both, a unique and a global definition of entries and the consistency and completeness of the necessary information.

- Beside this, the documents are currently not consistent and some gaps are still open and for further study.

Nevertheless, the concept of the CCITT-DS-model can be seen as a stable basis for further work. Appropriate extensions will fulfil the stated requirements. The VERDI study [DS-D-86] suggests such extensions as like as

- a category-concept for the Directory objects

- typing of attributes

- modeling of responsibilities.

The category concept can be compared with the scheme-definitions in data base descriptions. A category is the framework of a class of entries, which defines the structure, the components (attributes) and responsibilities for maintaining the entries.

The categories themselves with the basic structure and building rules for instances of entries will be objects in the Directory System. Therefore, each creation of a new entry at any domain of the Distributed Directory System relies on the relating category and will be supported independent of the place (domain) or the expected user. More details to this extensions are found in [DS-D-86].

7 Conclusion

The integration of an extended CCITT/ISO-oriented Directory System to a distributed system or network with various applications or services may bring some advantages:

- easy access to the services

- unique naming concept for persons, endsystems and services

- intensified usage of services

- support of application management

- reduction of administration work

The consequences of a Directory integration are as follows and can be disadvantegeous in some cases:

- provision and administration of information by the Directory System (only)

- unique data structure and semantics

- usage of the same Directory information for an information service, for the operation of applications and for the support of the system operator

- integration of DUAs in the applications and services (changes in existing implementations)

- charging directory usage within the applications

- dependency on the availability and reliability of the Directory System.

8 References

[DS-C-Xdsn]   CCITT STUDY GROUP VII: Directory Systems (Q35 / VII), Draft Recommendation X.ds1 - X.ds7, (Version 3), January 1986

[DS-CI-Xds/DP]   CCITT/ISO Directory Convergence Document, Part 1-8, June 1986

[DS-D-86]   VERDI - A Distributed Directory System for the German Research Network, Deutsches Forschungsnetz (DFN), Report, Eds.: H. Santo (GMD), M. Tschichholz (HMI), March 1986

[DS-E-32]   ECMA/TC32/85/32: OSI Directory Access Service and Protocol, December 1985

[DS-I-N97]   ISO/TC97/SC21/WG4/N97: OSI Directory Access Service Definition, December 1985

[DS-I-N98]   ISO/TC97/SC21/WG4/N98: OSI Directory Access Protocol Specification, December 1985

[LANT-86]   Lantz K.A., Edighoffer, J.L., Hitson B.L., Towards a Universal Directory Service, in: 4th PODC Conference Proceedings, ACM 1985, Departments of Computer Science and Electrical Engineering, Stanford University

[MHS-C-X4xx]   CCITT STUDY GROUP VII: Message Handling Systems, Recommendations X.400, X.401, X.408, X.409, X.410, X.411, X.420, X.430, Report R 38 and R 39, Malaga-Torremolinos, 1984

[MHS-D-85b]   DFN, Zentrale Projektleitung Berlin, The DFN Message Handling System, Specifications for Realization, May 1985

9 Annex: Examples of Directory Objects

Use	Information (* mandatory)	Example
Parent CE Class:	*Country	Deutschland
	*Organization	HMI
	Organizational Unit	Bereich D/M
attributes for naming use:	*Common Name	Michael Tschichholz
	Locality	Raum DV 308
	Telephone Number	8009 2570
	Telex Number	0185763
	Facsimile Telephone Number	8009 2999
Object specific attributes:	O/R-Name	EAN: tschichholz@vax.hmi.dfn
	Access rights for PAD	World
	Access rights for MHS	World
	Access rights for FT	Domain
	Access rights for RJE	Country
	Interests	Distributed Systems, DS
	Further Information (ffs)	
Relations:	Residential Person	M.T. D-1000/62 ...
	Organizational Roles	EAN Administrator, ...
	Organizational Groups	EAN-Adms, VERDI, ...
	End Systems	VAX: DMT, ...
DS internal use:	Objecttype (CE-Class)	Organizational Person
	Distinguished Name	
	Further Information (ffs)	

Table 1 : Example of an Organizational Person Object

Use	Information (* mandatory)	Example
Parent CE Class:	*Country	Deutschland
	*Organization	HMI
	Organizational Unit	Bereich D/M
attributes for naming use:	*End System Name	DVAX3
	Telematic address	49-30-1234
Object specific attributes:	Network address	DTX-P: 45 3000 217 17
	Network access	WAN, LAN
	Locality	Raum DV 308
	Manufacturer	DEC
	End system type	VAX 11/780
	Operating system	VMS
	OS-version	4.2
	PAD parameter pre-settings	1:0, 2:1, ...
	File name stucture	[<dir>]<fn>.<ext>;<vers>
	Command interpreter language	DCL
	RJE end system id	B_HMI17
	Job command language inform.	
	Further information (ffs)	
Relations:	Responsible Organizational Person	A.D. Ministrator
	Available OSI-applications	PAD, FT, EAN, PSIMAIL
	Available devices	Laser-Printer, Plotter...
DS internal use:	Objecttype (CE-Class)	End System
	Distinguished Name	
	Further Information (ffs)	

Table 2 : Example of an End System Object

Database Access in Open Systems

S. Pappe, W. Effelsberg, W. Lamersdorf

European Networking Center
IBM Germany GmbH

Abstract

After a short introduction into centralized and distributed database systems and communication in open systems, the integration of a Remote Database Access facility into the ISO Reference Model is described. Such a facility provides a communication path between an application program and a remote database system in a heterogeneous network. The current standard proposals of ECMA are introduced. Finally, the status of a prototype implementation is reported.

1. Introduction

Today's applications in data processing are characterized by the continuously increasing use of the possibilities of remote data processing. For many years there exists distributed computer networks. The majority of these systems, based on special manufacturers conventions, have the disadvantage that they are closed systems, i.e. they are restricted to closed organization units and unique manufacturer-dependent systems without access to public networks. These systems can be characterized by the following attributes:

- The application networks are not universal,
- they are restricted to a special user group,
- they often use manufacturer standards.

In the last years the development of open computer networks has come up. Open systems that enable data communication via standardized protocols and interfaces independent of manufacturers' architectures can be realized by using international standards. The idea of open systems advances continously in the daily bussiness and private environment. Examples of this are the worldwide data exchange via packet switching networks, the automization in administrative organizations, and the access of distributed and remote databases (**Remote Database Access - RDA**), i.e. access of databases installed on different computers of various manufacturers (**heterogeneous systems**) accessable via open systems. The last example is the topic of this paper.

In the following chapter the state of the art in database research is presented. The third chapter describes the communication in open computer networks based on the ISO reference model. The application-oriented layers are discussed in more detail, because remote database access is based on these. Chapter 4 describes the task and functionality of access to remote databases in open systems. Moreover the problems with the integration of RDA in the reference model are explained. This is reflected in two different standardization proposals, which are illustrated extensively. In chapter five a prototype implementation based on the older RDA standard proposal is explained. The software engineering environment, the current state of realization, and the ongoing research is presented.

2. Centralized and Distributed Databases

Today centralized databases are a cornerstone in the world of commercial data processing. They are generally seen as a vital resource of a company. The stored data can be shared by various applications. Transaction-oriented multi-user systems allowing parallel online access to applications are common. The access to remote systems from a terminal happens only via terminal networks. The application program runs on the same computer as the database management system (**DBMS**). There exists only one copy of the database management system and one copy of the data.

Current research in the field of centralized databases concentrates on modelling and storing of complex objects, such as graphical data or model data in engineering databases. Storage, query, and transaction techniques are developed for these objects (NF2 relational model, long transactions etc.). Another research topic in the area of centralized databases is the integration of databases and expert systems.

In contrast to centralized databases, the data in distributed databases is stored in several systems. The goal of maintainig copies to improve reliability, of increasing the performance by distribution on multi-processors or of integration of existing autonomous databases (with the special case of heterogeneous distributed systems with different local data models) depends on the purpose of the distributed database system. In any case, it is the aim of distributed databases to reach distribution transparency. The user does not need to know where the data is stored physically [CEPE84, ELRS].

The local database management systems of a distributed database must cooperate to execute global queries and to secure the global consistency of data. Procedures for this have been research topics for many years, for example the decomposition of global queries into local subqueries, and global transaction management with commitment, concurrency control and recovery. Although algorithms for these problems are known and prototypes have been developed, distributed database systems are only slowly finding acceptance in the commercial data processing. This is partly due to the great complexity of such systems, which makes maintenance difficult and causes high costs.

This paper discusses remote database access in computer networks (RDA), i.e., the communication between an application program in one system (e.g. a workstation) and a database management system in another system (e.g. a mainframe). It is not relevant for the dialogue between an application program and a database management system whether the database management system is centralized or distributed (transparency of distribution).

There should be a standard for the access to remote databases. It is an open systems problem because both participating systems can be different in hardware and software. The current state of research in this area is discussed in the following chapter.

3. Open Systems

An open system is defined as a set of

- one or more computers,
- the associated software, peripherals, terminals, human operators, physical processes, information transfer means, etc.,

forming an autonomous unit for information processing. Another part of a system is an application process that executes the information processing for oné application. Examples for application processes are:

- a person operating a banking terminal,
- a Pascal program accessing a database,
- a program controling technical processes.

A system is called an open system when application processes of various systems plan to communicate and the single systems are connected physically. In addition to a standardized physical connection other standards are necessary to realize the cooperation between the single systems. The implementation of these standards is part of the communication systems (see Figure 1).

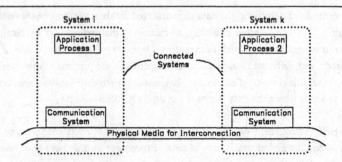

Figure 1. **Systems in an open environment**

Interaction of open systems is accomplished by standardized procedures for the information exchange (standardized protocols). If every system uses these protocols then all systems will be able to communicate with each other, i.e. compatibility between the single systems will be realized.

A framework for the standardization of communication protocols is the **ISO Reference Model** for open systems.

The ISO Reference Model for Open Systems

Since the mid-1970's, the development and use of computer networks is getting more and more attention. When it was realized that an open worldwide communication depends on the agreement and observance of communication rules, the technical committee 97 (TC97) of the International Organization for Standardization (ISO) founded a new subcommittee (SC16) for the standardization of communication in open systems (Open Systems Interconnection - OSI). SC 16's task was to develop a reference model that would provide an architecture to serve as a basis for all future development of standards for worldwide distributed information systems. Since spring 1983, the reference model is standardized as international standard IS 7498 [ISO84]; it describes functionally the data exchange between open systems. The reference model is independent of internal hardware and operating system properties.

In the reference model SC 16 took the way of a modular approach and hierarchical arrangement in layers. A communication system is subdivided into the well-known seven layers (see Figure 2). The seven layers can be grouped into two categories: the transport-oriented layers 1 to 4 (which are dependent on the characteristics of the transfer technology used) and the application oriented layers 5 to 7 which are problem-oriented [EFFL86, IEEE83].

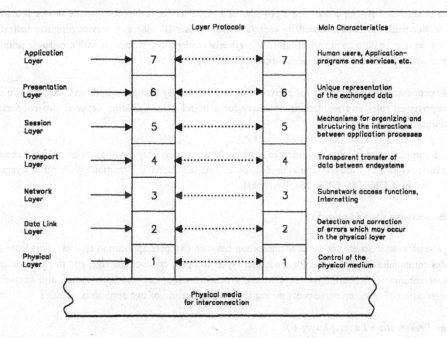

Figure 2. The Seven-layer OSI Architecture

Meanwhile several standards have been standardized internationally. Besides ISO, especially CCITT (Comite Consultatif International Telegraphique et Telephonique) and ECMA (European Computer Manufacturer Association) are actively developing new standards. Agreement has been reached in the standard drafts of the first five layers up to now, but layer 6 and especially some applications of layer 7 are currently discussed intensively.

The Principles of the Structuring

The ISO reference model describes a communication system in terms of a hierarchy of communication layers. In this approach, each layer (except the upper-most layer) provides a set of services to the layer above based on a different set of services provided by the layer below. So, the functionality of a specific communication layer can be viewed as the functional difference of those services provided by the instances (entities) of the layers above and below. An entity of layer N is called (N)-entity. An (N)-entity communicates directly with layers $(N+1)$ and (N-1) and indirectly with a corresponding peer (N)-entity at the communication partner's side. This last indirect communication is always preceded by a direct communication with layer *(N-1)*.

The entities within a given layer are based on the services of the entities in the layer below. Here, a service always represents the functionality of a layer *(N)* as provided to the entities of layer *(N+1)*. So, the entities of a layer N 'upgrade' the given services of layer *(N-1)*, and then offer these improved services to instances at layer $(N+1)$.

The interaction between a user and a provider of a given service is described by the **service primitives**. A service primitive **request** issued by user A arrives at user B's side as a service primitive **indication**. Some services require an acknowledgment (*confirmed service*); user B answers with a **response** primitive which arrives at user A's side as a **confirmation** primitive.

The cooperation of peer entities of a given layer is governed by that layer's **protocol**. So, the protocol comprises all rules and regulations necessary for a useful communication between two cooperating entities.

With respect to the (application-oriented) context of this paper, only the application oriented communication layers 5 to 7 are considered in more detail here; additional information on the other layers can be found in [GGHST85, GKSST85, TANE81].

The Session Layer (Layer 5)

The session layer structures the communication between two peer application entities. This layer provides communication primitives for the higher layers to open, conduct, manage, and to close a session. These communication primitives support the synchronization, i.e. the initialization and continuous preservation of common agreements during the communication of two application entities.

The Presentation Layer (Layer 6)

Beside common forms of communication behaviour as defined in ISO/OSI-layers 1 to 5, both peer application entities are required to speak the 'same language' [ISO85c, ISO85d, ISO85e]. In order to provide a common transfer language, the presentation layer has to transform all data exchanged between communication partners into a commonly agreed **transfer syntax**. Prior to exchanging data, both application entities have to negotiate a common language (syntax) for the representation of the data to be transmitted. By choosing a common transfer syntax, both sides agree on the data types and data structures to be sent. The **abstract transfer syntax** is constituted from all data type definitions of all application protocol data units, without bitwise defining the coding of the data. An abstract transfer syntax is chosen by specifying a name of certain syntax definition as commonly known to both sides. The **concrete transfer syntax** then defines the mapping of all abstract syntax elements into a sequence of bits. As, in general, for a given abstract syntax there may be several possible concrete syntaxes, a major task of layer 6 is to negotiate which transfer syntax is to be used. The result of this negotiation between application and presentation entities are a concrete and an abstract transfer syntax, both of which together are referred to as **presentation context**. In order to make even the first negotiation possible (without a previously given common language), there is always an implicitly pre-defined default context.

Besides these explicit presentation services, there are other services which make the session service primitives accessible as presentation service primitives.

The Application Layer (Layer 7)

The application layer supports the communication between application processes. Because most standards for the application layer are still in an early stage, there are still different approaches to standardizing this layer. An evolving common understanding seems to be that different applications of open computer networks have many (sub-)/problems in common, e.g. synchronization of parallel processes, error recovery, etc.. This leads to a substructuring of layer 7 into a sub-layer providing such generally available, commonly usable service elements (Common Application Service Elements - **CASE**) as, e.g, proposed by ISO. This sub-layer can then be jointly used by applications as, e.g., Remote Database Access(**RDA**) etc.. So, besides the commonly available service elements (CASE) there are additional application specific service elements (Specific Application Service Elements - **SASE**) which complete the application layer.

The **Common Application Service Elements (CASE)** provide a basic resource of services for the implementation of various applications. CASE can be divided into two categories:

1. Association Control Service Elements (ACSE)
2. Commitment, Concurrency and Recovery (CCR)

The **Association Control Service Elements (ACSE)** are used for application association control. They provide services for the establishment and termination of application associations. An association is the cooperative relationship between two application entities, formed by the exchange of application protocol control information, in this case by the exchange of ACSE-Application Protocol Data Units (**APDUs**). The Association Control Protocol Machine (**ACPM**) within the ACSE communicates with its service user by means of primitives defined in ISO 8649/2 [ISO86a]. Each invocation of the ACPM controls a single association. The set of all ACSE-services can be divided into three groups:

1. Association establishment (A_ASSOCIATE)
   The association establishment procedure is used to establish an association between two application entities. With the association establishment, the application entities negotiate the quality of

service to be used and the qualities of the application association to be available. Also a connection is established at the presentation layer.

2. Normal release of an association (A_RELEASE)
   This procedure is used for the normal release of an association by an application entity without loss of information in transit.

3. Abnormal release of an association (A_ABORT, A_P_ABORT)
   The abnormal release procedure can be used at any time to force the abrupt release of the association by a requestor in either application entity, by either ACPM or by the presentation service provider.

The **CCR services** coordinate the communication of several distributed application services [ISO85a, ISO85b]. They allow an atomic view of their communication with one or several partners.

A transaction is a unit of commitment accepted by the communication partners. Through that, it is guaranteed that changes in a distributed environment are consistent and could not be influenced by other processes. Each participating entity needs an unique identifier which is known to and administrated by the coordinating process to achieve a consistent commitment in distributed transactions.

The requirement for **concurrency** is that the effect of transactions which overlap in time should be the same as it would have been if they had been executed one after the other, i.e. the transactions are serialized.

Recovery is the ability to guarantee that once commitments are completed they will stay even after system failures. If a transaction could not be successfully completed then it should have no effect.

In addition to these ACSE and CCR services, there is another category of service elements (see figure 3) that could be called common service elements but it is not treated as a part of CASE in ISO. This is the **Remote Operations Service (ROS)** [CCITTb, ECMA85b], a method to specify external interactions of distributed applications. Its functionality is comparable to Remote Procedure Call.

To invoke remote operations, two ROS users have to establish a ROS association. When a ROS association is established, a ROS user can request the execution of operations from another ROS user, or can execute operations for the other user. In an answer to a ROS request, a ROS user can send the requesting partner the result, an error or a refusing message.

A ROS association is established by the **BIND** procedure and terminated by the **UNBIND** procedure. With the **OPERATION** procedure a certain operation is requested from a remote entity. It is not necessary for the initiator of an operation to wait for a result before he requests another operation (asynchronous mode).

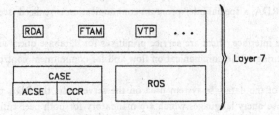

RDA : Remote Database Access
FTAM : File Transfer, Access and Management
VTP : Virtual Terminal Protocol

Figure 3. Structure of an Application Entity

4. Access to Remote Databases in Open Systems

Besides the input and output to terminals, access to data is the most common system call used by an application programmer. The access to remote databases [ECMA85a, ECMA86] is for this reason one of the most important areas of research in computer networks. With RDA, one has the ability to query or to manipulate from one computer a database on another computer system. (see Figure 4).

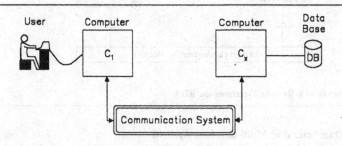

Figure 4. Simplified Representation of Remote Database Access

4.1. Functionality of Remote Database Access

When data is accessed in a remote database, the two communicating entities take over different functions.

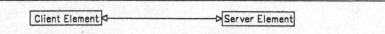

Figure 5. RDA Communicating Elements

At the query site (client), there exists an application program which queries the data available at the server site. Thus, in RDA, a special inter-program communication is realized (see figure 5).

At the RDA service interface, there are service primitives for database query and manipulation functions, for the structuring of the communication flow and for commitment control.

To be independent of the database system used on the server side, the RDA protocol supports only standardized database query languages which are mandatory for both peer entities. On the database side, the server process translates the received RDA messages into data retrieval and manipulation procedure calls of the database management system.

The following figure shows the use of RDA services:

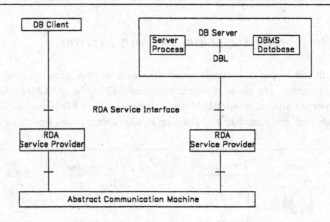

Figure 6. Access to a Remote Database via RDA

Distributed Databases and Multi-database Systems

In a distributed database the data is distributed over many single databases, which all have their own DBMS. They cooperate to form a Distributed Database Management System (**DDBMS**).

A user of the RDA service does not notice whether there is a single or a distributed database in the server element. The coordination of the different databases is executed through the DDBMS and is hidden from the user.

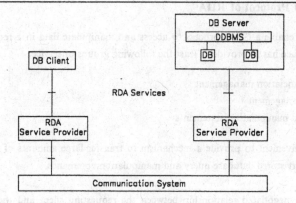

Figure 7. **Distributed Database/Single Connection**

The communication over RDA with a distributed database is not different from that with a single database (see figure 7).

When a client communicates with many separate databases through RDA, the client is himself responsible for the correct handling of a 2-phase commit mechanism. This presumes that the participating databases support such a procedure (see Figure 8).

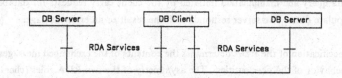

Figure 8. **Distributed Database/Multiple Connections to Separate Databases**

4.2. Integration into the ISO Reference Model

As described in the above sections, remote database access is one of the typical applications of communication in open systems. Because the work in standardization of the application layer is not yet finished, there is no final standard available for RDA. In this chapter, two alternative standard proposals are presented.

Some basic functions of the upper ISO layer are used by many applications. Therefore the work of standardization in layer 7 follows two different directions. On one hand, layer 7 is seen as one unit - like the other layers - which is specialized for one application. One the other hand, layer 7 is split into sublayers, which provide more generally usable generic functions. The advantage of this is that these common services (ACSE, CCR, ROS) can be used by different applications as sub-standards.

4.3. Services and Protocol of RDA

The RDA **services** enable a Client process to access and manipulate data in a remote database. An RDA service interface has to provide at least the following groups of service:

- Connection/association management
- Transaction management
- Data query and manipulation statements

Moreover, it is convenient to provide a mechanism to transfer large amounts of data (e.g. complete tables) and to repeat stored database query and manipulation commands.

An **association** is a negotiated relationship between the requesting client and the responding server which (in contrast to a connection in the lower layers) cannot be terminated without the agreement of both parties. If a lower layer connection is accidentally lost, the association is not terminated. In this case, a new connection must be established to restart the association.

A **transaction** is a logically complete unit of processing as determined by the application process. Every client may execute only one transaction on one association. A server may process many transactions concurrently on behalf of different clients. It is the server's responsibility to guarantee transaction serializability.

Services for **data query and manipulation** initiated by the client carry requests for database actions, to query or manipulate data. The server responds with the result or an error message.

The **protocol** specification of the RDA determines the contents of the exchanged messages and describes the expected behavior of the peer entities. The asymmetry of the two RDA roles (client and server) is reflected in two different protocol machines. In addition to the protocol machines, the protocol document describes the mapping of the RDA services to the lower layer services.

Another important aspect is the behavior in case of a failure. As usual, it is defined by state transition tables.

4.4. Status of the Standard

Today two different standard proposals for RDA are discussed in the European Computer Manufacturers Association (ECMA). The first proposal was published in 1985 by the Technical Committee 22 of the ECMA. This version of the **Remote Database Access Service and Protocol** document has the state of a Technical Report (TR 30) [ECMA85a].

This standard proposal uses the services of the Presentation Layer (layer 6) directly and assumes a Session Layer which supports the services of the Basic Synchronized Subset (BSS) and full duplex data exchange.

The service interface provides the following facilities:

- Single or grouped database statements
- Reading and writing of large data sets (bulk transfer)
- Commitment control
- Transaction management
- Definition and execution of 'macros'

A macro is a user defined sequence of data manipulation functions that can be invoked in a single service element.

A simplified model showing the principal states of an RDA connection (without the effects of interruption and reconnection) is described in the figure 9:

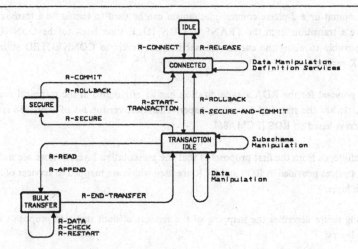

Figure 9. Simplified State/Transition Diagram of RDA (without error handling)

Starting with the IDLE state, a connection is established using the R-CONNECT service (R is an identifier for the RDA services). An association identifier, used for restarting an earlier association, and parameters for the connection establishment of the presentation layer must be given as arguments. A connection will be terminated normally with R-RELEASE, with a parameter if the association should survive.

In the CONNECTED state, data manipulation definition services can be requested to define temporary tables, cursors, and macros; which will be valid for all following transactions.

With the R-START-TRANSACTION service a transition to the TRANSACTION IDLE state is made. Now a transaction is in progress, and data manipulation functions are allowed (R-RDL-DO). As parameters, the client submits a specific database command (e.g. SELECT, INSERT, DECLARE CURSOR, etc.) and possibly user data. The server responds with a return code and, depending on the statement type, with the requested data. The response can carry only one row. To exchange more than one row, a cursor has to be opened or a transition to the BULK-TRANSFER state (with R-READ or R-APPEND) must be made. In this state, complete tables can be transferred to the client process, or a set of rows can be appended to a table in the database, uninterrupted by data manipulation commands. The service R-CHECK structures the bulk transfer with sequentially numbered checkpoints. In case of a failure, a communication entity can request a checkpoint number with the R-RESTART service at which data transfer is to restart. The bulk transfer is terminated with R-END-TRANSFER, and a transition back to the state TRANSACTION-IDLE is made.

Subschema management service elements allow the user to modify information concerning the database tables, database views, user privileges and macros. In the current standard proposal no facilities to access or alter the subschema data are provided. User privileges are identified by authority identifiers. Macro definitions can be declared and deleted by RDA macro handling functions.

A 1-phase commit or a 2-phase commit mechanism can be used to terminate a transaction normally and to make a transition from the TRANSACTION IDLE state back to the CONNECTED state. Also it is possible to abort the current transaction and revert to CONNECTED state with the R-ROLLBACK service.

The **second proposal** for the RDA standardization has its origin also in the technical committee 22 of the ECMA. It has the status of a working paper. The latest version 6.1 of July 1986 is titled **Remote Database Access based on ROS** [ECMA86].

The main difference from the first proposal is that the presentation layer services are not used directly but generic services provided in ROS and CCR are used which are mapped to services of ACSE and the presentation layer.

The following figure describes the mapping of the services of both standard proposals to lower communication layers:

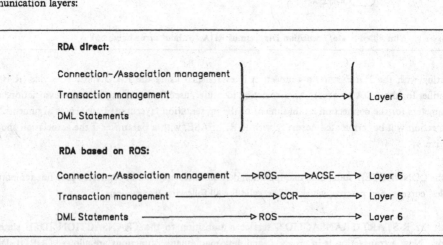

With the use of the ROS and CCR services, considerable functionality is moved to generic application layer elements which could also be used by other applications. This has the advantage that the complexity of the RDA protocol automaton gets would become lower. The protocol can be defined only with the mapping of the RDA service to the lower layer services in ROS, CCR and ACSE. The behaviour in case of a failure is not contained in the second version of the standard proposal.

The services of RDA based on ROS and the possible state transitions are shown in the following simplified figure (figure 10):

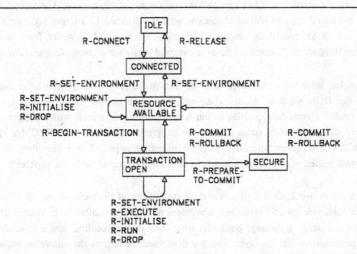

Figure 10. Simplified State/Transition Diagram of RDA based on ROS (without error handling)

Starting in the initial state (IDLE), a connection is established between client and server with the R-CONNECT service. The successor state is CONNECTED. Here it is possible to get access to data base resources. With R-RELEASE a connection is terminated, and a transition to the connectionless state IDLE is made.

When the first resource is acquired, a transition to the state RESOURCE AVAILABLE is made. Other resources can be defined, released, locked, etc. with the service R-SET-ENVIRONMENT. To validate and store or to nullify a certain database command, the service elements R-INITIALISE und R-DROP can be used. The command can then be can be executed as often as needed by an application by specifying a command identifier.

To inititiate a transaction, the service R-BEGIN-TRANSACTION has to be invoked. Then the current state is TRANSACTION OPEN in which the execution of database management commands is allowed. Such commands are invoked as single statements with R-EXECUTE. In this state, it is also possible to store, execute and erase certain database commands.

A transaction is terminated with a 1-phase commit or a 2-phase commit (via state SECURE) mechanism. To abort the current transaction and revert to the TRANSACTION IDLE state the R-ROLL-BACK service is used.

5. Implementation of a Prototype

The implementation of our RDA prototype is based on the first version of the standard proposal, using the presentation layer directly. There are several reasons for this: The first version is more thoroughly elaborated than the second, and has survived several rounds of discussions. The second version, based on ROS, is more elegant, and will be consistent with the planned ISO upper layer architecture when ROS, CCR and ACSE eventually become an international standard; however, for a prototype implementation, we consider the current status of the second version too incomplete and ambiguous.

We had a session layer implementation and the framework for a presentation layer available on our system, but not ROS, CCR or ACSE. Implementing these components as a prerequisite for RDA would have caused a considerable delay in our work, and our experience with the proposed RDA services and protocols would then come too late for integration into the work of ECMA TC 22. We had experience with the implementation of OSI protocol machines [PAPPE86]; therefore, the implementation of the well-defined protocol machine of the first version caused no major problems.

A major concern in our layer design was a clear structuring of the various aspects of communication software. We subdivide an OSI layer into a protocol machine, a number of predicates and actions (as defined in the standard), a message buffer for inter-layer communication, and a scheduler, controlling the parallel processes within the layer. The protocol machine defines the allowable sequences of messages sent and received and of internal operations. Internal operations are those computing the predicates and actions; they read and modify layer-internal variables. In order to allow asynchronous communication with higher and lower layers, a message buffer is maintained at each layer interface. Since an OSI layer can maintain several parallel connections to peer entities, we decided to provide layer-internal leight-weight processes, one for each connection, and implemented a scheduler to control these parallel processes. The processor is passed on to another light-weight process whenever the currently active process is waiting for an incoming event.

Currently, we are implementing the RDA protocol machine. In order to allow flexibility for the future, the protocol machine code on the server and the client side is identical. Their different behavior is controlled via parameters. In later versions of the RDA layer, when multi-site distributed transactions will be allowed, a server can become a client when it starts a subtransaction on a remote node. We will then still be able to use the same code in all nodes of the distributed system.

From the software engineering point of view, we have made a effort to produce high-quality code efficiently. Since we consider direct coding from the standard document too large a step for maintainability, we are using a specification technique called PASS (Parallel Activity Specification Scheme, [AFHHK84a, AFHHK84b]). A PASS specification is detailed enough to be translated automatically into programming language code. We have developed an automatic code generator, reading a PASS specification and producing PASCAL code. First experience shows considerable time savings in the coding and testing process, as well as reduced error rates in the resulting PASCAL code [PAPPE86].

Our prototype is implemented under IBM's operating system VM/CMS. We use PASCAL for both the generator and the layer itself. In our current test environment, all communicating RDA entities run in the same virtual machine under VM. Our current presentation layer offers pass-through services for the session layer (layer 5), with the added functionality of the presentation layer (e.g., negotiation of presentation contexts) under development. The session layer we are using is a complete ISO session with all functional units; it was developed here at IBM's European Networking Center. We use a simulated transport layer, running in the same virtual machine as the higher layers.

Currently, our RDA server layer only simulates the database behavior; it sends simulated rows back in response to simple relational queries. In the next phase of our project, we will connect the RDA server to a SQL/DS database management system, allowing access to a relational database.

Conclusion

After a short introduction into the world of OSI, we discussed the current status of the ISO upper layer architecture proposals, including ROS, CCR and ACSE. We then presented ECMA's two standard proposals for remote database access and explained the different protocols. The status of a prototype implementation at IBM's European Networking Center was then outlined.

We participate actively in ECMA TC 22 where the work on RDA standardization is done. We are also investigating a possible cooperation with another major manufacturer for practical testing of our prototype over a wide-area network.

A second version of a prototype is under design where we will separate ROS, CCR and ACSE service elements into separate components. These can then be replaced by ROS, CCR and ACSE software once it becomes available.

Acknowledgments

We would like to thank P. Pistor from the AIM team of IBM's Scientific Center Heidelberg for his great help in the interpretation of the ECMA proposals. We also thank G. Wulferding for the production of the figures.

References

[CEPE84]   St. Ceri, G. Pelagatti: Distributed Databases - Principles and Systems. Mc Graw-Hill, 1984.

[ECMA85a]   ECMA: Remote Database Access Service and Protocol, Technical Report 30, December 1985.

[ECMA85b]   ECMA: Remote Operations - Concepts, Notation and Connection-Oriented Mappings. ECMA Technical Report 31, December 1985.

[ECMA86]   ECMA: Remote Database Access based on ROS, Working Paper for a Standard, Version 6.1, July 1986.

[EFFL86]   W. Effelsberg, A. Fleischmann: Das ISO-Referenzmodell für Offene Systeme und seine sieben Schichten. Informatik Spektrum Vol. 9, No. 5, 1986

[ELRS]   H. Eckhardt, W. Lamersdorf, K. Reinhardt, J.W. Schmidt: Datenbankprogrammierung in Rechnernetzen. GI-Jahrestagung, Berlin, Springer Verlag, 1986.

[GGHST85]   E. Giese, K. Görgen, E. Hinsch, G. Schulze, K. Truöl: Dienste und Protokolle in Kommunikationssystemen, Springer Verlag Berlin (Heidelberg, New York, Tokio), 1985.

[GKSST85]   G. Görgen, H. Koch, G. Schulze, B. Struif, K. Truöl: Grundlagen der Kommunikationstechnologie; ISO-Architektur offener Kommunikationssysteme. Springer Verlag Berlin, 1985.

[IEEE83]   Proceedings of the IEEE, Vol. 71, No. 12, December 1983. Open Systems Interconnection.

[ISO84]   ISO: International Standard 7498. Open Systems Interconnection - Basic Reference Model, 1984

[ISO85a]   ISO: International Standard 8649/3. Open Systems Interconnection - Definition of Common Application Service Elements - Part 3: Commitment, Concurrency and Recovery, 1985

[ISO85b]   ISO: International Standard 8650/3. Open Systems Interconnection - Specification of Protocols for Common Application Service Elements - Part 3: Commitment, Concurrency and Recovery, 1985

[ISO85c]   ISO: International Standard 8822. Open Systems Interconnection - Connection Oriented Presentation Service Definition, June 1985.

[ISO85d]   ISO: International Standard 8823. Open Systems Interconnection - Connection Oriented Presentation Protocol Specification, June 1985.

[ISO86a]   ISO: International Standard 8649/2. Open Systems Interconnection - Service Definition for Common Application Service Elements - Part 2 : Association Control, 1986.

[ISO86b]   ISO: International Standard 8650/2. Open Systems Interconnection - Protocol Specification for Common Application Service Elements - Part 2 : Association Control, 1986.

[PAPPE86]   Anwendung einer Implementierungstechnik für Kommunikationsprotokolle auf das ISO-Kommunikationssteuerungs-Protokoll. Master Thesis, University of Karlsruhe, March 1986.

[TANE81]   A. Tanenbaum: Computer Networks. Prentice Hall, Englewood Cliffs, 1981.

MANUFACTURING AUTOMATION PROTOCOL (MAP)
OSI FOR FACTORY COMMUNICATIONS

Mark Adler
General Motors Corporation
30300 Mound Road, A/MD-39
Warren, Michigan 48090-9040

Only 15 percent of the 40,000 programmable tools, instruments, controls, and systems already installed at General Motors facilities are able to communicate with one another. When such communication does occur, it is costly, accounting for up to 50 percent of the total expense of automation because of the special wiring and the custom hardware and software interfaces needed.

Wiring costs are incurred whenever new systems are installed and again each time a production process changes. In the automotive business, where retooling for new models occurs annually, the rewiring costs are significant. Custom interfaces are necessitated by supplier-unique communication methods. Custom software is usually required to interface two process applications. To make matters worse, incompatible software performing similar functions may exist for different process applications.

With installed programmable equipment in GM plants expected to increase by 400 to 500 percent over the next five years, the communications problem would rapidly get out of hand if no solution were found.

Fortunately there is a solution -- the Manufacturing Automation Protocol. The origins of MAP date back several years, when a few employees at GM realized that the incompatibility of computer systems was the largest single roadblock to the future automation of plants. In assessing the situation, they concluded that there were three possible ways to provide a compatible electronic link from the design through the manufacturing processes:

- Conduct business as usual by continuing to buy stand-alone machines from a variety of suppliers and develop custom solutions as needed. Because of the appreciable costs involved, the proliferation of automation equipment, and growing communications needs, this approach became less feasible daily.

- Buy all equipment from one supplier. But this was not possible, since no single vendor sells all the types of programmable equipment -- mainframe computers, minicomputers, programmable controllers, numerically controlled machine tools, robots -- needed on the plant floor.
- Develop a standardized approach to communications on the plant floor.

It was obvious that only the third alternative was feasible. A task force was formed in 1980 to investigate use of the Open Systems Interconnection (OSI) Reference Model developed by the International Organization for Standardization (ISO) as the basis for a standardized approach. Representatives from seven GM Divisions served on the task force to ensure that a broad spectrum of communications requirements were considered.

In 1981 the GM Task Force began to explore communications possibilities with the Digital Equipment Corporation, Hewlett-Packard, and IBM and proposed solving the problem through a local-area network using the OSI Model. In line with this, GM decided in 1982 to act as the catalyst for developing the set of communications protocols known as the Manufacturing Automation Protocol (MAP). The task force selected specific existing standards that would be appropriate for some of the layers of the OSI Model and recommended interim "standards" for other levels until national and/or international standards were developed (See Figure 1).

Once GM decided that MAP was the way to go the questions became, "How do you get there?" It was apparent the other companies were struggling in the same jungle of factory automation communications. So GM asked McDonnell Douglas to join it in forming a MAP Users Group to convince suppliers to build MAP-compatible products. The group also was charged with testing and other aspects that would make MAP a reality. McDonnell Douglas hosted the first meeting of small MAP Users Group in St. Louis, Missouri, on March 29, 1984. Sixty people representing 36 companies attended.

From that modest beginning, interest has soared. North American meetings now typically attract 600 people representing 300 companies. The secretariat for the U.S. MAP Users Group is the Society of Manufacturing Engineers in Dearborn, Michigan. MAP Users Groups have been established in Canada and Europe, and soon groups will be formed in Australia and Japan. A worldwide loosely knit federation of MAP Users Groups representing four geographical areas has been formed to ensure that the MAP protocol is applied in a standard way.

LAYER	MAP VERSION 2.1 PROTOCOLS
7	ISO FTAM (DP) 8571 File Transfer Protocol Manufacturing Messaging Format Standard (MMFS) Common Application Service Element (CASE)
6	NULL * (ASCII and binary encoding)
5	ISO Session (IS) 8327 Session Kernel, full duplex
4	ISO Transport (IS) 8073 Class 4
3	ISO Internet (DIS) 8473 Connectionless and for X.25 - Subnetwork dependent
2	ISO Logical Link Control (DIS) 8802/3 (IEEE 802.2) Type 1, Class 1
1	ISO Token-Passing Bus (DIS) 8802/4 (IEEE 802.4) Token-Passing-Bus media access control

* A null layer provides no additional services but exists only to provide a logical path for the flow of network data and control.

MAP NETWORK ARCHITECTURES
Figure 1

FIGURE 1 INFORMATION

The MAP Task Force adopted the OSI model, specifying standards that are appropriate for each layer (See Figure 1). MAP specifies IEEE 802.4 (token-passing bus access method for local-area networks) for Layer 1 and IEEE 802.2 for Layer 2. Recently MAP Layers 3,4, and 5 became accepted ISO standards. Layer 6 has not been completely specified and parts of Layer 7 are in the Draft International Standard stage.

The rationale for selecting broadband cable for Layer 1 was:

- Broadband allows multiple networks to exist simultaneously on the same media, minimizing wiring modifications while effecting a smooth and orderly transition to MAP.

- Broadband will not only support the high-speed data requirements of a local-area network, but will handle voice and video requirements for applications such as security surveillance, closed-circuit television, and teleconferencing.

- Broadband technology is part of the IEEE 802.4 communications standard and under study for CSMA/CD (IEEE 802.3 - 10BROAD86).

- The initial investment has already been made to install broadband systems in many GM locations.

Token-passing on a bus configuration (IEEE 802.4) was chosen for these reasons:

- Token-passing is the only Data Link protocol presently supported on broadband by IEEE 802.

- Many programmable device suppliers were already providing token-bus-based equipment.

- Token-passing supports a message priority scheme.

- Unless a physical failure occurs, high-priority messages will be delivered within a specified and calculable time limit.

To help establish a timetable for MAP implementation, GM sponsored a demonstration of a working MAP network at the July 1984 National Computer Conference in Las Vegas, Nevada. In a similar vein, the National Bureau of Standards and the Boeing Computer Company sponsored aa OSI booth for office communications. The NBS-Boeing effort has since evolved into the Technical and Office Protocols (TOP) system.

The MAP demonstration at the computer conference was critical to its success. GM had yet to convince other manufacturers, as well as suppliers, that MAP was technically feasible. Seven companies agreed to develop and demonstrate equipment at the conference: Allen-Bradley (Vistanet data gateway and PLC-3 programmable controllers); Concord Data Systems (Token/Net interface module); Digital Equipment Corporation (VAX-11/750 computer); Gould (Concept 32/2705 and Modicon 584 programmable controller); Hewlett-Packard (HP 1000); IBM (Series 1); and Motorola (VME/10).

Equipment was demonstrated using protocols for four of the seven layers that make up MAP. The exhibit showed that files and standard message could be transmitted between computers and other programmable equipment.

The next milestone was a more elaborate demonstration at the Autofact '85 automation conference in Detroit. It involved 21 suppliers and more complex capabilities (See Figure 2). It proved that MAP and TOP could be the basis for a factory-wide communication system.

PROTOCOL LAYER OR APPLICATION	1984 NCC DEMONSTRATION	MAP		TOP
7 APPLICATION	LIMITED FILE TRANSFER	NETWORK MANAGEMENT DIRECTORY SERVICE MMFS SUBSET FTAM ISO CASE KERNAL		FTAM
6 PRESENTATION	NULL	NULL		NULL
5 SESSION	NULL	ISO SESSION KERNAL		ISO SESSION KERNAL
4 TRANSPORT	NBS CLASS 4	ISO TRANSPORT CLASS 4		ISO TRANSPORT CLASS 4
3 NETWORK	NULL	ISO CLNS	ISO CLNS	ISO CLNS
2 DATALINK	IEEE 802.2	IEEE 802.2 LINK LEVEL CONTROL CLASS 1	IEEE 802.2 LINK LEVEL CONTROL CLASS 1	IEEE 802.2 LINK LEVEL CONTROL CLASS 1	IEEE 802.2 LINK LEVEL CONTROL CLASS 1
1 PHYSICA	IEEE 802.4	IEEE 802.4 TOKEN ACCESS ON BROADBAND MEDIA	IEEE 802.4 TOKEN ACCESS ON BROADBAND MEDIA	IEEE 802.3 CSMA/CD BASEBAND MEDIA	IEEE 802.3 CSMA/CD BASEBAND MEDIA

AUTOFACT '85 OSI IMPLEMENTATIONS
Figure 2

Computer networks may be arranged in several configurations. MAP is based on a bus configuration, in which network stations are wired together with a common bus so all can receive all messages. Any station, when sending a message, designates in the address portion of that message where it should go. All other stations receive the transmitted message and examine its address field (a code for the particular station to which the message is being sent). If a station decodes its own address, it takes the appropriate action; otherwise it ignores the message.

With many messages passing through the network, priorities must be established to avoid chaos. MAP uses what is called a token-passing technique to ensure that all stations have an opportunity to send messages. It involves passing a token (special bit patterns or packet) from one station to another. The station that holds the token has momentary control over the medium. The token is passed by all the stations tied to the bus. As the token is passed from station to station, a logical ring is formed. The station transmitting a message may do so for only a preset period of time. The maximum time that any network station must wait before it receives the token and can then transmit is also precisely determined. This is essential for a network handling a large quantity of messages in real time.

Because different types of equipment carrying a variety of signals-- voice, data, and video-- must share the same transmission medium, MAP calls for the use of a broadband cable like that used in cable television systems. The cable eliminates the need for costly individual wires and connectors.

The MAP broadband signaling technique is AM-phase-shift-keying, specified in the IEEE 802.4 standard. Directional matching taps are used to access the cable. The MAP specification calls for a bus speed of 10 megabits per second, which requires two adjacent 6-megahertz channels.

The physical medium and configuration were chosen after an exhaustive debate within communication and factory automation circles. On the factory floor the critical factor in process control is time. The token technique ensures that all signals are transmitted or received within a prescribed time limit that is compatible with process needs. This criterion excluded IEEE 802.3 or contentionbased Ethernet types of networks that cannot guarantee a network component will be able to transmit a message within a given amount of time. Another now-popular configuration called the token ring, which could be applicable on the factory floor, was still under development when the MAP decision was made.

An alternative medium for MAP that may be considered at a future date is fiber optics. At this time, however, the inability to tap easily into fiber-optic cable precludes it from serious consideration for wide use on the plant floor. The lack of recognized standards in this area further dampens the near-term prospects for fiber optic use in MAP.

Early on, a five-step implementation plan was adopted to install MAP in GM facilities. Progress in meeting the goals is close to target. Two recently installed pilot systems in GM plants covering Steps 1 and 2 will soon be evaluated. Step 3, application services, is slightly behind the original schedule but should be in place by the end of this year. Many MAP-compatible products have been introduced by equipment suppliers, and more are expected. Chip makers are developing MAP chips to be introduced this year. A steady stream of MAP-compatible products and MAP chips is expected next year. (See Figures 3 and 4)

For suppliers, MAP development programs have evolved to the point of making products available for network implementation. Users should begin planning for MAP or inquiring about MAP products from computer and communications suppliers. At GM, the first plant-wide implementation of MAP will take place this year at five truck and bus plants and at a factory-of-the-future project in Saginaw, Michigan.

Further MAP development is needed for real-time communication between host computers and intelligent devices on the plant floor. To accomplish this, MAP's broadband communications will have to be linked to work cells communicating via MAP on the carrierband. This development will be needed by GM plants in early 1987.

In the past there was a misconception that MAP was only for the backbone communications network, linked to small, proprietary networks performing local control functions. This would be only a partial solution to the factory communications problem. MAP is necessary throughout the factory at all levels of controls: providing management information at the factory level, coordinating multiple stations at the work cell or work center level, coordinating systems of real-time devices at the station level, and controlling real-time devices at the process level.

On the carrierband (baseband), MAP uses the IEEE 802.4 phase-coherent signaling specification running at 5 megabits per second. The medium is a 75 ohm coaxial cable with nondirectional

A Five-Step Migration Strategy has been followed of which we recently completed Step 3. Major functions and pilot dates for each step are outlined below.

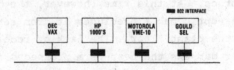

STEP 2: LOCAL AREA NETWORK

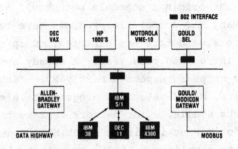

*3 QTR '84

STEP 2: LOCAL AREA NETWORK AND GATEWAYS

STEP 1:   **CENTRALIZED NETWORK**

- Multi-vendor connections via a centralized computer node.
- Terminal emulation via a centralized computer node.
- Completed 2Q84

STEP 2:   **LOCAL AREA NETWORK (LAN)**

- Multi-vendor connetion via a distributed LAN.
- Gateway to selected Programmable Controllers.
- Completed 3Q84

STEP 3:   **APPLICATION SERVICES**

- Enhance Step 2 Lan with additional application services.
- Gateway to wide area network.
- Completed 4Q85

GM IMPLEMENTATION PLAN
Figure 3

173

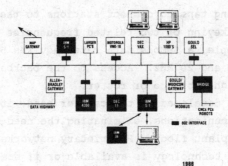

STEP 4: LOW COST HARDWARE

- Reduce ISO Layers 1-4 to hardware.
- Expanded vendor-base
- Proliferate Step 3 on multiple processors.
- Complete ISO Layer 5.
- Add ISO Layer 6.
- 1986

STEP 5: COMPLETE NETWORK UTILITY

- 'Plug compatibility' by majority suppliers.
- 1988

GM IMPLEMENTATION PLAN
Figure 4

impedance-matching taps to connect stations to the trunk cable.
Frequency-shift keying of two related frequencies encodes data on the
cable. For example, a 5-Mb/s data rate requires 5 and 10 MHz; a 10 Mb/s
rate requires 10 and 20 MHz. Access to the backbone MAP communications
network is through a bridge or router.

When MAP is developed on the carrier band, it will provide rapid
peer-to-peer communications, eliminating the need for proprietary
networks on the plant floor. Proprietary networks will be used at GM
only if no other technology is available or if some special situation
or circumstance requires it. Having MAP tied to proprietary networks
via gateways will be discouraged because gateways are expensive, slow,
and cumbersome.

To accommodate all industries interested in using MAP for plant
communications, mini-MAP and Enhanced Performance Architecture have
been developed. Mini-MAP implements on the first two layers of the
specification and is targeted for communications between MAP-based
systems such as bar-code readers, smart sensors, and vision systems.
The Enhanced Performance Architecture is mini-MAP coupled with the
seven-layer OSI architecture. It is designed to assure verification of
the receipt of messages between systems in less than 20 milliseconds.

MAP Version 2.0 products are being debugged at GM plants and MAP
Version 2.1 systems are about to come on stream. The numerical
designations were chosen to freeze technology at certain points in
development so implementation could begin. Later versions of MAP will
build up MAP 2.0 and 2.1 in such a way that MAP 2.0 and 2.1 products
are not made obsolete.

For additional information on the MAP adoption process in the
United States, contact Paul Borawski, Executive Director of Computer
and Automated Systems Association (CASA), Society of Manufacturing
Engineers, One SME Drive, Dearborn, Michigan 48121; telephone number
(313) 271-1500. In Canada, contact Robin Haighton, Canadian Standards
Organization, 178 Rexdale Boulevard, Rexdale, Ontario, Canada M9W 1R3;
telephone (416) 747-4017. In Europe, contact Tony Scarr, Cranfield
Institute of Technology, College of Manufacturing, Cranfield, Bedford,
England MK43 OAL; telephone 44 (1) 920-2260.

SME has available a new videotape, "The Story of MAP and TOP."
The two-tape set examines MAP and TOP specifications local-area
networks and their components, the OSI Model, and the Autofact '85
demonstration. A "graceful migration" plan is introduced that provides
guidelines for companies unable to wait for a full system to be

installed. "The Story of MAP and TOP" is available for purchase or rental. Contact the Society of Manufacturing Engineers, Video Communications Department, One SME Drive, P.O. Box 930, Dearborn, Michigan 48121.

Copies of the MAP 2.1 specification are available upon written request from General Motors Technical Center, Manufacturing Building, A/MD-39, 30300 Mound Road, Warren, Michigan 48090-9040, ATTN: MAP Chairman.

MAP and TOP Users Group meetings, formerly held separately, are now combined. The next MAP and TOP Users Group meeting is scheduled for September in Ann Arbor, Michigan, is open to all who are interested. For additional information, contact the Society of Manufacturing Engineers at the above address.

COMMUNICATION SUPPORT FOR DISTRIBUTED PROCESSING:
DESIGN AND IMPLEMENTATION ISSUES

Liba Svobodova
IBM Zurich Research Laboratory
8803 Rüschlikon, Switzerland

Abstract

This paper presents a brief survey of the various arguments and results concerning design, implementation, and use of special and general-purpose protocols to support distributed processing, especially in high-speed local-area networks. After some examples of special protocols, it focuses on the problem of implementing general-purpose layered protocols to achieve good performance.

1. Introduction

Standardization of communication protocols to support information processing in a distributed environment has reached a great extent and also a wide acceptance. Yet, special protocols, incompatible with the existing and emerging standards, are still being proposed and applied. Why? Performance and complexity are the two primary reasons. The standard protocols are intended to be very general: they must be able to support many different applications in any kind of network environment. That generality usually implies greater complexity is a well-known fact. More man-power and resources are needed to implement and maintain such protocols. And the complexity has a direct impact on the computational overhead of a protocol implementation. The communication speed offered by modern networks, especially local-area networks, becomes easily obliterated by the overhead of higher-level protocols.

The standardization of communication protocols follows the OSI Reference Model [OSIRM], which defines a communication architecture consisting of seven layers, some of which are further divided into sublayers. Layer N builds on services of layer N-1, progressively hiding more and more of the idiosyncrasies of remote communication in what might be a very heterogeneous environment. Standard protocols have been or are being specified for each of the layers. Together, the suite of OSI protocols should be able to support a wide range of distributed applications on top of an internet consisting of many subnetworks differing widely in their topology, access protocol, and performance.

But, each layer represents an added overhead. First, each layer has its own internal protocol, which may generate additional messages, not directly related to the communication at the application level. As part of its protocol, each layer has to construct appropriate headers on the sending side, and analyze these headers on the receiving side. Last but not least, each layer has to communicate with its neighbors; these interlayer interfaces can have a significant impact on performance.

There are two approaches to this problem. The first, applied frequently in distributed systems for local-area networks, is to simplify the communication architecture by reducing the number of layers and streamlining the actual protocols. The second approach is to improve the implementation strategies and techniques for general layered communication architectures. Both of these approaches will be examined.

The paper is organized as follows. Section 2 gives a brief overview of the concepts and terms needed in the following discussion, in particular, the notion and characteristics of connection-mode and connectionless-mode services and protocols. Section 3 summarizes various arguments concerning use of special and general-purpose protocols in local-area networks. Section 4 presents examples of special protocols. Section 5 discusses various implementation issues with emphasis on the techniques for layered communication architectures. Section 6 summarizes the various observations and conclusions of the preceding discussions.

2. Connection-mode and connectionless-mode service and protocols

One of the basic decisions in designing or choosing a communication service is whether the service should be connection-oriented or connectionless. Both modes are covered by the OSI Reference Model [OSIRM, OSIRMa].

In the OSI Reference Model, a connection, or, more precisely, (N)-connection, is defined to be an association established by layer N (the service provider) for the transfer of data between two or more peer entities at layer $N+1$ (the service users). Connection-mode service and protocols have three distinct phases: connection establishment, data transfer, and connection release. While a connection release may be initiated by either one of the users or the service provider, a connection establishment takes place only at the explicit connect request from a user. Connection establishment sets up a logical channel between the users, represented at each end by a record containing data-transfer parameters and connection-state information. Various protocol options and parameter values can be negotiated during this phase, thus allowing certain tailoring of the connection characteristics to a particular environment and situation. During the data-transfer phase, the connection information in the connection record is used to control the addressing, formatting, sequencing, and flow of data units on the connection. Connection release destroys the connection records at both ends.

During the data-transfer phase, a connection maintains the sequence of the data units exchanged between the users, that is, successive Service Data Units (SDU's) submitted by one user are delivered in the same sequence to the receiving user (expedited data being an exception), after being possibly segmented and reassembled by the service provider. Internally, the connection-mode protocol in layer N either associates sequence identifiers with its internal Protocol Data Units (PDU's) or relies on the layers below to ensure proper sequencing. In the former case, the protocol in layer N can (may have to) also include error detection and recovery mechanisms to handle duplicated, misordered, and lost PDU's. Finally, a connection provides means for regulating the flow of data between the two ends.

A connectionless-mode service provides for transfer of SDU's without a prior connection establishment. Each SDU is treated as a self-contained message which bears no relation to SDU's sent to the same destination before or after. The service provider again may have to segment SDU's internally and reassemble them at the receiving side. In fact, a protocol supporting a connectionless-mode service could include similar flow and error-control mechanisms as a connection-mode protocol, but such mechanisms can be effectively applied only for each SDU individually.

Connectionless service and the corresponding protocols are in general simpler and have lower computational overhead than the connection-oriented counterpart. First, the connection establishment and release procedures have been eliminated. Second, in spite of the remark above, the data-transfer protocol is often rather simple. Connection-less service is normally not assumed to be reliable. Thus, even when SDU's are segmented within the layer providing the service, no recovery action is called for when some of the pieces get lost. On the other hand, connectionless mode usually leads to a greater transmission overhead: since each SDU or SDU segment must be completely self-identifying, the headers may end up substantially longer than the corresponding headers in connection-mode protocols.

Finally, let us look at yet another mode, sometimes called 'light-weight connection', which has been used in various special protocols. A connection of this kind usually spans a single logical operation such as a file transfer. Similar to the connectionless-mode, light-weight connection service and protocols do not have explicit separate connection-establishment and release phases. A connection is set-up implicitly by the service provider with the first service call from a user. Subsequent interactions with the user are considered to be logically related until some indication of the end of data transfer or a failure, which is related to the users and which closes the connection. During the data transfer, the protocol provides some degree of flow and error control. This mode thus appears to offer more extensive service than the connectionless-mode service defined in the OSI Reference Model; however, it might be just a matter of a proper interpretation of what constitutes the service specification as opposed to the service interface implementation.

3. Use of special and general protocols in LAN's

In many distributed systems based on local-area networks (LAN's), there is a general tendency to use the simplest protocols possible, in particular, to minimize the number of layers and the use of connected mode. First, a connectionless-mode protocol is used whenever possible. Second, when a connection must be established, it is maintained only for the time needed to satisfy a request. Consequently, connection establishment and release must be inexpensive operations; this requirement leads to the light-weight connection mode discussed above. The desire to take advantage of the low delays and high throughput of LAN's is the primary factor in the drive to simplify the protocols. Short-term connections reflect well the dynamic nature of many distributed applications, and, in particular, of distributed system control.

There have been many arguments concerning the role of LAN's in relation to OSI [BURG84]. On one hand, the whole local-area network with all of the stations attached to it can be viewed as a single system, which may use special, highly-optimized protocols for local communication and resource sharing. To make it an "open system", a shared communication server (gateway) supporting the right OSI protocols can be employed [BRAD83, CHERI83a]. An opposite extreme is to regard each node on a local-area network as an open system, using the OSI protocols for communication within the LAN environment as well [MAP85]. Finally, a node could support both local and OSI protocols, choosing the local ones when the desired partner is on the same LAN, and OSI protocols otherwise. The transport service, being the lowest-level service with end-to-end (process-to-process) significance, plays an important role in the choice between and actual implementations of these different scenarios.

The role of the transport service is to assure reliable exchange of data between processes in different nodes reachable via the underlying network. The transport service is provided by a combination of the transport, network, and data-link layers; the actual mechanisms needed to achieve end-to-end reliable data delivery can be distributed in different ways across these layers. In LAN's, it is common to use connectionless protocols in the data-link and network layers. Consequently, the mechanisms necessary to provide a *general-purpose transport service*, that is, the mechanisms for error detection, recovery, and flow control, need to be provided in the transport layer itself. In contrast, in various special protocols such mechanisms may appear first and only in the application layer, tailored to a particular application.

A very good analysis of the features, advantages, and drawbacks of special-purpose and general-purpose end-to-end transport mechanisms was published by Watson and Mamrak [WATS84]. However, while Watson and Mamrak argue in favor of general-purpose protocols, they are concerned only with the generality with respect to the networks on which such a protocol can run and applications it can support. In this paper, the general protocols of interest are those that ensure also wide compatibility, that is, protocols which are accepted as standards.

Current standardization efforts for local-area networks [MAP85] require use of the ISO/OSI transport protocol class 4, the so-called 'error detection and recovery class', which includes mechanisms for detecting and recovering from lost TPDU's (Transport Protocol Data Units) [OSITP]. The TCP protocol developed for DARPA [POST81a] provides essentially the same mechanisms.[1] This transport protocol, which together with IP (DARPA Internet Protocol [POST81b]) has been implemented on many different systems and operated in complex internetwork environments [LEIN85], has been finding its way to local-area networks as well [BALK85].[2]

---

[1] A good overview of transport protocol functions and mechanisms and a comparison of the ISO/OSI transport and TCP protocols can be found in [STAL84].

[2] TCP was initially used also in Andrew, the ITC/CMU project providing extensive resource sharing over interconnected LAN's [MORR86]. It is being replaced with an internally-designed protocol, to a large degree for reasons not directly attributable to TCP itself, but to the particular TCP implementation on hand.

4. Special protocols: examples

Special protocol designs have two aspects: exploiting some special features of the underlying layers, and tailoring to a specific class of applications. An example of the former is the Basic Block Protocol devised for the Cambridge ring [WILK79], which, using special features of the MAC (Media Access Control) layer, provides an effective end-to-end flow control. Two rather different classes of special protocols of the second type are described below.

4.1. RPC protocols

A great number of specialized protocols have been developed in support of so-called Remote Procedure Call (RPC). Many RPC protocols build directly on top of MAC or a connectionless internetwork service, while incorporating, in a single 'layer', functions, which according to the OSI Reference Model belong to the transport, session, and even application[3] layers [BIRR84].

One of the major problems in designing distributed systems is how to deal with the uncertainties caused by component failures and imperfect communication. Reliable end-to-end delivery of data, such as provided by the transport service, is not sufficient to ensure reliable operation at the application level. Synchronization mechanisms are needed to ensure proper coordination. In the OSI Reference Model, such mechanisms are provided by the session and the application layer, which, however, rely on the reliable communication at the transport layer. In special RPC protocols, the same mechanism, namely, a timeout associated with an application-level request, is used to handle recovery from both unreliable communication and node and application failures.

When a timeout expires, the requestor has two options: abandon the request, or retry. When the request was sent as a datagram, not supported by a reliable connection at any level below, retrying at the request level is the only possible recovery from packet losses due to transmission errors or lack of buffers. However, if the request is still executing or the reply was lost, a retransmission creates another problem, and that is duplicate requests. How to handle duplicate requests has been one of the major problems in the design of RPC; a more detailed discussion of these problems was given in [SVOB84, SVOB85]. In an

---

[3]   In the OSI sense, the application layer does not represent the final application, but service elements to be used by such applications.

RPC based on a reliable transport, it is possible to eliminate the timeout-driven retransmission of a request, and with that also the problems of how to detect and suppress potential duplicates.

4.2. 'Blast' protocols

'Blast' protocols are special protocols for high-speed bulk-data transfer. Usually, they are of the type of light-weight connection, that is, there is no explicit connection establishment phase. Although they build on top of a connectionless-mode service, they use few or no acknowledgements during the data transfer phase. Three specific examples, two file-transfer protocols and one transport protocol, are described below.

The Cambridge File Server of the Cambridge Distributed System employs an extremely simple file-transfer protocol [NEED82]. A file is transferred as a set of basic blocks, without any acknowledgements. At the end of the transfer, the receiving side checks only whether the right amount of data arrived, and whether the checksum on this total data is correct. When not, the whole batch has to be retransmitted. Retransmissions at this level have proven to be extremely rare, because of the low error rate and simple but effective flow-control mechanisms of the Cambridge ring.

A more robust type of blast protocol, actually called BLAST, was designed by David Reed and his colleagues at MIT as part of their investigations of the feasibility and benefits of so-called non-FIFO protocols [REED82]. BLAST [COOP83], similar to the Cambridge file-transfer protocol, postpones recovery from errors and lost packets to the end of a file transfer. However, rather than simply signaling an error and forcing a retransmission of the whole file, the receiver returns a map indicating precisely which blocks need to be retransmitted. The retransmission step may have to be repeated several times if some of the retransmitted blocks become lost or damaged again, but, unless communication is so bad that nothing is received correctly, this procedure clearly converges. BLAST does not require that file blocks have to be transmitted in their sequential order within the file. Thus, one could also optimize the disk access, fetching (and transmitting) file blocks in an order that minimizes the disk seek and rotational delays.

Although blast protocols have been developed mostly for local-area network environments, they could be used in more general environments, especially the latter kind, with selective retransmission; the first kind, relying on full retransmission, behaves poorly even for relatively moderate error rates [ZWAE85].

In fact, the blast mode can be employed in a general transport-level protocol,[4] but, first, it is necessary to solve the problem of having to buffer possibly huge amounts of data for retransmission and resequencing. An important feature of the protocols described above is that they rely on the ability of the application to reproduce the blocks that need to be retransmitted (by rereading them from the disk), and, in case of BLAST, also to receive them out of order. This property must be preserved, but made application-independent. NETBLT (Network Block Transfer) protocol [CLAR86] solves this problem. NETBLT is a connection-mode transport protocol aimed at high-speed transfer of data supplied to it in potentially very large blocks. The buffers to hold these blocks must be provided by the users; the protocol ensures the availability of a receive buffer before a block transfer begins. A block is segmented so that each segment fits into a single network packet. The segments of a whole block are then transmitted in a blast mode with selective retransmission as in BLAST. The buffer is returned to the user only after the entire block has been successfully received or the transfer was aborted by a connection release. Thus, this protocol specifies also the *interface flow control*, which is normally considered to be 'local implementation matter' and as such left out of the service and protocol specifications.

Finally, a bulk-data transfer protocol needs also some internal flow-control mechanisms, especially when built on top of a connectionless-mode service. In NETBLT (and BLAST), flow control is based on a negotiated maximum transmission rate which specifies how many packets the sender may transmit in a given time interval. The transmission rate can be re-negotiated between block transfers. How well this form of flow control can adapt to dynamically changing conditions in the network and the receiver node is an open question [ZHAN86].

5. General protocols: implementation issues

Performance of a communication subsystem depends not just on the protocols used, but to a large extent on their implementation. In fact, it has been stated by various experienced protocol designers and implementers that implementation has more profound impact on performance than protocol design [WATS84, COLE86] (assuming that the different designs provide similar functionality). Recent experimental projects have demonstrated that good performance can be obtained with general protocols, even when implemented on personal computers [SALT85]. Further, OSI protocols (adhering to MAP specifications) are appearing on front-end boards. A detailed simulation study from NBS predicts that, with

---

[4]   In Zwaenepoel's study, blast protocols are treated as transport protocols [ZWAE85].

current technology, these front-end boards can provide 1.5 Mbit/sec throughput and one-way delay between 6 and 10 msec [MILL86].

Fast front-end systems are clearly a very attractive solution to whether and how to support general-purpose protocols. The throughput figures quoted above are better than many results reported for special, problem-oriented designs. In fact, most applications today cannot even utilize such high throughput, in particular, when access to other devices such as a disk or a display is part of the application [MEIS85, SALT85]. However, the figures quoted above are for the case where both the sender and receiver are restricted to a single connection; with multiple connections per station, the throughput per connection will be significantly less. Further, real-time applications and tightly-coupled distributed applications require low delays; here special protocols have demonstrated significantly better results [POPE81, CHERI83b]. Hence, studies that will provide new insights into the implementation issues and development of improved implementation methods remain important research subjects.

The implementation issues fall into two categories: those pertaining to an implementation of a single protocol, and those pertaining to how to build a complete communication subsystem. The latter comprises interlayer interfaces and an incorporation of communication protocols into the local system. Both of these aspects will be discussed in more detail.

5.1. Implementation considerations for individual protocols

Many of the optimizations possible in a single layer are protocol-dependent. However, there are some common problems and implementation rules.

A protocol consists of the actual protocol logic (a finite-state machine) and PDU (Protocol Data Unit) formatting. In an actual implementation, headers have to be composed by the sending entity and decomposed and analyzed by the receiving entity. Processing of headers can be time-consuming, especially on the receiving side [CLAR82]. Here the programming language can have a significant impact, specifically, the data structures available and the efficiency of the operations on these data structures. It might be possible to use (partially filled) header templates on the sending side [SALT85], but the checking of the individual fields on the receiving side cannot be skipped without weakening the robustness of the implementation.

The protocol itself might have many states and mechanisms needed to handle exceptional cases. The general rule is to design the internal structures and procedures so that they favor the normal cases. Although this sounds very common-sense, Clark has assembled a whole list of examples to the contrary [CLAR82].

Finally, there might be mechanisms in a protocol the use of which is not specified rigorously, thus leaving room for optimizations. A typical example is a use of acknowledgements. Since an acknowledgement of the k-th PDU acknowledges all previous ones, it is not necessary to send an explicit acknowledgement for each PDU. The acknowledgement-accumulation strategy can improve throughput significantly, as demonstrated, for example, in [MEIS85]. Another example is the management of timers. The ISO specifications of the transport protocol class 4 require that a timer be associated with each TPDU sent out, but this "association" may be only logical, that is, one timer can be used to protect several TPDU's. This strategy reduces the number of timers that have to be maintained, and thus the cost of setting and cancelling timers.

5.2. Implementation considerations for communication subsystems

The term 'communication subsystem' will be used here to mean a package comprising one or more protocol layers that provides, on one end, access to one or more networks, and on the other end, a well-defined service interface for the user programs. Such a subsystem can be embedded in a particular operating system, implemented as a user process, or off-loaded into a front-end processor. In all three cases, the designer has to solve the problem of interlayer interfaces and the support of the package in the given environment.

It should be kept in mind that although a layered architecture prescribes specific layers and protocols, it does not prescribe that individual layers must be clearly separated in an implementation. Above all, it is not required that each layer provides a service interface that could be used by arbitrary programs. Consequently, various optimizations are possible inside such a communication package. On the other hand, it is also desirable to be able to separate the individual layers, so that they could be replaced by different protocols (protocol classes) if necessary, or ported to various environments which require either different protocols in one of the other layers or a different packaging.

The portability is an important design issue. The portability goal might be set for a communication subsystem as a whole, or, as already discussed above, for the protocol layers individually. In both cases, services expected from the local operating system must be well defined, and mappable to a variety of environments. The possible restriction on the interfaces and the mapping from abstract to local operating system services reduce most likely the efficiency. Unfortunately, although portable implementations have been developed, no data is available on how the portability impacts performance.

The rest of this section will discuss the internal structure of a communication subsystem and related design issues from the point of view of efficiency rather than portability.

5.2.1. 'Softening' layer boundaries

A general implementation model of a layered architecture is to view each layer as an asynchronously running entity, communicating with each of the neighboring layers via a pair of message queues. A layer will remove a message from one of its input queues, process it, and possibly place a message into one or both of the output queues. Given a message-passing multitasking system, a natural solution is to implement each layer as a separate task (process). However, this kind of modularization has a number of drawbacks. The most frequently cited one is the number of context switches imposed by such a communication subsystem in the course of sending and receiving messages on behalf of its users. Although in many multitasking systems the overhead for context (task) switching is very low, it remains significantly higher than typical overhead of a procedure call. Building and buffering of messages needed to pass data and commands (even without data) between layers may have even more significant cost [CLAR85]. Finally, this model makes it difficult to perform optimizations requiring cooperation between layers.

An alternative implementation model was suggested several years ago by David Clark; it has since been applied in several experimental implementations of layered protocols at MIT [SALT82, SALT85]. Clark's structuring method is based on two concepts: upcalls and multi-task modules [CLAR85]. A protocol layer implemented as a multi-task module is basically a collection of subroutines that may be executed concurrently by several tasks. The shared state variables of a layer are protected by monitor locks. The processing of an incoming or outgoing message through multiple layers is handled, as far as possible, within a single task. The communication between layers takes form of procedure calls, where 'upward' procedure calls, named *upcalls*, do the bulk of the work. An upcall, as the name indicates, is a call from a lower layer to an upper layer, that is, from a service provider to a user. Upcalls are claimed to be a natural mechanism not only for passing messages from the network towards the application, but for the opposite direction as well. 'Downcalls', that is, calls from a user to a service provider, are used mostly only as 'arming' calls, to indicate to the provider that a certain service is desired. When the provider is ready to perform it, it will do it via an upcall. Essentially, the objective of this upcall-downcall strategy is to match the natural flow of control in the system. In addition, the upcall mechanism can be used to 'soften' the boundaries between layers: via an upcall, layer N may be able to get information about layer N + 1 that would help it to optimize its own behavior.

An important source of context switches which cannot be eliminated with the sort of structuring outlined above is the arrival of data from the network. Thus, it is desirable to minimize the number of packets that need to be exchanged to perform the service required. First, when a higher-level protocol such as a character-oriented virtual-terminal protocol generates a large number of very small data units, these ought to be concatenated into larger ones whenever possible.[5] The second area for optimization is the use of acknowledgements. Acknowledgment accumulation is one useful strategy. Another one is so-called 'piggy-backing', where an acknowledgement generated by layer N is attached to a data or acknowledgement PDU generated by layer N + 1. However, this latter technique is rather difficult to implement, since layer N alone does not know whether layer N + 1 will have anything to send and thus whether it is advisable to wait with the acknowledgement. The upcall mechanism provides a good solution to this problem [CLAR85].

The above-described structuring approach has yielded encouraging results. However, programming in this way poses many problems and requires many clever tricks. More work is needed to determine whether and how it can be developed into a coherent software-engineering discipline.

Cooper approached the problem of facilitating better cooperation between layers from a different angle [COOP83]. In his proposal, a functional specification of a layer is to be augmented with a *usage model* for which the layer is optimized. The user layer does not have to obey this model, but if it does, the service will be more efficient. An implementation optimized for a specific usage model can also be substantially simpler than a general-purpose implementation [SALT85]. Unfortunately, such an implementation may perform very poorly if it is used in a way very different from the expected usage model. Thus, following this path might necessitate support of several special-purpose implementations[6] of the same protocol in a given system, but even this approach has been

---

[5]   There are various other, and possibly more important reasons for performing such concatenations [NAGL84].

[6]   Special-purpose implementations must not be confused with special-purpose protocols. First, they are *legal* implementations of a given specification, optimized for a particular application. Second, such a specialized implementation does not require the same on the partner's side, although it may not be possible to realize the expected performance benefit without help from the partner.

considered as a potential solution to improving the overall efficiency of a communication subsystem [CLAR82].

5.2.2. Buffer management

Copying data to and from buffers between protocol layers can be a major source of overhead; it has been one of the most widely publicized 'don'ts' in a design of multi-layer communication architectures. Of course, whether data buffers can be passed between layers by reference depends on the nature of the layer boundary. However, even when all layers run in a common address space, the problem of copying cannot always be avoided. For outbound traffic, it is necessary to prefix a header to the data passed from layer $N+1$ to layer N. The space required for the combined headers can either be preallocated in the same buffer that contains the ultimate data, or the headers can be constructed in separate (non-contiguous) memory segments and 'gathered' into one packet by the network adapter before transmission [LASK84]. The former approach works only up to the point when segmentation needs to be performed. At such a point, the data segments either have to be copied into individual buffers with enough space for the rest of the headers, or the 'gathering' approach can be used to pull out these segments directly from the original buffer. For incoming messages, 'scattering', the reverse to 'gathering', is more difficult if possible at all, since the size of the individual headers and thus the beginning of the actual data is not known at the network interface; a special mark would have to be used to facilitate such a separation at this level [LEAC83]. To reassemble segmented data units, the simplest solution is to copy the data segments into a buffer of suitable size.

It would be an oversimplification to concentrate only on the techniques for assembling and disassembling packets without copying. Passing data between layers without copying implies also passing the responsibility for the buffers. A buffer allocated by one layer might be freed by another layer. Or, it might be returned to the layer that allocated it, but as an asynchronous step. This complicates not just the buffer management but also the interlayer interfaces. On the buffer management side, a very important issue is to ensure that the communication subsystem does not end up in a deadlock because it has no free buffers to send or receive a message that would lead to a release of some of the allocated buffers.

Unfortunately, very little information can be found in the literature on these different aspects of buffer management in layered communication subsystems. Upcalls seem to be a good mechanism for buffer handling across layer boundaries. However, a thorough analysis of the trade-offs between copying and not copying and the underlying buffer-management schemes is still missing.

6. Conclusions

This paper surveyed a variety of issues concerning protocol design and implementation. On the design side, it focused on the comparison of special and general-purpose protocols and the arguments which kind to use in high-speed local-area networks. Whether or not use of special protocols is justified, depends to a large extent on the type of distributed system the protocols are to support. Distributed systems that require very close cooperation between the nodes with the goal to provide a single-system image are best treated as such with respect to the outside world. On the other hand, if a LAN is just a part of what ought to be a rather open resource-sharing system (usually concentrating shared resources in a relatively few centralized servers), general-purpose protocols are a better solution. Although it seems feasible to 'open' a LAN-based system by providing gateways to outside networks, this task may turn out to be rather hard for general cases.

The underlying motive of this paper is a belief that it is more beneficial to concentrate on implementation techniques for communication subsystems utilizing standardized general-purpose protocols than on design of special protocols, which, while inherently more efficient, do not contribute towards the ultimate goal of uniform communication in a heterogeneous environment. Besides operational compatibility, wide acceptance of certain standards means that these standards are likely to become available as common components for a large variety of systems. The provision of the MAP protocols on front-end boards demonstrates clearly this trend.

But, to achieve good performance with general-purpose layered protocols is still a challenging design and implementation problem. Although various guidelines for protocol implementers have been compiled, a coherent software-engineering methodology for layered communication architectures that would produce correct and also efficient implementations is still a matter for further research. Such research should include more than the sort of structuring techniques discussed in this paper: generation of protocol code matching the desired interlayer structures directly from protocol specifications and testing are important components of such a methodology.

Acknowledgements

I wish to acknowledge contributions of my colleagues from the Interconnected Systems Group in our laboratory who helped me to 'discover' and understand many of the implementation issues. Thanks also to Harry Rudin, Rainer Hauser, and Robin Williamson for their comments on various technical points and the presentation of this material.

References

BALK85   Balkovich, E., Lerman, S., and Parmelee, R., "Computing in Higher Education: The Athena Experience," *Comm. ACM 28*, 11 (Nov. 1985), pp. 1214-1224.

BIRR84   Bifrell, A.D. and Nelson, B.J., "Implementing Remote Procedure Calls," *ACM Trans. Computer Systems 2*, 1 (Feb. 1984), pp.39-59.

BRAD83   Braden, R., Cole, R., Higginson, P., and Lloyd, P., "A Distributed Approach to the Interconnection of Heterogeneous Computer Networks," *Proc. ACM SIGCOMM'83 Symposium on Communications Architectures and Protocols*, Austin, Texas, March 1983, pp. 254-259.

BURG84   Burg, F.M., Chen, C.T., and Folts, H.C., "Of Local Networks, Protocols, and the OSI Reference Model," *Data Communications*, Nov. 1984, pp. 129-150.

CHERI83a   Cheriton, D.R., "Local Networking and Internetworking in the V-System," *Proc. 8th Data Communications Symposium*, Cape Cod, Massachusetts, Oct. 1983, pp. 9-16.

CHERI83b   Cheriton, D.R. and Zwaenepoel, W., "The Distributed V Kernel and its Performance for Diskless Workstations," *Proc. of the 9th ACM SIGOPS Symposium on Operating Systems Principles*, Bretton Woods, New Hampshire, Oct. 1983, pp. 129-140.

CLAR82   Clark, D., "Modularity and Efficiency in Protocol Implementation," Internet Protocol Implementation Guide, Network Information Center, SRI International, Menlo Park, California, August 1982.

CLAR85   Clark, D., "The Structuring of Systems Using Upcalls," *Proc. of 10th ACM SIGOPS Symposium on Operating Systems Principles*, Orcas Island, Washington, Dec. 1985, pp. 171-180.

CLAR86   Clark, D., Lambert, M., and Zhang, L., "NETBLT: A Bulk Data Transfer Protocol," Network Working Group, RFC No. 969, Laboratory for Computer Science, Massachusetts Institute of Technology, Cambridge, Mass., January 1986.

COLE86   Cole, R. and Lloyd, P., "OSI Transport Protocol -- User Experience," *Proc. of Open Systems '86*, Online Publications, 1986, pp. 33-43.

COOP83   Cooper, G.H., "An Argument for Soft Layering of Protocols," MIT/LCS/TR-300, Laboratory for Computer Science, Massachusetts Institute of Technology, Cambridge, Mass., May 1983.

LASK84   Lasker, V., Lien, M., and Benhamou, E., "An Architecture for High Performance Protocol Implementations," *Proc. IEEE INFOCOM 84*, San Francisco, Calif., April 1984, pp. 156-164.

LEAC83   Leach, P.J., Levine, P.H., Douros, B.P., Hamilton, J.A., Nelson, D.L., and Stumpf, B.L., "The Architecture of an Integrated Local Network," *IEEE Journal on Selected Areas in Communications SAC-1*, 5 (Nov. 1983), pp. 842-857.

LEIN85   Leiner, B.M., Cole, R., Postel, J., and Mills, D., "The DARPA Internet protocol suite," *IEEE Communications Magazine 23*, 3 (March 1985), pp. 29-34.

MAP85   Manufacturing Automation Protocol, Version 2.1, General Motors Corp., Warren, Michigan, March 1985.

MEIS85   Meister, B., Janson, P., and Svobodova, L., "Connection-Oriented Versus Connectionless Protocols: A Performance Study," *IEEE Trans. on Computers C-34*, 12 (Dec. 1985), pp. 1164-1173.

MILL86   Mills, K., Wheatley, M., and Heatley, S., "Prediction of Transport Protocol Performance Through Simulation," *Proc. of ACM SIGCOM'86 Symposium on Communications Architectures and Protocols*, Stowe, Vermont, Aug. 1986.

MORR86   Morris, J.H., Satyanarayanan, M., Conner, M., Howard, J.D., Rosenthal, D.S.H., and Smith, F.D., "Andrew: A Distributed Personal Computing Environment," *Comm. ACM 29* 3 (March 1986), pp. 184-201.

NAGL84   Nagle, J., "Congestion Control in IP/TCP Internetworks," *Computer Communication Review 14*, 4 (Oct. 1984), pp. 11-17.

NEED82   Needham, R.M. and Herbert, A.J. *The Cambridge Distributed Computing System*, Addison-Wesley Publ. Co., Reading, Massachusetts, 1982.

OSIRM   "Information Processing Systems -- Open Systems Interconnection -- Basic Reference Model," International Organization for Standardization, International Standard 7498, 1984.

OSIRMa   "Information Processing Systems -- Open Systems Interconnection -- Addendum Covering Connectionless-mode Transmission," International Organization for Standardization, International Standard 7498/AD 1, 1984.

OSITP   "Information Processing Systems -- Open Systems Interconnection -- Connection Oriented Transport Protocol Specification," International Organization for Standardization, International Standard 8073, 1984.

POPE81   Popek, G., Walker, B., Chow, J., Edwards, D., Kline, C., Rudisin, G., and Thiel, G. "LOCUS: A Network Transparent, High Reliability Distributed System," *Proc. 8th ACM SIGOPS Symposium on Operating Systems Principles*, Asilomar, California, Dec. 1981, pp. 169-177.

POST81a   Postel, J.B., Internet Protocol, DARPA Internet Program Protocol Specification, Sep. 1981.

POST81b   Postel, J.B., Transmission Control Protocol, DARPA Internet Program Protocol Specification, Sep. 1981.

REED82   Reed, D.P., Computer Systems Structures, Annual Report 1981-1982, Laboratory for Computer Science, Massachusetts Institute of Technology, Cambridge, Mass., 1982.

SALT82   J. H. Saltzer, and D. D. Clark, Computer Systems and Communications, Annual Report 1981-1982, Laboratory for Computer Science, Massachusetts Institute of Technology, Cambridge, Mass., 1982.

SALT85   Saltzer, J.H., Clark, D.D., Romkey, J.L., and Gramlich, W.L., "The Desktop Computer as Network Host," *IEEE Journal on Selected Areas in Communications SAC-3*, 3 (May 1985), pp. 468-478.

STAL84   Stallings. W., "A Primer: Understanding Transport Protocols," *Data Communications*, Nov. 1984, pp. 201-215.

SVOB84   Svobodova, L. "File Servers for Network-Based Distributed Systems," *Computing Syrveys 16*, 4 (Dec. 1984), pp. 353-398.

SVOB85   Svobodova, L., "Client/Server Model of Distributed Processing," *Kommunikation in verteilten Systemen I*, (Krüger, G., Spaniol, O., and Zorn, W., Editors) Springer-Verlag, 1985, pp. 485-498.

WATS84   Watson, R.W. and Mamrak, S., "Special or General Purpose End-to-End Mechanisms in Distributed Systems: One View," *Proc. 4th International Conference on Distributed Computing Systems*, San Francisco, Calif., May 1984, pp. 154-165.

WILK79   Wilkes, M.V., Wheeler, D.J., "The Cambridge Digital Communications Ring", *Proc. of the NBS Local Area Communications Network Symposium*, Boston, Massachusetts, May 1979.

ZHAN86   Zhang, L., "Why TCP Timers Don't Work Well," *Proc. of ACM SIGCOM'86 Symposium on Communications Architectures and Protocols*, Stowe, Vermont, Aug. 1986.

ZWAE85   Zwaenepoel, W., "Protocols for Large Data Transfers Over Local Networks," *Proc. 9th Data Communications Symposium*, Whistler Mountain, British Columbia, Sept. 1985, pp. 22-32.

COMMITMENT, CONCURRENCY, AND RECOVERY - ISO-CASE AND IBM LU.62

J Larmouth
Director
Information Technology Institute
University of Salford
Salford M5 4WT
England

This paper describes the Commitment, Concurrency and Recovery (CCR) features found in the ISO Common Application Service Element (CASE) for CCR and the SYNCPT verb of IBM's LU6.2. The International Standard for CCR is contained in ISO 8649 Part 3 and ISO 8650 Part 3.

1.   History and objectives

The use of "two-phase commitment" to handle "simultaneous" updates of two or more separate resources has been known for many years in database work, and has become a common feature of "transaction" protocols.

Much of the early impetus in ISO came from the needs of the JTM work (see Chapter 14), where the distributed nature of JTM activity led to a requirement for commitment. The JTM work also emphasised the need for well-defined protection against crashes of systems, and the interaction of this with "commitment" handshakes.

Finally, the crucial interaction with concurrency controls was recognised and the term "CCR" - Commitment, Concurrency, and Recovery control was born. CCR was absent from SNA protocols until the introduction of LU6.2. The IBM and ISO approaches to CCR were developed independently of each other, but there was a very strong similarity of function and mechanism between the SYNCPT verb of LU6.2 and the ISO CCR Standard. This similarity was increased when, at a late stage in the development of ISO CCR, a detailed comparison with the LU6.2 SYNCPT verb functions was undertaken. This showed some features of LU6.2 which could usefully be added to the ISO Standard. These were

        a)   heuristic commitment; and

        b)   a single protocol round-trip exchange to end one atomic action and start the next one.

These have been incorporated in the final International Standard in a way which enhances the current LU6.2 features.

The result permits a program interface to be developed for transaction processing which could be supported by either the LU6.2 protocol or the ISO CCR protocol.

CCR is designed to ensure successful coordination and completion of activities distributed across several open systems, taking into account the possibility of network failures and system crashes.

CCR is often described as coordinating across several connections synchronisation of communications activity which, for a single connection, is handled in ISO by the session layer. A dual, but perhaps more useful, focus, is to regard it as coordinating the information processing at the nodes linked by the connections.

The ISO work neared stability at the end of 1983, but at this time there was a growing interest in use of CCR for other applications - transaction processing, remote database access, directory updates, which introduced elements of instablility.

At the time of writing this paper (July 1986) the final form of the initial ISO CCR Standard is not certain. There is also instability in the precise way the ISO CCR specifications are to be included in other standardisation work, revolving around the so-called "cooperating main service" issue.

This paper describes the main features of CCR, present in the second ISO Draft International Standard, and broadly present also in the LU6.2 SYNCPT verb. Where differences exist, the ISO approach is described first, and LU6.2 is covered in later sections.

2.   Application of CCR facilities

CCR is primarily concerned with exception cases. It worries about what happens if a crash (loss of information, release of concurrency controls) occurs at critical points in an activity, or if one update succeeds but a related update cannot be done because the network has gone down.

This means that almost any application using CCR can be run without CCR, and will work a lot of the time. Nonetheless, for almost all applications, use of CCR gives a degree of reliability which is generally highly desirable.

The following are the main areas where use of CCR, or some CCR-like handshake, is highly desirable:

- no-loss, no-duplication transfer of material from system A to system B, without requiring an indefinite retention of knowledge of completed transfers or the intervention of human intelligence on crashes.

- up-dates to two different systems by a third party, where it is essential for neither or both updates to occur before other users access the affected information.

- any remote operation where guarantees are required of precisely one performance of the operation, without requiring an indefinite retention of knowledge of completed operations or the intervention of human intelligence on crashes.

Within ISO, the CCR Standards are used (mandatory) for Job Transfer and Manipulation and (optionally) for File Transfer, Access, and Management, and have been proposed for use in International Standards being developed for Remote Database Access, Reliable Transfer Service, Transaction Processing, and Common Directory Services.

3.   What is Commitment, Concurrency, and Recovery?

In a simple protocol, an initiator requests an action and a responder either performs the action and acknowledges it or refuses to do it while providing an appropriate diagnostic. This simple protocol suffers from two serious flaws for general use. These flaws arise firstly from crashes of one or another system and secondly from the need for an initiator to work with several responders simultaneously (e.g. to debit one bank account and credit another).

The first flaw, then, is concerned with crashes. Suppose the initiator receives no reply in the simple protocol outlined above. In connectionless operation, this usually means no reply in some finite time; in connection-oriented operation, this usually means some form of reset or disconnect or provider-generated abort. If the crash was due to a network failure, certain classes of connection-oriented transport service in ISO and lower level protocols in SNA permit recovery. If, however, one or another end-system or end-application crashed (described in the ISO CCR Standard as an application-failure), the lower level protocols offer no help. In this case, the initiator does not know whether the action was performed but the crash lost the acknowledgement or whether the action was refused but the diagnostic was lost, or whether the action request was lost. (Illustration 1 shows these options.)

Illustration 1 : Problems in a simple protocol

Normal case:

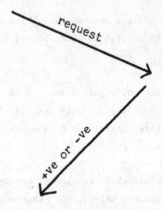

request

+ve or -ve

Diagnosis: Action done if response +ve
Action not done if response -ve

Exceptional cases:

   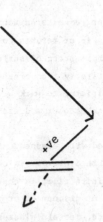   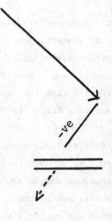

+ve   *-ve*

Not done   Done   Not done

Diagnosis: ????

A simple "recovery" protocol puts an identifier on the request. Following the failure case, the initiator retries the action with the same identifier, and the responder who has remembered the identifier (on disc) detects a duplicate request and reacts accordingly. However, for how long must the identifier be remembered? One protocol says "until a different identifier comes from the same source". This works fairly well unless you want multiple simultaneous activities between the two, but still involves holding the information indefinitely following the "last" operation.

"Indefinitely" needs qualifying. What we really require is "for a period which is long compared with the expected recovery time of the initiator". Now suppose we guess wrong, and it is too short. The "reliability" breaks down and we are back to humans sorting the mess out. Depending on the application, the risk of loss of reliability may need to be low or very low, but in almost all cases it is important for the breakdown to be reliably detected and signalled to a human. Some of these problems will be clearer when heuristic commitment is discussed.

Many experts believe that the occasional duplication of an action is inherent in protocol design. This is not so. The ISO CCR Standard, and the LU6.2 SYNCPT verb, if incorporated in an application, can be used to provide "no loss, no duplication" operation. The following actions are ones where risk of duplication ranges from "no problem" to intolerable:

> Reading a file (no problem)
> Writing a file (provided the duplicate action occurs before
>   the file has been seen and, perhaps, edited ...)
> Sending mail to a human (humans are good at spotting duplicates)
> Running a batch job (duplication is often accepted, but has
>   nasty side-effects)
> Appending to a file (Ugh!)
> Debiting a bank account (nasty)

In general, application-specific mechanisms can often be used to determine whether an action has occurred or not. Unfortunately, this becomes arbitrarily complex if, before communications can be resumed between initiator and responder (e.g. the initiator system requires serious engineering attention, and is not running again for two days) some other initiators generate actions at the responder. This relates to concurrency (concurrent access) aspects. The ISO CCR Standard recognises the need to prevent access by other users when distributed resources are in an inconsistent state, and LU6.2 has the concept of a "protected resource".

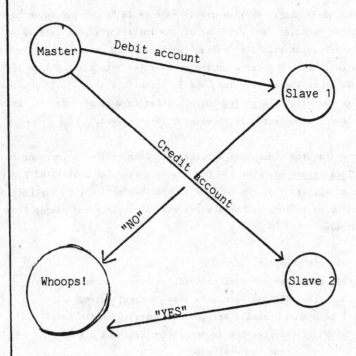

Illustration 2 : Updating two slaves

Now   assume a network crash before Slave 2
can be debited back to its original state.
And another user withdraws the cash!   The
reverse situation can be almost as bad for
large sums - thousands of pounds of
interest are lost!

The second flaw in the simple protocol is exemplified by one system (often called the master or superior) initiating changes on two (or more) other systems (often called the slaves or subordinates), such that, for consistent operation, either both changes must occur simultaneously or neither must occur. (Simultaneously means "before other users access the resource" - concurrency again.) A typical example is debiting an account at System A and crediting that money to an account at System B. In the simple protocol, the master "commits" himself (to the possibility that the action will happen) when he makes the request. (See illustration 2.)

The ISO CCR protocol and the LU6.2 SYNCPT verb ensure that the master retains control. In what is called "Phase I" of the CCR handshake, the superior initiates the action and receives a commitment by both subordinates to perform it (or refusal by one or more of them). At this stage, concurrency controls are in place to prevent the changes being aparent to other users. The subordinate undertakes to monitor actions by other users to ensure that the changes to which it committed will continue to be possible, and that "rollback" to the initial state also remains possible (because the master has not yet committed to the action). It is important here to note that these requirements do not in general, mean that other users are completely locked out of the resource. Considerable flexibility can be adopted over permitting the use of uncommitted data - particularly if the second use is subject to CCR and the custodian of the data (the subordinate) can still ensure rollback of the second use if the expected commitment of the first change is not forthcoming. The precise means of concurrency control will in general depend on the nature of the protected resource and implementation decisions.

Again, a good example comes from debiting and crediting a bank account, where the unaffected balance can be available for other users at all times (see 7.). Another example comes from commitment to a credit card debit, when a hotel verifies your credit card when you check in. For example:

> On checking into a hotel, the hotel (the commitment master) asks the credit card company "Will you commit to making a debt for up to this amount?" The credit card company says "Yes" and provides an "atomic action identifier" (authorization code). At the end of your stay, the hotel commits the debit by sending in the bank copy with the "atomic action identifier" on it. Until this action is committed (to the same or a lesser amount), the credit card company avoids any additional credit commitments that would place you over your credit limit.

If one or more subordinates refuse an action, the master (using the ISO CCR C-ROLLBACK primitive or the LU6.2 BACKOUT verb) will (in what is called "Phase II") order rollback, restoring all resources to their initial state, and releasing

cocurrency controls. If, on the other hand, all subordinates have offered commitment, the master will, in Phase II, "order commitment", that is to say, require the
action to occur and concurrency controls to be released. (In ISO, C-ROLLBACK is
only issued by the master, but it is requested by a subordinate using C-REFUSE. In
LU6.2 the subordinate as well as the master can issue BACKOUT. The implications of
this difference are described later.)

The extra handshakes (protocol) needed to mend each of the two flaws in the simple
protocol, and their interactions with concurrency controls, are the same for each
flaw. Thus Commitment, Concurrency, and Recovery is a single set of procedures.
Separation of the three features is not meaningful.

The CCR mechanisms (or their equivalent) are useful (some would say essential)
whenever more than two parties communicate. These mechansisms are also useful (some
would say essential) in the two-party case if application-independent recovery is
required. Application-dependent recovery is often extremely difficult (due to
access by other users) - even when using human intelligence.

4.   Terminology

The following sections introduce some of the terminology used in the CCR work. It
is as well for the reader to have available at this point a list of broad
equivalences between the notation and terminology of ISO CCR and that of LU6.2.

ISO CCR	LU6.2
service primitive	verb
atomic action	unit of work
bound data	protected resource
	recoverable resource
rollback	backout
master	initiator

5.   The ISO CCR exchanges

The ISO CCR service and protocol operates on each of the two-way communications that
form part of some tree of activity. The master at the top of the tree communicates
with subordinates, each of which can act as a superior to further subordinates, and
so on. The ISO CCR Standard defines the protocol (and corresponding local actions)
between a superior and a subordinate. The local actions for a superior specify

Illustration 3 : A CCR tree

The distributed application in this example involves the application-
entity instances A, B, C, D, E, F, G, H and J.   These may all be
different application-entities on different open systems, or may have
some degree of overlap.

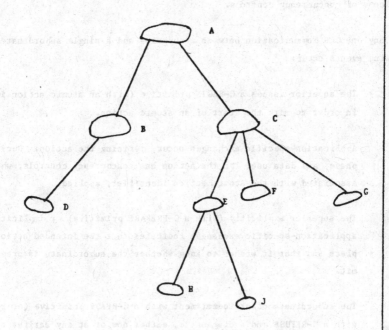

Master: A

Subordinate	Superior
B, C	A
D	B
E, F, G	C
H, J	E

rules that link the exchanges with all its subordinates (and its own superior, if any), to ensure that the master retains control at all times. The CCR atomic action tree is shown in illustration 3.

The term atomic action describes the piece of work performed during one invocation of CCR. (In LU6.2, it is called a unit of work.) The action is called atomic because, to an outside observer, either all the changes (on all systems) required by the action occur, or none of them occur (the action is rolled-back by the master). Ensuring that an action is atomic is dependent on the coordinated application and release of concurrency controls.

On any one CCR communication between a superior and a single subordinate, the following events occur:

1.　　　The superior issues a C-BEGIN primitive (with an atomic action identifier) in order to mark the start of an atomic action.

2.　　　Application-specific exchanges occur, defining the action. During this phase, any data used for the action has concurrency controls, which are associated with the atomic action identifier, applied.

3.　　　The superior explicitly (with a C-PREPARE primitive) or implicitly (by an application-specific exchange) indicates that the intended action is complete and that it wishes to know whether the subordinate is prepared to commit.

4.　　　The subordinate offers commitment with a C-READY primitive (or refuses it with a C-REFUSE and a diagnostic, either now or at any earlier time). From this point on, the subordinate is required to ensure that both commitment to the action and rollback continue to be possible. This is achieved by applying concurrency controls to all affected resources.

5.　　　The superior orders commitment by a C-COMMIT primitive, or rollback (restoration of the initial state of resources) by a C-ROLLBACK primitive.

6.　　　The C-COMMIT or C-ROLLBACK is a confirmed service, the response/confirm providing an acknowledgement by the subordinate to the superior that all concurrency controls have been released, and the subordinate has "forgotten" the action. When this is received, the superior can also "forget" the action and, if it wishes, can even re-use the atomic action identifier in a subsequent action. Note the critical nature of the final message in telling the master his responsibilities are at an end.

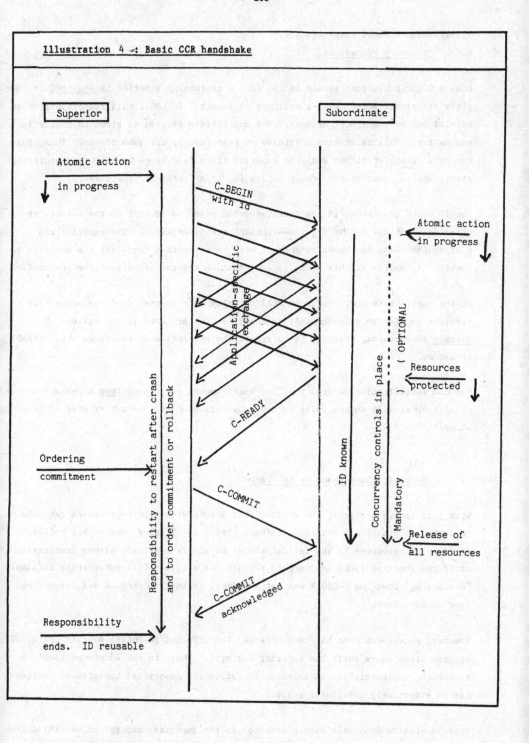

Illustration 4 —: Basic CCR handshake

Illustration 4 shows this sequence.

6.   Recovery Procedures

Once a C-BEGIN has been issued in ISO CCR, a conforming superior is <u>required</u> to complete the atomic action with a rollback or commit. In LU6.2 this responsibility is
carried out by the SYNCPT Manager. The application program is given no choice in
the matter. This is to ensure release of resources by the subordinate. Thus, following a crash (of either end) the superior tries (and keeps trying) to restart the
atomic action, quoting the atomic action ID, in a C-RESTART primitive.

The ID would be unknown if the crash occurred prior to receipt of the C-BEGIN or
after issue of the C-COMMIT (or C-ROLLBACK) acknowledgement. The superior can
distinguish these two cases because it will have recorded (on disk) its decision to
commit. An update at this point is essential to the correctness of the protocol.

At the subordinate end, resources will normally be protected (and the protection
recorded on disk to guard against crashes) as they are used in the action. The
<u>minimum</u> requirement, however, is to record the protection of resources when C-READY
is issued.

The CCR specification ensures that its requirements are the <u>minimum</u> necessary number
of disk updates to ensure fully reliable operation of CCR, no matter when crashes
occur.

7.   Heuristic Commitment or Rollback

With this simple protocol, the actions that a subordinate implementation can take
following a "crash" are very constrained. Prior to offering commitment, rollback
(release of resources in the initial state) is always possible, either immediately
or if the superior takes too long to restart the action. This subordinate is simply
"pretending" that the C-BEGIN was not received, or that a C-REFUSE was issued, but
lost in the crash.

However, once commitment has been offered, the affected resources must be protected
against other users until the superior restarts. Thus, in the simple protocol, a
conforming implementation is expected to retain the associated concurrency controls
for an effectively indefinite period.

What is missing from this simple protocol is the important concept of heuristic commitment. This concept was introduced into the ISO work only following a study of
LU6.2, where the concept was seen as vital for operation in the real world. The

precise handling of heuristic commitment varies between different IBM operating
systems incorporating LU6.2. The ISO Standard includes enough flexibility to cover
all current IBM approaches, but also allows limited negotiation of heuristic commitment parameters - a feature missing from the current LU6.2.

Heuristic commitment (or rollback) permits a subordinate who has offered commitment,
and who subsequently loses communication with its superior or crashes, to decide
unilaterally either to commit - heuristic commitment - (or to rollback - heuristic
rollback) if the superior fails to restart soon enough.

This is called heuristic commitment or rollback because the subordinate must guess
whether:

      a)   the superior is more likely to order commitment than rollback; and

      b)   whether guessing wrong does more damage in the commitment or the
         rollback case.

The decision is normally expected to be taken by a human being!

The disadvantage of a heuristic decision is that the whole CCR service fails if the
guess is wrong - the atomic nature of the action is lost, and distributed resources
can be left in an inconsistent state (e.g. a bank account has been debited, but no
account has been credited), and loss or duplication of actions can arise.

The need for a heuristic decision arises from the undesirability of applying
concurrency controls to a resource for an indefinite period (especially if "locking"
is used - see 8.). In general, the need for heuristic action will depend on the
demands of other users to access the resource and the importance of meeting those
demands.

In the full OSI CCR protocol, the superior, on C-BEGIN:

      a)   can say how long it requires the subordinate to wait (after a
         C-READY) before indulging in heuristic commitment or rollback; the subor-
         dinate accepts this constraint by issuing C-READY - if the contraint is
         unacceptable, it issues C-REFUSE; and

      b)   can say "please make the heuristic action (if taken) commitment", or
      "please make the heuristic action (if taken) rollback", or "you can choose".

This level of user control over heuristic commitment is missing from the current
LU6.2.

It is important to distinguish release of resources (concurrency control) on
heuristic commitment or rollback from the complete amnesia which occurs on a normal
commitment or rollback. The subordinate who takes a heuristic decision has a
responsibility to retain knowledge of the atomic action, and the decision it took,
until the master has attempted restart, been told a heuristic occurred, <u>and told the
subordinate it has received that</u> information. Only then can the subordinate forget
the identifier, and only when the master has been told this has been done does the
master's responsibilities end. Thus heuristic commitment adds significant complication and extra hand-shakes to the restart procedures.

8.   Concurrency Controls

A very crude approach to protecting resources to permit later orders to commit or
rollback is to apply "locks" which prevent access by all other users. ISO CCR
recognises that this is too simplistic an approach. Consider, for example, debiting
an account. Provided there is still money left, other debits should clearly be
allowed "simultaneously" (while commitment is offered), and certainly credits should
be allowed!

ISO CCR defines "the period of use of a datum" by an atomic action, which is the
time from first to last reference to the datum. The CCR requirements are simple:

>   a)   an implementation is required to rollback an atomic action if, dur-
>   ing the period of use of a datum by the atomic action, the datum is changed
>   by other actions; and

>   b)   an implementation is required to prevent an atomic action from com-
>   mitting if it has used any datum that has been changed by an atomic action
>   that has not yet committed.

These requirements can be satisfied in many ways. For example, if a datum is locked
(preventing change by other actomic actions) from first use to the end of Phase I,
a) above is automatically satisfied. It can be equally satisfied by issuing a
C-REFUSE (BACKOUT in LU6.2) during an action if a) is found to have been violated.

In the case of b), locking (preventing another action from accessing the resource
once commitment has been offered) is one possibility, but another would be to allow
other actions to proceed on an assumption that commitment will occur, and to delay

the offer of commitment for these actions until the earlier commitment arrives. Yet a third variant covers the case of debiting an account. Here the data (the balance) can be regarded as two separate pieces of data; one is the residue after debit, which is unaffected by the atomic action, and the other is the bit being (potentially) removed, which disappears on commitment or is reinstated on rollback. Any subsequent debits that can be supported by operations on the stable residue part (the account has enough money in it for both debits) can now proceed to commitment.

9.   Additional features

Three aspects of the CCR operation are worth discussing but are not covered by the initial CCR Standard (or by LU6.2).

The first aspect is (global) checkpointing, which covers the totality of the actions up to some point in time within the entire atomic action tree. (This should not be confused with checkpointing of a one-way flow of data between two parties; it is both more powerful and more "expensive" in round-trip exchanges.) This is important for long atomic actions in order to prevent crashes from having to be recovered by starting the entire atomic action again from the beginning.

The second aspect is that of nested atomic actions. Consider, for example, a series of atomic actions ordering a list of goods from a series of warehouses, and perhaps involving negotiating with a transport contractor (for each one). The ordering of each item in the list involves a commitment handshake to handle the correlation of the transport contractor commitment and the warehouse commitment, and hence can usefully be structures as an atomic action. However, the customer may wish to abandon the entire list unless he gets about 80%-of it satisfied from this supplier. Thus he wishes to have an outer atomic action wrapped around the series of inner ones.

The final aspect involves <u>reopening</u> Phase I of an atomic action. The point here is best illustrated by the hotel and credit card example. In Phase I, the credit card company commits to a certain size of debit. The actual debit is, however, less. In the real world this is handled by indicating the lesser debit when ordering commitment in Phase II.

With OSI CCR and LU6.2, however, the choice is to either:

> a)   commit the greater debit, then do another atomic action to add a
> credit for the difference; or

b)   rollback the first atomic action, then do the actual debit as a new atomic action, risking other users getting in inbetween, and hence failure of the second debit.

What is required is the opportunity to renegotiate a commitment offer (to the lesser debit) while retaining the possibility of committing the earlier offer.

10.   CCR Service Primitives and Parameters

This section provides a more detailed discussion of ISO CCR exchanges. At this level of detail the differing architectures of ISO and SNA make a detailed comparison difficult.

The CCR service primitives are listed below:

Primitive	Issued by	Confirmed?
C-BEGIN	Superior	No
C-PREPARE (optional)	Superior	No
C-READY	Subordinate	No
C-REFUSE	Subordinate	No
C-COMMIT	Superior	Yes
C-ROLLBACK	Superior	Yes
C-RESTART	Superior or Subordinate (only if communications have not failed)	Yes

The parameters of C-BEGIN are:

atomic action identifier - an application-entity-title that unambiguously identifies the master, and a suffix that is a character string unambiguously identifying the action among all current actions with the same master.

branch identifier - the application-entity-title that unambiguously identifies the superior (and which is carried in the Association Control CASE protocol) together with a suffix that is a character string unambiguously identifying this branch of the atomic action tree among all branches of the same atomic action with the same superior.

atomic action timer - (optional and advisory) - a signed integer value (N) that warns the subordinate that the superior intends to rollback the atomic action if it is not completed in 2**N seconds.

heuristic timer - (optional) if present, specifies the time the subordinate is required to wait after offering commitment before it is permitted to perform heuristic commitment or rollback. It is either "indefinite" or a value in seconds obtained like the atomic action timer. (Note that the timer is permissive. Heuristic commitment is not required to occur after its expiry.)

heuristic-decision - (optional) if present, constrains any heuristic decision to be the one it specifies (i.e. either COMMIT or ROLLBACK).

user data - carries information determined by the SASE with which CCR is incorporated.

e C-PREPARE, C-READY, and C-REFUSE primitives carry only user data. The user data C-REFUSE is expected to carry structured diagnostic information.

e C-COMMIT and C-ROLLBACK primitives carry no parameters.

e C-RESTART request and indication carries the following parameters:

atomic action identifier - the same as C-BEGIN.

branch identifier - the same as C-BEGIN.

restart timer - (optional) the same form as the atomic action timer; the superior intends to break the association (and attempt the restart later) if a reply is not received within the stated time.

resumption point - ACTION if the superior has not issued (prior to the crash) a C-COMMIT or C-ROLLBACK, otherwise it is COMMIT or ROLLBACK respectively.

> **user data**

The C-RESTART response and confirm carries only:

> **resumption point** - DONE, RETRYLATER, REFUSED, COMMITTED, ROLLEDBACK, MIXED
> or ACTION.

A response of DONE indicates either that the C-BEGIN was not received or a
C-ROLLBACK or C-COMMIT response has been issued (atomic action ID unknown), or that
the C-RESTART ordered COMMITT or ROLLBACK and commitment or rollback (respectively)
has occurred (either now or by an earlier heuristic).

The response RETRYLATER indicates that restart is not possible at this time.

A response of REFUSED indicates that the subordinate has previously issued a
C-REFUSE for this atomic action.

A response of COMMITTED (or ROLLEDBACK) means that ROLLBACK (or COMMIT) had been
ordered, but that a previous heuristic commitment (or rollback) decision had been
taken (respectively).

A response of MIXED means that some subordinates of the subordinate have
heuristically committed and others have heuristically rolled back, or that a partial
heuristic has been taken, with some resources still available for commitment or rollback.

Following a COMMITTED (or ROLLEDBACK) response, the superior issues a C-COMMIT (or
C-ROLLBACK), respectively). (This final handshake is needed to ensure that the
superior has received notification of the incorrect heuristic decision before the
subordinate forgets the atomic action.)

Following a MIXED response, the superior issues either a C-COMMIT or C-ROLLBACK to
commit or rollback any remaining resources, but with a "MIXED" parameter present to
acknowledge receipt of the information that heuristics have occurred.

Following a response of ACTION, the atomic action is restarted, either from the
beginning or from an application-specific checkpoint, or from a CCR global checkpoint.

11.   The ISO CCR Protocol

The CCR specifications contain the following features:

a)   the syntax and encoding of a set of messages capable of carrying the information on the CCR service primitives;

b)   a specification of the semantics and local actions associated with these messages, and the order in which they can be sent;

c)   a means of transferring these messages using the Presentation service.

The correspondence between the messages for each service primitive and the presentation service primitive used to carry them is shown below:

CCR Service	Presentation Service
C-BEGIN	P-SYNC-MAJOR
C-PREPARE	P-TYPED-DATA
C-READY	P-TYPED-DATA
C-REFUSE	P-RESYNCHRONIZE (ABANDON)
C-ROLLBACK	P-RESYNCHRONIZE (ABANDON)
C-COMMIT	P-SYNC-MAJOR
C-RESTART	P-RESYNCHRONIZE (RESTART)

12.   Cooperating Main Services

The inclusion of c) above in the CCR Standard has been a matter of controversy, because:

1)   the ISO FTAM protocol conveys CCR messages as part of its own PDUs (in order to avoid a round-trip time for FTAM and for CCR, and in order to make minimum use of session features)

2)   extending protocols based on use of S-ACTIVITY (as most early CCITT
protocols were) by adding the CCR messages as further (optional) fields in
the existing protocol provides an easy upgrade path, but adding extra
S-SYNC-MAJORs is more or less impossible;

3)   one view of Upper Layer Architecture is that CASE Standards have no
business specifying the means of transfer; it is for the designer of the
SASE making use of the CASE to put together the total protocol in whatever
way he think is best, using all available tools.

In partial recognition of these points, an annex to the CCR Standard describes the
use of CCR by "a cooperating main service", which carries CCR messages in fields of
its own PDUs, as FTAM does.

The cooperating main service approach is also fundamental if any vendor-specific
protocol was to migrate its CCR features to the International Standard, without having to adopt the rest of the ISO architecture and protocols.

13.   The LU6.2 approach

Terminology differences apart, the CCR services in LU6.2 are very similar to ISO
CCR, but some differences do exist at the current time (July-86). This section
explores the differences.

13.1   Internal architecture

The most obvious difference relates to the nature of vendor-specific protocol
specifications versus those of ISO, and has little to do with the approach to CCR.
In vendor-specific specifications in general, and SNA LU6.2 in particular, aspects
of the implementation architecture are described (the existence of a so-called synch
point manager in each system, and of a resources manager), as well as aspects of the
program interface between a transaction program (TP) and those parts of the operating system providing support for LU6.2. All ISO specifications (and CCR in
particular) try to avoid giving internal implementation detail. The specifications
are designed to be the minimum necessary to ensure interworking between conforming
systems. Thus while the functionality of IBMs synch point manager and resources
manager and the user's TP are required to conform to the ISO Standard, the
partitioning of these functions is not specified by ISO. Other vendors might choose
to leave the entire functionality in the TP. This would give greater flexibility in
driving the protocol, but would produce a more complex TP interface. It could also
make conformance to the Standard depend not on correctness of the vendor-supplied

software, but on correctness of the user's TP - an unsatisfactory solution.

An example of the flexibility issue arises if an initiator (master) system has a number of CCR subordinates. In ISO CCR, if one subordinate refuses to commit, the initiator can attempt to complete the action by introducing some other subordinate activity (which may be a local action). In IBM LU6.2 implementations, this flexibility is not exploited because the synch point manager intercepts the refusal and __automatically__ rolls back the entire atomic action. This is an internal architecture/implementation decision. It does not affect the protocol or inter-working.

The IBM LU6.2 implementations of CCR provide the TP with a single interface to the atomic action. In terms of the CCR tree described earlier, every node in the tree is a synch point manager. The TP is either a subordinate of or the master of the local synch point manager. Only the synch point manager is aware (at the level of CCR exchanges) that there are multiple subordinates. A single SYNCPT verb (the TP interface) can produce protocol exchanges with all subordinates.

13.2   Optimisation

One notable difference between the ISO CCR emphasis and that of IBM LU6.2 is the sort of atomic action they are optimised for.

The ISO CCR tends to be optimised for relatively long, isolated, atomic actions where repetition of the action in the event of loss of an eventual error message may be more costly than an extra round-trip in rollback/backout cases.

On the the other hand, IBM LU6.2 is optimised for a __sequence__ of short atomic actions with the same partners, and therefore puts a lot of attention on optimisation of round-trip messages.

There are four main areas of optimisation in LU6.2, one of which has been incorporated in ISO CCR, and three of which have been discussed and rejected.

The first concerns clean starts and ends of atomic actions. When an atomic action is embedded in a two-way simultaneous general conversation, a round-trip is needed at its start and end to delineate it. When, however, one atomic action immediately follows another with the same partner, the two round-trips can be "piggybacked" into a single round-trip. This feature was added to ISO CCR as a result of comparison with IBM LU6.2.

The second concerns the so-called "last agent". If a CCR superior has already checked local resources and has received commitment offers from all subordinates except
one, the superior can choose to effectively pass control of the commitment to the
"last" subordinate, who directly orders commit or backout. This saves a round-trip
time (for the last agent only), but at the expense of considerable extra complexity
in the recovery procedures. The gains were not thought worth the extra complexity
in the ISO discussions.

The third relates to atomic actions which are "read-only" (no change to protected
resources at the subordinate). In this case there is no difference between rollback
and commitment and the "prepare" can be immediately followed by a "done" ("forget")
message from the subordinate instead of a "ready" ("request commit"). Adoption of
this optimisation increases the number of disc updates (forced writes) which the
master must employ, so that there is a trade-off situation.

The final area relates to the issue of rollback/backout by a subordinate. This is
allowed in IBM LU6.2, but not in ISO CCR, making a difference of half a round-trip
for communication with the agent generating the refusal. The cost is repetition of
the whole action with all agents if the diagnostic is lost, and on balance ISO
rejected this optimisation.

The above describes optimisations concerned with round-trips. There is, however,
one further area to consider. This is the number of disk updates (forced writes)
required to correctly handle failures during an atomic action. ISO CCR requires the
theoretical minimum number, but allows implementations to do more to reduce recovery
time for some cases of failure. IBM LU6.2 has intrinsically more (performed by the
synch point manager) to handle some of the optimisations that it permits.

13.3   Heuristic control

The heuristic mechanisms were absent from ISO CCR prior to comparison with LU6.2.
As a result of this study, heuristics were added. The opportunity was taken,
however, to provide marginal improvements on the LU6.2 mechanisms. The improvements
permit the atomic action master to influence (should he so wish) the direction a
heuristic decision a subordinate takes (commit or backout), and to specify a time
interval before the expiry of which the heuristic mechanism is disabled.

13.4   Summary of differences

The above are the only major differences identified to-date. They are largely unimportant for any interface definition, and show a broad level of agreement between
the two protocols. Most differences are minor ones relating to the view taken on

the relative merits of certain optimisations against their costs. It must be added, however, that the ISO Standard benefitted considerably from the detailed comparison with LU6.2.

14.   Analysing "reliable" protocols

This part of the text introduces the reader to the technique of analysing protocols which claim to be "reliable". It is not essential to an understanding of CCR, and can be omitted on a first reading.

The reader will encounter, and may become involved in designing, "reliable" protocols.

It is important to be capable of analysing such protocols to see if the stated goals are being achieved.

Analysis is relatively simple provided a suitable methodology is followed. The following steps are needed

    a)   identify clearly the incoming protocol events which produce changes on disk (memory which survives crashes), and any resulting outgoing events;

    b)   draw a time sequence diagram similar to illustration 4, marking these points clearly. (The six horizontal arrows are the points for the ISO CCR protocol.)

    c)   consider crashes (cessation of activity and reversion to the last horizontal arrow) at all possible timelines, and see what the protocol does.

Illustration 5 shows the time-lines (dashed lines) for ISO CCR, marked "A" to "E".

It is important to note that for ISO CCR all actions depend only on the main CCR protocol. Other reliable protocols will place semantics on lower layer exchanges, such as P-DISCONNECT or P-RELEASE messages in OSI. If so, these must be included in the time sequence diagram.

A crucial point in any anaylsis is examination of the <u>last</u> message. This message always has the property that if the crash occurs after its issue and before its receipt (line "E") this is undetected by the sender of it. Moreover, as he has no remaining memory (if he has, your time sequence diagram is not complete), any information on it is <u>lost</u> and cannot be repeated following a restart. This leads to an interesting "principle" or "law" for reliable protocol endings - "A correct reliable protocol cannot carry any information on its last message". This often provides a very quick rule of thumb for saying "protocol x does not work", even without detailed analysis.

Let us now examine ISO CCR. (Note, we are only outlining the "proof". A more rigorous treatment would examine the restart exchanges and the heuristic commitment exchanges also.) Referring to illustration 5:

At point "A":   The superior issues C-RESTART (ACTION); the subordinate says "DONE" (the id is unknown); the superior starts the action again with C-BEGIN. O.K.

At point "B":   The superior issues C-RESTART (ACTION); the subordinate replies C-RESTART (ACTION), and the action begins again (from the start or from an application-specification checkpoint). O.K.

At point "C":   The superior issues C-RESTART (ACTION); the subordinate repeats any refusal or restarts the action, typically from a checkpoint near or at the end. O.K.

At point "D":   The superior issues C-RESTART (COMMIT) or (ROLLBACK); the subordinate performs commitment or rollback as if it were a C-COMMIT or C-ROLLBACK. O.K.

At point "E":   The superior issues C-RESTART (COMMIT) or (ROLLBACK); the subordinate says "DONE" (the id is unknown); the superior has recorded his order to commit or rollback, so can distinguish this from case "A", and tidies up. O.K.

A further complication in CCR (ISO or LU.2) is to consider the records which need to be kept by intermediate nodes in the CCR tree. The absence of "OFFERED" on ISO C-RESTART responses is to avoid forcing further disc updates in "normal" (non-crash) cases. (Inefficiency in the restart case is several orders of magnitude less important.)

The above analysis was for a protocol which works. Now let us try one that has some slight flaws!

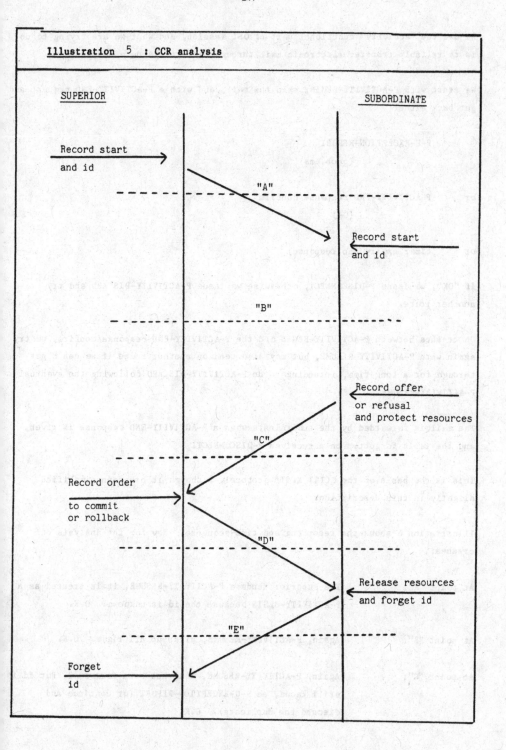

Illustration 5 : CCR analysis

We use the "activity" functional unit of OSI session, and what we are trying to do is to reliably transfer electronic mail through mail relays to a destination.

We start with P-ACTIVITY-BEGIN, ship the mail, end with a P-ACTIVITY-END request and get back either:

> P-U-EXCEPTION-REPORT
> > (problems)

or   P-ACTIVITY-END response/confirm
> > (OK)

or   timer expiry (no response)

If "OK", we issue P-DISCONNECT, otherwise we issue P-ACTIVITY-DISCARD and try another route.

On crashes between P-ACTIVITY-BEGIN and the P-ACTIVITY-END response/confirm, we try again with P-ACTIVITY-RESUME, but may also send on another route if we can't get through for a long time, intending to do P-ACTIVITY-DISCARD following the eventual P-ACTIVITY-RESUME.

The mail is forwarded by the subordinate when a P-ACTIVITY-END response is given, and the id is forgotten on a received P-DISCONNECT.

This is the basis of the CCITT X.410 protocol, although it has been simplified slightly in this description.

Illustration 6 shows the recording and time-sequence. Now for the analysis of crashes:

At point "A":   The superior sends a P-ACTIVITY-RESUME, it is treated as a P-ACTIVITY-BEGIN because the id is unknown. O.K.

At point "B":   Again, P-ACTIVITY-RESUME, picks up all right. O.K.

At point "C":   Again, P-ACTIVITY-RESUME, mail has been forwarded, But id is still known, so S-U-EXCEPTION-REPORT (or continue and discard the duplicate). O.K.

At point "D":   Id is left dangling - has to be remembered indefinitely, or deleted by other means. (Note that the subordinate cannot distinguish "D" from "C".) BAD (but not, of course, disastrous).

Now consider again the crash at "C" (no response for a long time, say), and the issue of a P-ACTIVITY-DISCARD by the superior (now, perhaps colliding with the late P-ACTIVITY-END response, or later), with the mail now being shipped on an alternative route. We now have duplication of the mail.

Comparing this with the CCR handshake we can see the cause of the problems. The first is the release of concurrency controls (sending the mail onward) at point "X" instead of point "Y" (the master lost control, and so potential duplication was introduced), and the second is the absence of confirmation at point "Y", leading to the occasional "dangling id".

One final remark is relevant. We have seen here the occasional dangling id, and the occasional duplication. For some purposes, even occasional duplication in a supposedly "reliable" protocol would be wholly unacceptable. For human mail, it is clearly all right. Thus formal analysis must be tempered by "fitness for purpose" considerations.

The above examples should now enable the reader to examine any "reliable" protocol, to find if it has any flaws. As a quick check:

- Does it have the double handshake (keeping control with the master) before release of the material?

- Does it have the one with restart responsibility forgetting the id last?

- Does it have information on the last message which affects the reliable protocol?

If you get YES, YES, NO, there is a good chance it works! Otherwise - start looking for the flaws!

In LU6.2, the superior (the initiator) normally "forgets" last, and has restart responsibility. The "last agent" optimisation, however, reverses these roles. If crashes occur when this switch is occurring, both ends may attempt restart. LU6.2 partially solves this by having the agent (subordinate) wait for a period before attempting the restart. "Restart collisions" are, however, still possible.

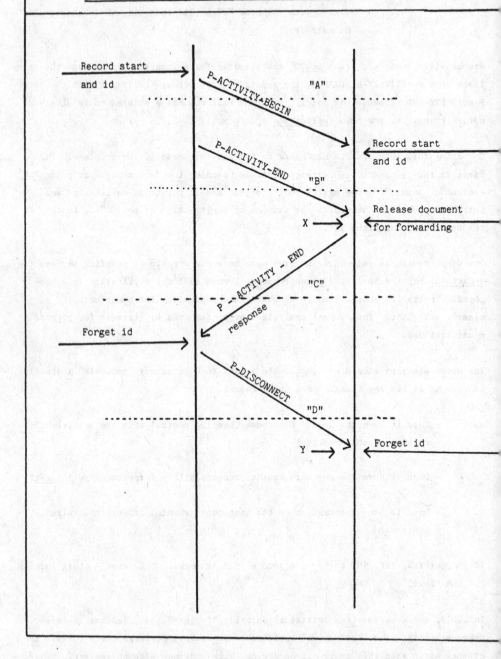

Illustration 6 : Another "reliable" protocol

15. Implementation

A CCR protocol is very simple to implement from the communications point of view. All the problems are concerned with

      a)   getting reliable storage of information which must be retained across crashes; and

      b)   managing the protected resources which are to be committed or rolled-back.

The first problem stems from the fact that the master <u>must</u> have a disc record of his responsibilities to clear the action before the subordinate ties up resources by making <u>his</u> disk record. On most 1980's operating systems, this involves writing to a disk file, closing the file, <u>and ensuring operating systems buffers are written up</u> before sending C-BEGIN. (In IBM LU6.2, this is called a <u>forced write</u>.) <u>Updating</u> this stored information by an intermediate node in the CCR atomic action tree requires updates of records for different connections to be made as an atomic action - all or none. This often requires considerable care, and use of cyclic version numbers for file-names holding the records. There are no insuperable difficulties, but understanding and care is needed if the crucial reliability features of CCR are not to be lost by an incorrect implementation.

The second problem is much more difficult to handle unless support for CCR is built deeply into the operating system (which it is, in the case of LU6.2). It requires in the simplest case mechanisms to reinstate "locks" following recovery from crashes; in the case of flexible application of concurrency controls (see 7. and the bank balance discussion) it can require highly resource and application-dependent gate-keeping of access which can become extremely complex, particularly if use of uncommitted resources is to be permitted.

The implementation of CCR is considerably eased, however, if it is being considered in relation to a specific-application which is the only accessor of the relevant resources. In this case support of the CCR controls can be built into the application, and some of the problems evaporate.

16. Conclusion

The CCR mechanisms provide a very important tool for application designers.

Similarities between the LU6.2 and ISO approaches to CCR considerably exceed the differences.

COMMUNICATION PRIMITIVES IN PROGRAMMING

AND SPECIFICATION LANGUAGES

M Hennessy

University of Sussex

Presented at : Distributed Computing in Open Systems
               IBM Europe Institute 1986
               Oberlech, August 18-22, 1986

Abstract: After a review of semaphores and monitors the principles
underlying the programming language OCCAM are outlined. We show how
it can also be used as a specification language. We also examine in
detail the language CCS, principely by giving examples of processes
or specifications. We end with some remarks on the issues raised by
these modern languages which on the one hand may be viewed as programming
languages for highly parallel architectures and on the other languages
for specifying the required behaviour of systems.

§ A program written in a language such as PASCAL specifies the
sequential execution of a list of commands or statements. During its
execution there is only one locus of control, and the program is often
said to specify a unique process. A concurrent programming language,
on the other hand, allows the programmer to define two or more such
processes, all of which can be active and executing at any one time.
The task of the programmer is to organise the execution of these
processes so that they each contribute to some common aim and the
overall behaviour of the system is consistent and coherent.

The usual model of a sequential process is that of an algorithm
which acts on a store, i.e. a set of variable bindings; each statement
of the process gives a transformation of the store, and these statements
are executed sequentially, one at a time. When more than one process
is allowed to be active, a straightforward method of allowing them to
interact or communicate is to have a common store shared between
processes. This is a very popular framework which provides the basis
for many programming languages. The store memory acts as a kind of
shared knowledge which is common to all; it can be interrogated and
changed by all processes and therefore acts as the medium for
transferring information from one process to another. However, care
must be taken to ensure that the memory remains consistent and uncorrupted.
A typical problem arises when two processes (writers) want to update a
variable simultaneously. At the hardware level this cannot happen, of
course, and the outcome will be nondeterministic; it will depend on
exactly how the various fetch and put commands which are used to implement
an update are interleaved. It is possible that these are interleaved
in such a way that the resulting store does not reflect the updates
requested by the writers. (See [1] for an example.)

The general situation is as follows: the common store is a resource shared by all active processes. To avoid abuse of the resource processes may require, from time to time, exclusive use of it. A typical example is when a writer wishes to update a variable he needs exclusive access to it for the duration of the update. These important segments are often called critical regions. The problem is to ensure that a process can only enter a critical region when it has exclusive use of the resources it requires.

Many software solutions have been suggested and perhaps the most famous is that of semaphores. These are distinguished integer or boolean-valued variables on which one can perform only two operations:

$$V \quad : \quad \text{increment}$$
$$P \quad : \quad \text{decrement} \quad \text{if possible.}$$

The crucial aspect of these operations is that P is indivisible: it can only be performed if the semaphore is non-zero and when it can be performed the testing and decrementing cannot be interrupted or interleaved. In general, resource problems are solved by using one (or more) semaphores as sentries whose task is to guard access to the resource. Usually access can only be gained by performing a P operation and when the resource is relinquished a V operation must be performed. Thus if the semaphore is boolean-valued this means that only one process can have access at any one time.

Figure 1 gives a typical use of semaphores to implement a bounded buffer between a producer process and a consumer process. Three different semaphores are used to ensure that processes will only add messages when the buffer is not full; the consumer will only take messages when it is not empty and only one producer has access to the buffer at any one time.

Producers/Consumers using a Bounded Buffer

k producers P_i, $1 \leq i \leq k$

1 consumer C

messages are transferred from producers to consumers using a

bounded buffer   BUFFER.

Problem: to limit access to the buffer to ensure integrity of messages

SEMAPHORE :   nm :=0 ; number of messages in the buffer

                       bf :=1 ; is the buffer free?

                       nep := bound; the number of empty spaces in the buffer

PROCESS P_i :   do _forever_

                   generate (message);

                   P(nep);

                   P(bf);

                   add(message, BUFFER);

                   V(bf);

                   V(nm)   od

PROCESS C :   do _forever_

               P(nm);

               take(message , BUFFER)

               V(nep)   od

Figure 1: Using Semaphores

This example underlines the inadequacy of semaphores; they are too primitive. It requires three different semaphores to ensure the consistency of the buffer. The dynamic relationships between these semaphores are non-trivial and it is not immediate that the scheme actually works. In general, as the number of semaphores increases, this problem of ensuring correctness gets worse.

A monitor can be viewed as a sophisticated semaphore. The latter has a simple data structure (a variable) on which only two operations can be performed, P and V; moreover, only one of these can be active at any one time. Similarly a monitor has

      i) an internal private collection of data-structures

     ii) a set of operations or procedures for
           accessing/modifying these data-structures.

As with semaphores, only one procedure can be active at any one time. Since these procedures are the only way of accessing the internal data-structures this restriction ensures the consistency of the internal data.

An example of a monitor is given in Figure 2. Here the bounded buffer is inside the monitor and can only be accessed either by calling one of the associated procedures add or take. The result is that only one process, either a producer or a consumer, can have access to the buffer at any one time. Of course, having gained access to the monitor a process may realise that he cannot proceed; for example, a consumer may gain access only to find that there are no messages. To handle this situation monitors have associated with them objects called conditions. Executing wait.cond. suspends the process until some other process executes the corresponding command signal.cond. There are many

Producers/Consumers using a Bounded Buffer

```
MONITOR      BBUFFER

    begin    ds ;  declaration of a suitable data structure for holding
                      messages

             count : 0.. bound :=0 ; number of messages in the buffer

             nonempty,nonfull : condition




    procedure   add (x : message)

             begin     if count = bound then wait.nonfull ;

                       append (x,ds) ;

                       count : = count + 1;

                       signal.nonempty

             end


    procedure   take (result x : message)

             begin     if count = 0 then wait.nonempty

                       x := next(ds) ;

                       count:=count -1 ;

                       signal.nonfull

             end

END OF BBUFFER
```

The producer P_i calls BBUFFER.add(m) to add a message to the buffer

and the consumer C calls BBUFFER.take to receive a message from the buffer

Figure 2 : Using Monitors

variations of the precise details of how these conditions work; a good discussion can be found in [2] . For our purposes it is sufficient to have a general picture of how monitors work in languages such as [3] , [4].

In such languages there are at least two different kinds of objects, processes which are active and monitors which are passive. A process is an independent object which has a life of its own, so to speak. On the other hand, monitors are passive in that they remain unactivated until some active process calls them. What actually happens is the calling process is suspended, control is given to the monitor which is activated; when it is finished the task it has been assigned it hands back control to the calling process and remains idle until another call from a process brings it to life again. Another set of languages, such as [5], [6] abolish this distinction between active and passive objects; essentially the only kind of object around is a process. In this setting the monitor in Figure 2 can be viewed as a permanently active and independent process which periodically accepts requests from other processes either to add or take elements from the buffer. There is considerable variation in the mechanisms used for interaction between these processes. For example, the buffer monitor could be viewed as having two "communication ports" called add and take. Calling it at the first amounts to sending it a message to be added to the buffer, whereas calling it at the second amounts to receiving a message from it. In designing this message-passing framework various decisions must be made. For example

- does it take time to send a message?

- is the sending process blocked until the message has been received?

- how does the sender identify the receiver?

We examine one language where processes intercommunicate via message-passing in detail, OCCAM[5] Here communication is instantaneous via channels and the sender must wait for the receiver.

§2. OCCAM is conceptually very simple. It has some minimal features for the management of variables and channels and a small number of constructors for defining new processes in terms of existing ones. For example, if P,Q and R have already been defined as processes

```
          PAR
          P
          Q
          R
```

is a new process which consists of three subprocesses running in parallel, P,Q and R. These three processes are in turn defined in terms of constructors. Primitive processes are defined using the primitive statements

```
          x: = e
          chan?x
          chan!e.
```

The first is the usual assignment statement; the second is an input statement; it represents a process which <u>inputs</u> a value from a channel named 'chan' and assigns it to the variable x. chan!e is a process which outputs the value of the expression e in the present environment along the channel named 'chan'. Of course these statements only make sense in contexts where the variable x has been defined or declared. Channel names are treated in much the same way as variables; they must be declared and their declaration has scope exactly as variable declarations. A simple example of a process is

```
          WHILE true
          Var x
          SEQ
             in?x
             out!x²
```

Two new constructors are used here, SEQ and WHILE.* Their effect is
as one would expect: SEQ is the sequential composition of the processes
it governs, in this case in?x and out!x^2, and WHILE has its usual meaning.
It is assumed that the two channel names in and out have already been
declared. In this case the process can be viewed as:

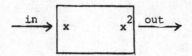

It never terminates and is always ready to accept input from channel in
and to output its square at the channel out. Each time it performs one
of these actions it is assumed that some other process simultaneously
performs the complementary action. For example, it can only input the
value 7 from the channel in if some other process simultaneously outputs
the value 7 to the same channel. (Of course, the external use may be
such a process.) This is the only way that information can flow between
processes. In particular they are not allowed to common variables. An
example of a process which uses this form of communication is given in
Figure 3. The first subprocess outputs to the channel comms and the
second inputs from the same channel. The processes described in
Figure 4 is essentially the same, except that the channel comms is
declared. The effect of the declaration is to limit the scope of the channel
to the process defined; in other words, comms can only be used for
communication between the subprocesses. On the other hand, in Figure 3 it
is still in scope outside the process being defined. So it could be used
by other processes.

In Figure 5 we give an example which underlines the difference between
the two processes. The first one introduces nondeterminism into the
overall system; when 6 is input at channel in we don't know if 37 will be
output at channel nout or 1296 at channel out. If the process in Figure 4
is used, the resulting system is completely deterministic and the extra
subprocess is never activated. It can never input on comms because the
channel it refers to as comms is different than that referred to by the
other two subprocesses; they refer to different definitions.

---

*In general the scope of constructors is determined by
 identation.

```
PAR
  WHILE true
    VAR x:
    SEQ
      in?x
      comms!x*x
  WHILE true
    VAR y;
    SEQ
      comms?y
      out!y*y
```

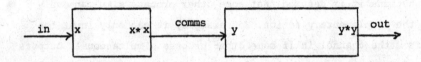

<u>Figure 3</u> : Computing x $\longrightarrow$ x^4

```
CHAN comms
PAR
  WHILE true
    VAR x:
    SEQ
      in?x
      comms!x*x
  WHILE true
    VAR y:
    SEQ
      comms?y
      out!y*y
```

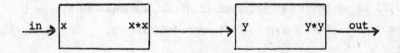

<u>Figure 4</u>: Computing x $\longrightarrow$ x^4, again

```
CHAN comms
PAR
  P - process from Figure 3
  WHILE true
    VAR z:
    comms?z
    nout!(z + 1)
```

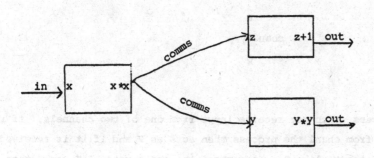

```
CHAN comms
PAR
  Q - process from Figure 4
  WHILE true
    Var z
    comms?z
    nout!(z+1)
```

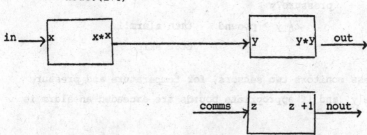

Figure 5: Scope of channels

It is only fair to point out that there are many syntactic restrictions imposed on the use of these constructs. With these restrictions the definitions given in Figure 5 are actually disallowed. However, because the restrictions are fairly arbitrary and have been imposed primarily to aid implementation we choose to ignore them.

There is a form of nondeterminism allowed:

```
ALT
    chan1?x
        P
    chan2?y
        Q
```

is a process which may receive input from one of two channels; if it is received from chan1 the process then acts as P and if it is received at chan2 it acts like Q. In general the constructor ALT can act on guarded processes. In the above example the process P is guarded by chan1?x whereas Q is guarded by chan2?y. This constructor can be very useful for defining a process which has to react to different stimuli which can occur in any order. The following is a typical example

```
WHILE true
    ALT
        temp?x
                if x > tbound    then alarm!0
                                 else skip

        pressure?y
                if y > pbound    then alarm!1
                                 else skip
```

This process monitors two sensors, for temperature and pressure respectively, and if appropriate bounds are exceeded an alarm is sounded.

OCCAM has a very powerful notation for the duplication of
parameterised processes. For example, if n is some instantiated variable

    ALT [i = 0 to n]

        P(i)

is a shorthand for

    ALT

        P(0)

        P(1)

        ⋮

        P(n)

where P(i) is a definition of a guarded process parameterised on i.
It can be applied to all the constructors of the language. As an
exmaple of its use we define a sieve for enumerating prime numbers.
To print the first n primes we need n identical filters which are
arranged as in Figure 6. The pump process merely pumps out all the
natural numbers in ascending order, starting at 2. The system is designed
so that the first number input by the kth filter will be the kth prime.
It is immediately printed and stored permanently. Subsequent numbers input
from the left are passed to the right only if they are not divisible by the
stored number, i.e. the kth prime. Before the actual system is configured
a number is read in which is taken to be the number of primes required.
The system designed as in Figure 7 will then output the first n primes at
the channel chan[n] followed by the infinite sequence of natural numbers
which are not divisible by any of these numbers.

We have not described all of the features of OCCAM but instead have
restricted ourselves to a general outline. It takes an extreme view of
processes:they are capable of essentially nothing other than communication.

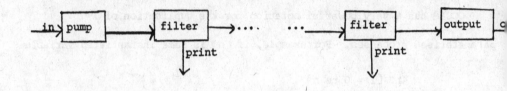

```
CHAN chan [i for 0 to n]

VAR n,k

SEQ

in?n                          - no. of primes to be output

k:=2                          - first prime

PAR

    WHILE true
                                                    ⎫
      SEQ                                           ⎬  pump
        chan[0]!k                                   ⎭
        k:= k+1


    PAR i = [1 to n]                                ⎫
                                                    ⎪
    VAR perm,temp                                   ⎪
                                                    ⎪
      SEQ                                           ⎪
                                                    ⎪
      chan [i-1]? perm                              ⎪
                                                    ⎬  filter
      print!perm                                    ⎪
                                                    ⎪
      WHILE true                                    ⎪
                                                    ⎪
        chan [i-1]? temp                            ⎪
                                                    ⎪
          if div(perm,temp) then skip               ⎪
                            else chan [i]!temp       ⎭


      WHILE true                                    ⎫
                                                    ⎪
        VAR x                                       ⎬  output
                                                    ⎪
        chan [n]?x                                  ⎪
                                                    ⎪
        out!x                                       ⎭
```

Figure 7 : Printing prime numbers

It also emphasises their structure; in fact, a process can only be defined by saying how it is to be constructed from simpler components. This highly structural approach is ideal for supporting a hierarchical view of the design of systems. As a completely trivial example, consider the problem of designing a process which will

> input x at channel in
>
> output x^6 at channel out.

The OCCAM program

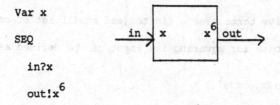

```
Var x
SEQ
    in?x
    out!x^6
```

can be viewed as a <u>specification</u> of the required process. It can be <u>implemented</u> using simpler components and one possibility of this is:

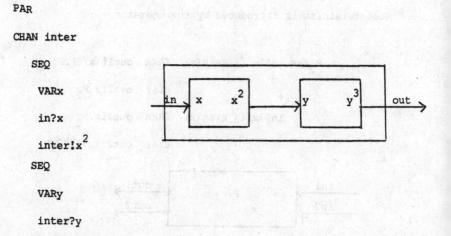

```
PAR
CHAN inter
    SEQ
    VARx
    in?x
    inter!x^2
    SEQ
    VARy
    inter?y
    out!y^3
```

These two simpler components can also be viewed in turn as specifications to be implemented by even simpler components.

§3. OCCAM was designed as a programming language and this imposes constraints which make it very difficult to use it effectively and simply as a specification language. Here we examine a less constrained language CCS, a Calculus for Communicating Systems,[6]. It is very similar in design to OCCAM but it is more abstract. It is an applicative language in that it lacks the notion of a store. Essentially one can give recursive definitions of processes using a small number of constructors or combinator Unlike OCCAM their application is unconstrained, so that processes are essentially recursive terms over a finite (and small) set of combinators. For example, a process for squaring its input can be defined as

$$P_1 \Leftarrow in?x.out!x^2.P.$$

Diagrammatically it can be represented as usual by :

$$\xrightarrow{\text{in}} \boxed{\text{x} \quad \text{P1} \quad x^2} \longrightarrow$$

Nondeterminism is introduced by the operator + :

$$
\begin{aligned}
P_2 \Leftarrow \ & in1?x.if \ even(x) \quad then \quad out1!(x/2).P_2 \\
& else \quad out2!x.P_2 \\
+ \ & \\
& in2?x.if \ even(x) \quad then \quad out1!x.P_2 \\
& else \quad out2!(x+1)/2.P_2
\end{aligned}
$$

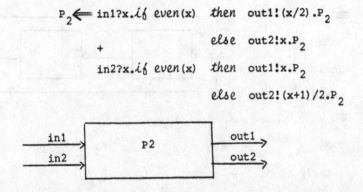

Communication is handled exactly as in OCCAM: a communication is the
simultaneous occurrence of an input and output along some channel. Thus
processes cannot identify other processes directly by name. It is part
of the programmer's duty to ensure that channel names are used sensibly
so as to enable the required communication to occur. In OCCAM the
usual scoping mechanism for declarations gives a method for protecting
channels from interference by unwanted processes. In CCS this is
effected by another combinator <u>restriction</u>, \chan. An example is given
in Figure 7.   A semaphore is represented by :

$$\text{SEM} \Longleftarrow \text{p!v?SEM}$$

Here we suppress reference to any values passed in communications since
their value is unimportant. A reader and writer is represented by,
respectively:

$$\text{R} \Longleftarrow \text{p? beginr.endr.v!.R}$$
$$\text{W} \Longleftarrow \text{p? beginw.endr.v!.W}$$

Here we use beginr, etc. for actions which the processes can perform and
which are not associated directly with communication.

The three processes running in parallel are represented by

$$\text{R|SEM|W.}$$

The symbol | represents a binary combinator and P|Q means the processes
P and Q running in parallel. We would expect this operator to be associative
so we can write R|SEM|V rather than (R|SEM)|V or R|(SEM|V). In these
processes one can see, at least intuitively, no pair of complementary
actions, such as beginr, endr, can be interleaved by an action performed by
the other process, such as beginw, endw. However, here the channels p and v

are still public in that they are still accessible to the outside world.
To restrict their use to R and W only we write

$$(R|SEM|W)\backslash p,v$$

In general $X\backslash c$ is a process

A semaphore   : $SEM \Longleftarrow p?.v!.SEM$

A reader   : $R \Longleftarrow p!.beginr.endr.v?.R$

A writer   : $W \Longleftarrow p!beginw.endw.v?.W$

The System $(R|SEM|W)\backslash p,v$

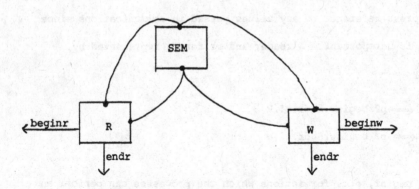

Figure 8 : Readers and Writers

---

which acts like X except that it cannot use channel c for external
communications. However, it can be used internally. In our example p and
v are therefore the private links between SEM and the competing R and W.
The uniformity of the language lends it considerable power. For example,
a simple unbounded counter can be defined by

$$C \Longleftarrow up.(C|down.NIL)$$

Here NIL is a process which can do nothing, like skip in many
programming languages. So whenever an up action is performed a
corresponding down action can be performed: C|down.NIL acts just like
C except that it also can perform a down action.

Another powerful advantage of the language is that derived
operators or "macros" can be defined and one can program directly
in them. Before we see an example of this we introduce the final
constructor of CCS called renaming. A renaming R is simply a function
from channel names to channel names. The term

$$P[R]$$

now represents a process very similar to P except that the channels
are renamed via R. .As an example of its use consider two processes

$$P \Longleftarrow l?x.r!x^2.P$$
$$Q \Longleftarrow l?x.r!x^3.Q$$

and suppose we want to join the output of P to the input of Q to form
a new process to compute $x \longrightarrow x^6$. A solution is given in Figure 9.
The output channel of P, r, is renamed to int and the input channel
of Q, l, is renamed to the same channel int. So that when P thinks it is
using channel r it is actually using channel int. Similarly Q uses int
when it thinks it is using channel l. The net effect is that the
output from P is the input to Q. The restriction \int internalises
the communication along channel int. This scheme leads to a process
which computes the required function, $x \longrightarrow x^6$.

This approach can be generalised, as outlined in Figure 10. Here
we assume processes which have a set of left and right channels of
the form lx, ry respectively where x and y are arbitrary. The derived

operator or macro □ is defined in such a way that in the process
X□Y all channels of the form rx of the subprocess X are joined to
channels of Y of the form lx. To show the power of this operator

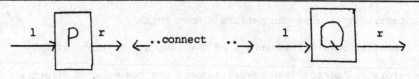

L : r ———→ int

 chan—→chan, otherwise

R : l ———→int

 chan—→chan, otherwise

combined process: :   (P[L] |Q[R]) \int

Figure 9 : Renaming

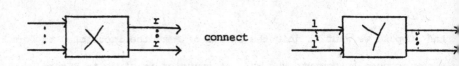

L : rx ———→ intx

 chan ———→ chan,otherwise

R : lx ———→ intx

 chan ———→ chan, otherwise

X □Y $=_{def}$ (X[L]|Y[R])\ INT

   where INT is set of channel names of the form intx.

Figure 10: Derived Operator □

we give a description of a highly parallel system of processes

of the form

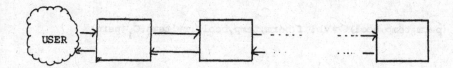

for recognising palindromes. The general idea is that the user

pumps in symbols at one end; after each symbol he demands a boolean

value, which is true if the word read in up to this point is a

palindrome and false otherwise.

Each cell, except the special last one, is identical :

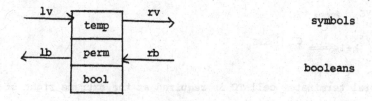

It has three states: the first one is an initial state and the

behaviour in this state is given by

$$C_0 \Longleftarrow \text{lv?perm.lb!true.}C_1 \text{(perm)}$$
$$+$$
$$\text{lb!true.}C_0$$

If the new state $C_1(x)$ is reached the value read in is stored in the

permanent memory of the cell. The behaviour in this state is deterministic:

$$C_1 \text{(perm)} \Longleftarrow \text{Pv?temp.rb?bool. } C_2 \text{(perm,temp,bool)}.$$

A new symbol temp is input from the left and a boolean value from the right. The new state is an output state:

$$C_2(\text{perm},\text{temp},\text{bool}) \Longleftarrow \text{lb!} \ f(\text{perm},\text{temp},\text{bool}).\text{rv!temp}.C_1(\text{perm})$$

where $f(\text{perm},\text{temp},\text{bool})$ =   true if perm = temp and bool is true

                                         =   false, otherwise

Here the temporary symbol is passed on to the right and a boolean is passed left; the exact value of the boolean is given by the function f.

A system with k functioning cells is defined formally by:

$$SYS_0 \Longleftarrow TC$$
$$SYS_{k+1} \Longleftarrow C_0 \ \square \ SYS_k$$

A special terminator cell TC is required at the extreme right of the machine.

$$TC \Longleftarrow \text{lb!true}.TC + \text{lv?x}.\text{lb!true}.TC$$

An example of the functioning of the machine for $k = 4$ is given in Figure 11. This machine will function properly until its storage bound has been exceeded. In general malfunctioning of a system with k cells will start after a word of length 2k has been read in. However the language is sufficiently powerful to describe a machine (presumably to be implemented in software) which will generate new storage as required:

$$SYS \Longleftarrow \text{lb!true}.SYS$$
$$+$$
$$\text{lv!perm}.C_1(\text{perm}) \square \ SYS,$$

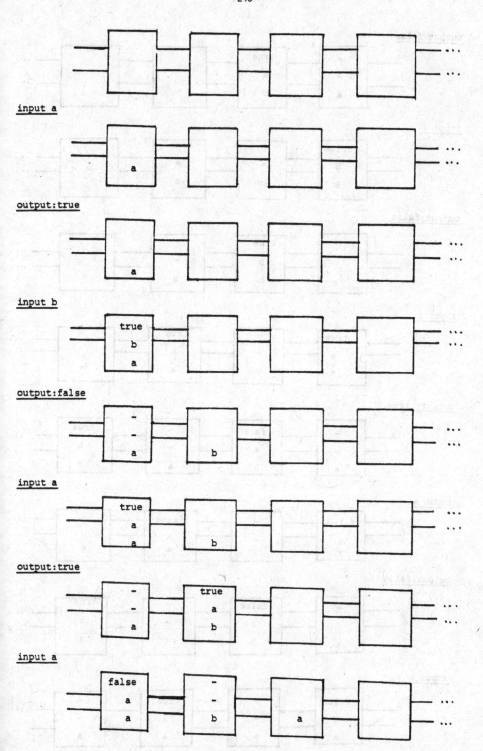

Figure 11: Recognising Palindromes

246

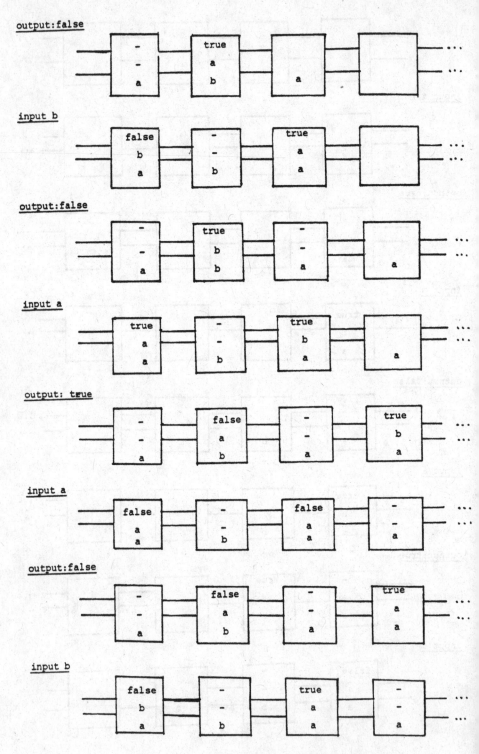

Figure 11 (contd.)

CCS is an example of one of a number of so-called algebraic process languages, [7], [8], [9] and later in the week we will see another one, LOTOS [10]. The last example shows that they can describe very complicated systems in a compact and precise manner. They can also be used to describe the specifications of actual or projected systems at various levels. For example, a natural description of a palindrome recogniser is:

$$SPFC \;\Longleftarrow\; lb!true.SPEC + SPEC(\varepsilon)$$
$$SPFC(w) \Longleftarrow lv?x.lb!ispalindrome(xw).SPEC(xw).$$

To understand this specification one needs to know what a palindrome is, i.e. have the predicate ispalindrome.
We will not pursue the topic of specifications any further. Instead we conclude with some remarks on the issues which these descriptive semantics raise.

The first is the problem of a semantics for such languages. The entire talk has been about syntax and we have left the semantics to the intuition of the reader. However, because of the complicated nature of the descriptions one can write, a formal semantics is essential for a proper understanding of how processes work. For example, the behaviour of SYS, given in Figure 11 is difficult to deduce from our intuitive description of the effect of each construct in the language. Considerable advances have been made in this area of semantics. There is now a well-developed theory of <u>operational semantics</u> [11], which is readily applicable to concurrent languages. It uses labelled transition systems to specify in a structured way the next possible moves that a process can perform. The rules for inferring these possible moves are in

general straightforward and look rather like PROLOG clauses. The
result is that prototype high-level interpreters or simulators can easily
be written for these languages.

The relationship between specifications and implementations is
also a topic of much interest in the research community. In general
these are two different descriptions of essentially the same semantic
object. This leads immediately to the question of what is the nature
of the objects which can be described in these languages. What is a
process? The objects described by sequential languages such as
PASCAL, etc. are simply functions from the space of inputs to the
space of outputs. For concurrent languages there is no straightforward
analogue and there is not even a consensus of what type of object it
should be. This philosophical question can be avoided by developing
instead a semantic equivalence between processes:

$$P \approx Q \text{ if P and Q have essentially the same behaviour.}$$

For example, one would expect the implementation of the palindrome recognis
to have essentially the same behaviour as the specification, i.e.

$$SPEC \approx SYS.$$

Similar examples occur in the domain of protocol specification. One
description would be of the requirements of a service at a certain
level and the other would be of the protocol, or more precisely, the
protocol together with the lower level services it uses. The equivalence
of these two descriptions amounts to a statement of correctness of the
protocol. There are essentially two well-developed theories of

semantic equivalence which may be found in [6]. [7], [12]. Both

are attempts at formalising the intuitive idea that processes are

equivalent if no user could ever distinguish between them. We will

not pursue how this intuitive notion is formalised except to say

that both theories give rise to elegant and well-behaved equivalence

relations. For example, they are preserved by the constructors used

in the language and they can be characterised by a set of equational

laws. It leads to a proof methodology based on syntactic transformations

and induction. (This explains the use of the word "calculus" in the

name CCS). These transformations can be used to give a formal proof

facts such as

$$SPEC \; = \; SYS$$

i.e. the implementation of the palindrome recogniser meets its

specification.

REFERENCES

[1] E. Dykstra, "The Hiararchical Design of Processes". Acta
     Informatica 1, pp.115-138, 1971.

[2] G. Andrews and F. Schneider, "Concepts and Notations for
     Concurrent Programming", Computing Surveys 15, no.1, pp.3-43,
     1983.

[3] P. Brinch Hansen, "The Programming Language Concurrent Pascal",
     IEEE Transactions on Software Engineering, vol.1(2), pp.199-207,
     1975.

[4] B. Lampson and D. Redell, "Experience with Processes and Monitors
     in Mesa", CACM vol.23(2), pp.105-117, 1980.

[5] Inmos Ltd., "OCCAM Programming Manual", Prentice-Hall, 1984.

[6] R. Milner, "A Calculus for Communicating Systems", Lecture
     Notes in Computer Science, vol.   , Springer-Verlag, 1980.

[7] C. Hoare, "Communicating Sequential Processes", Prentice-Hall,
     1985.

[8] D. Austry and G. Boudol, "Algèbre des Processus et Synchronisation",
     Theoret. Comp. Science 30, No.1, pp.91-132, 1984.

[9] J. Bergstra and J. Klop, "Process Algebra for Synchronous
     Communication", Information and Control 60, pp.109-137, 1984.

[10] 150 DP8807, "LOTOS - A Formal Description Technique based on
     Temporal Ordering of Observational Behaviour", 1985.

[11] G. Plotkin, "A Structural Approach to Operational Semantics",
University of Aarhus Notes, 1982.

[12] R. De Nicola and M. Hennessy, "Testing Equivalences for Processes",
Theor. Computer Science 34, pp.84-133, 1984.

Experiences with the Accent Network Operating System[1]

Richard F. Rashid
Computer Science Department
Carnegie-Mellon University
Pittsburgh, Pa. 15213

Abstract

This paper describes experiences gained during the design, implementation and use of the CMU Accent Network Operating System. It outlines the major design decisions on which the Accent kernel was based and how those decisions evolved from the experiences gain by the author with the University of Rochester's RIG system. Also discussed are some of the major issues in the implementation of message-based systems, the usage patterns observed with Accent over a three year period of extensive use at CMU and a timing analysis of various Accent functions.

1. Background

Accent is a communication-oriented operating system kernel built to support a large network of scientific personal computers. Accent was developed as part of the Carnegie-Mellon Spice Project [9] [17] which had as its goal a distributed programming environment for the support of computer science research.

Early in the design of Accent, it was felt that it would have to provide:

- fast interprocess communication (IPC) which could be transparently extended across a network without kernel intervention,

- multiple processes, each with a large (2↑32 byte), sparsely allocatable, paged address space, and

- the ability to communicate large (megabytes) amounts of information in messages efficiently between processes.

These and other major design decisions were partly based on the requirements of the Spice project itself (such as the requirement that large sparse address spaces be supported for large AI tasks) and partly on past experiences with a previous network operating system -- the Rochester RIG system -- with which several members of the project were directly familiar.

The first line of Accent code was written in April 1981. Accent is used at CMU in a network of over 150 PERQ workstations and the Accent network IPC facility is supported under a modified version of Berkeley UNIX on the Department's 120 VAX minicomputers.

This paper discusses what has been learned in the design, implementation and use of Accent, including:

- the evolution of Accent facilities from those in RIG based on experiences with the RIG system,

---

[1]This research was sponsored by the Defense Advanced Research Projects Agency (DOD), ARPA Order No. 3597, monitored by the Air Force Avionics Laboratory Under Contract F33615-81-K-1539. The views and conclusions contained in this document are those of the author and should not be interpreted as representing official policies, either expressed or implied, of the Defense Advanced Research Projects Agency or the U.S. Government.

- the key implementation issues inherent to the Accent design,

- the evolution of language support facilities for distributed programming within Accent, and

- the behavior of Accent vs more traditional systems.

Serious development work on Accent ended in 1985. Most of the critical features of Accent have since been incorporated into CMU's Mach multiprocessor operating system [1]. Mach currently runs on all VAX architecture machines (multiprocessors as well as uniprocessors) and is running on the IBM RT PC.

2. The Evolution of Accent from RIG

Implementation of RIG began in 1975 on an early version of the Data General Eclipse minicomputer. The first usable version of the system came on-line in the fall of 1976. Eventually the Rochester network included several RIG Eclipse nodes as network servers and a number of Xerox Altos acting as RIG client hosts. RIG provided clients network file services, ARPANET access, printing services and a variety of other functions. Active development continued well into the 1980's but obsolescence of its Data General Eclipse and Xerox Alto hardware base eventually dictated its demise in the Spring of 1986.

2.1. The RIG Design

The basic system structuring tool in RIG was an interprocess communication (IPC) facility which allowed RIG processes to communicate by sending packets of information between themselves. RIG's IPC facility was defined in terms of two basic abstractions: *messages* and *ports*.

A RIG port was defined to be a kernel-provided queue for messages and was referenced by a global identifier consisting of a dotted pair of integers *<process number.port number>*. A RIG port was protected in the sense that it could only be manipulated directly by the RIG kernel, but it was unprotected in the sense that any process could send a message to a port. A RIG port was tied directly to the RIG abstraction of a *process* -- a protected address space with a single thread of program control.

A RIG message was composed of a header followed by data. Messages were of limited size and could contain at most two scalar data items or two array objects. The type tagging of data in messages was limited to a small set of simple scalar and array data types. Port identifiers could be sent in messages only as simple integers which would then be interpreted by the destination process.

Due largely to the hardware on which it was implemented, RIG did not allow either a paged virtual memory or an address space larger than $2\uparrow16$ bytes. RIG did, however, use simple memory mapping techniques to move data [4]. The largest amount of data which could be transferred at a time was 2K bytes.

2.2. Problems with RIG

The RIG message passing architecture was originally intended more as a means for achieving modular decomposition (much like Brinch-Hansen's RC4000) rather than as the basis for a distributed system. It was discovered early on, though, that RIG's message passing facility could be adapted as the communication base for a network operating system. Unfortunately, as RIG became heavily used for networking work at Rochester a number of problems with the original design became apparent:

- **Protection**

  The fact that ports were represented as global identifiers which could be constructed and used by any process implied that a process could not limit the set of processes which could send it a message. To function correctly, each process had to be prepared to accept any possible message sent to it from any potential source. A single errant process could conceivably flood a process or even the entire system with incoherent messages.

- **Failure notification**

  Another difficulty with global identifiers was that they could be passed in messages as simple integers. It was therefore impossible to determine whether a given process was potentially dependent on another process. In principle any process could store in its data space a reference to any other process. The failure of a machine or a process could therefore not be signaled back to dependent processes automatically. Instead, a special process was invented which ran on each machine and was notified of process death events. Processes had to explicitly register their dependencies on other processes with this special "grim reaper" process in order to receive event-driven notifications.

- **Transparency of service**

  Because ports were tied explicitly to processes, a port defined service could not be moved from one process to another without notifying all parties. Transparent network communication was also compromised by this naming scheme. A port identifier was required to explicitly contain the network host identifier as part of its process number field. As the system expanded from one machine to one network to multiple interconnected networks this caused the port identifier to expand in size -- usually resulting in considerable reimplementation work.

- **Maximum message size**

  The limited size of messages in RIG resulted in a style of interprocess interaction in which large data objects (such as files) had to be broken up into chunks of 2K bytes or less. This constraint impacted on the efficiency of the system (by increasing the amount of message communication) and on the complexity of client/server interactions (e.g., by forcing servers to maintain state information about open files).

2.3. The evolution of RIG

CMU's Spice [9] distributed personal workstation project provided an opportunity to effectively "redo" a RIG-like system taking into account that system's limitations. The result was the Accent operating system kernel for the PERQ Systems Corporation PERQ computer.

The Accent solution to the problems present in the RIG design was based on two basic ideas:

1. **Define ports to be capabilities as well as communication objects.**

By providing processes with capabilities to ports rather than a global identifier for them[2], it was possible to solve at one time the problems of protection, failure notification and transparency:

- Protection in Accent is provided by allowing processes access only to those ports for which they have been given capabilities.

- Processes can be notified automatically when a port disappears on which those processes are dependent because the kernel now has complete knowledge of which processes have access to each port in the system. There is no hidden communication between processes.

- Transparency is complete because the ultimate destination of a message sent to a port is unknown by the sender. Thus transparent intermediary processes can be constructed which forward messages between groups of processes without their knowledge (either for the purpose of debugging and monitoring or for the purpose of transparent network communication).

2. Use virtual memory to overcome limitations in the handling of large objects.

The use of a large address space (a luxury not possible in the design of RIG) and copy-on-write memory mapping techniques permits processes to transmit objects as large as they can directly access themselves. This allows processes such as file servers to provide access to large objects (e.g., files) through a single message exchange -- drastically reducing the number of messages sent in the system [10].

The first line of Accent code was written in April 1981. Today Accent is used at CMU in a network of 150 PERQ workstations. In addition to network operating system functions such as distributed process and file management, window management and mail systems, several applications have been built using Accent. These include research systems for distributed signal processing [11], distributed speech understanding [7] and distributed transaction processing [18]. Four separate programming environments have been built -- CommonLisp, Pascal, C and Ada -- including language support for an object-oriented remote procedure call facility [13].

3. The Accent Design

Accent is organized around the notion of a protected, message-based interprocess communication facility integrated with copy-on-write virtual memory management. Access to all services and resources, including the process management and memory management services of the operating system kernel itself, are provided through Accent's communication facility. This allows completely uniform access to such resources throughout the network. It also implies that access to kernel provided services is indistinguishable from access to process provided resources (with the exception of the interprocess communication facility itself).

---

[2]An idea borrowed from Forest Baskett's DEMOS system [6].

3.1. Interprocess communication

The Accent interprocess communication facility is defined in terms of abstractions which, as in RIG, are called *ports* and *messages*.

The *port* is the basic transport abstraction provided by Accent. A port is a protected kernel object into which messages may be placed by processes and from which messages may be removed. A port is logically a finite length queue of messages sent by a process. Ports may have any number of senders but only one receiver. Access to a port is granted by receiving a message containing a port capability (to either send or receive).

Ports are used by processes to represent services or data structures. For example, the Accent window manager uses a port to represent a window on a bitmap display. Operations on a window are requested by a client process by sending a message to the port representing that window. The window manager process then receives that message and handles the request. Ports used in this way can be thought of as though they were capabilities to objects in a object oriented system(Jones78). The act of sending a message (and perhaps receiving a reply) corresponds to a cross-domain procedure call in a capability based system such as Hydra [3] or StarOS [12].

A *message* consists of a fixed length header and a variable size collection of typed data objects. Messages may contain both port capabilities and/or imbedded pointers as long as both are properly typed. A single message may transfer up to $2\uparrow32$ bytes of by-value data.

Messages may be sent and received either synchronously or asynchronously. A software interrupt mechanism allows a process to handle incoming messages outside the flow of normal program execution.

Figure 3-1 shows a typical message interaction. A process A sends a message to a port P2. Process A has send rights to P2 and receive rights to a port P1. At some later time, process B which has receive rights to port P2 receives that message which may in turn contain send rights to port P1 (for the purposes of sending a reply message back to process A). Process B then (optionally) replies by sending a message to P1.

Should port P2 have been full, process A would have had the option at the point of sending the message to: (1) be suspended until the port was no longer full, (2) have the message send operation return a port full error code, or (3) have the kernel accept the message for future transmission to port P2 with the proviso that no further message can be sent by that process to P2 until the kernel sends a message to A telling it the current message has been posted.

3.2. Virtual memory support

Accent provides a $2\uparrow32$ byte paged address space for each process in the system and a $2\uparrow32$ byte paged address space for the operating system kernel. Disk pages and physical memory can be addressed by the kernel as a portion of its $2\uparrow32$ byte address space. Accent maintains a virtual memory table for each user

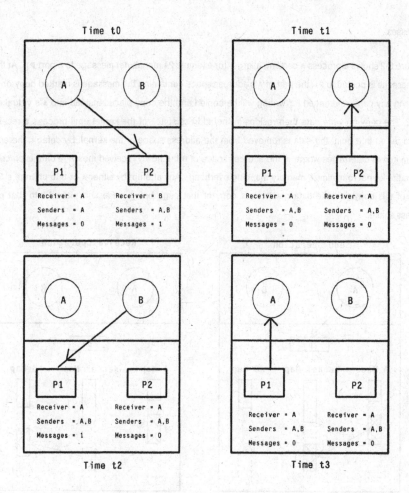

Figure 3-1: Typical message exchange

process and for the operating system kernel. The kernel's address space is paged and all user process maps are kept in paged kernel memory. Only the kernel virtual memory table, a small kernel stack, the PERQ screen, I/O memory and those PASCAL modules required for handling the simplest form of page fault need be locked in physical memory, although in practice parts of the kernel debugger and symbol tables for locked modules are also locked to allow analysis of system errors. The total amount of kernel code and symbol table information locked is 64K bytes [10].

Whenever large amounts of data (the threshold is a system compile-time constant normally set at 1K bytes) are transmitted in a message, Accent uses memory mapping techniques rather than data copying to move information from one process to another within the same machine. The semantics of message passing in Accent imply that all data sent in a message are logically copied from one address space to another. This can be optimized by the kernel by mapping the sent data copy-on-write in both the sending and receiving processes.

Figure 3-2 shows a process A sending a large (for example 24 megabyte) message to a port P1. At the point the message is posted to P1, the part of A's address space containing the message is marked copy-on-write -- meaning any page referenced for writing will be copied and the copy placed instead into A's virtual memory table. The copy-on-write data then resides in the address space of the kernel until process B receives the message. At that point the data is removed from the address space of the kernel. By default, the operating system kernel determines where in the address space of B the newly received message data is placed. This allows the kernel to minimize memory mapping overhead. Any attempt by either A or B to change a 512 byte page of this copy-on-write data results in a copy of that page being made and placed into that process' address space.

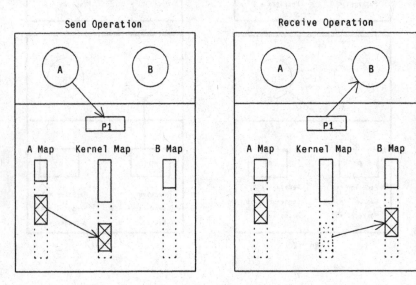

Figure 3-2: Memory mapping operations during message transfer

3.3. Network communication

The abstraction of communication through ports permits the distinction between access to local and remote resources to be completely invisible to a client process. In addition, Accent exploits the integration of memory management and IPC to provide a number of options in the handling of virtual memory, including the ability to allow memory to be sent copy-on-reference across a network. Each entry of an Accent virtual memory table maps a contiguous region of process virtual memory to a contiguous portion of an Accent *memory object.* A memory object is the basic unit of secondary storage in Accent. Memory objects can be contiguous physical memory (as used for the PERQ screen or I/O buffers) or a randomly addressed disk file. A memory object can also be backed not by disk or main memory, but by a process through a port. Initial references to a page of data mapped to a port are trapped by the kernel and a request for the necessary data is forwarded in a message on that port. This feature allows processes to provide the system with virtual memory that they themselves maintain (either locally or over a network connection to another machine). In this way network communication servers can provide copy-on-reference network transmission of pages in a large message.

4. Key Implementation Issues in Accent

Many of the implementation decisions made in Accent were based on experiences with RIG. Nevertheless, the addition of virtual memory and capability management to the RIG design made it unclear how the RIG experiences would extrapolate to the Accent environment.

4.1. IPC Implementation

The actual implementation of the message mechanism relied on several assumptions about the use of messages:

- the average number of messages outstanding at any given time per process would be small,

- the number of port capabilities needed by a process could vary from two to several hundred, and

- the use of simple messages (meaning messages which contained port capabilities only in their header and which contained less than a few kilobytes) would so dominate complex messages that simple messages would be an important special case.

Each of these assumptions had held true for RIG [5, 15]. It was hoped that although Accent provided a substantially different application environment than RIG, the RIG experiences would provide a reasonable prediction of Accent performance.

Given these expectations, the implementation was optimized for anticipated common cases, including:

- The assumption that there would seldom be more than one message waiting for a process at a time led to an implementation in which messages are queued in per-process rather than per-port queues.

- To allow large numbers of ports per process and fast lookup, port capabilities are represented as indexes into a global port record array stored in kernel virtual memory. Port access is protected through the use of a bitmap of process access rights kept per port (the number of processes is much less than the number of ports).

- The assumption that simple messages would be an important special case led to the addition of a field to the message header so that user processes can indicate whether or not a message is simple and thus allow special handling by the kernel.

These usage assumptions did in fact prove true for Accent. Table 4-1 demonstrates the properties of Accent message passing as measured during an active day of use.

1.01	Average probes to requested message
33.42	Average port rights held per process
14.38	Average ports owned per process
0.094	Ratio of complex to simple messages

Table 4-1: Message use statistics

4.2. Virtual Memory Implementation

The lack of sophisticated virtual memory management in RIG (and in fact in nearly all message-based systems of that era) meant that Accent could not benefit from previous experience with virtual memory use resulting from message operations. Instead, the design of Accent's virtual memory implementation grew out of simple assumptions based purely on intuition. These initial assumptions influenced the design of the Accent virtual memory implementation:

- process maps had to be compact, easy to manipulate and support sparse use of a process address space,

- the number of contiguously mapped regions of the address space would be reasonably small, and

- large amounts of memory would frequently be passed copy-on-write in messages.

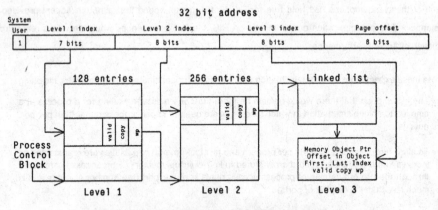

Figure 4-1: Mapping a virtual address in Accent

The Accent process virtual memory map is maintained as a two-level indirect table terminating in linked lists of entries (see Figure 4-1). Each entry on the linked list maps a contiguous portion of process virtual memory into contiguous regions of Accent memory objects. The map is organized so that large portions can be validated, invalidated or copied without having to modify the linked lists of map entries. This is accomplished by having valid, copy-on-write and write-protect bits at each level of the table. During lookup, these bits are "ored" together. Thus all of memory can be efficiently made copy-on-write by just setting the copy-on-write bits of valid entries in level one of the process map table. Figure 4-1 illustrates the translation of a virtual address to an offset within a memory object.

Physical memory in Accent is used as a cache of secondary storage. There are no special disk buffers. Access to all information (e.g., files) is through message passing (and subsequent page faulting if necessary).

This scheme is flexible enough to be used internally by the kernel to remap portions of its own address space. An entire process virtual memory map, for example, is copied in a fork operation without physically copying the map by using Accent's copy-on-write facility. To reduce map manipulation overheads, changes caused by copy-on-write updates are recorded first in a virtual to physical address translation table (kept in physical memory) and are not incorporated into a process map until the relevant page must be written out to secondary storage.

Copy-on-write access to memory objects is provided through the use of *shadow* memory objects which reflect page differences between a copied object and the object it shadows (which could in turn be a shadow). Disk space for newly created pages or pages written copy-on-write is allocated on an as-needed basis from a special paging area. No disk space is ever allocated to back up a process address space unless the paging algorithms need to flush a dirty page. See figure 4-2.

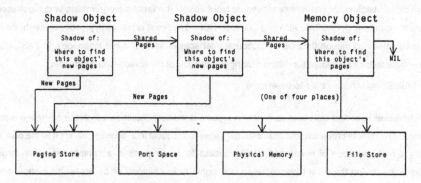

Figure 4-2: An example of memory object shadowing

Most shadow memory objects are small (under 32 pages). Most large shadows contain only a few pages of data different from the objects they shadow. These facts led to an allocation scheme in which small shadows are allocated contiguously from the paging store and larger shadows use a page map and are allocated as needed.

Overall, the basic assumptions about the use of process address space in Accent appear to hold true. The typical user process table:

- is between 1024 and 2048 bytes in size,

- contains 34-70 mapping entries, and

- maps a region of virtual memory approximately eight megabytes in extent (in PERQ PASCAL each separately compiled module occupies a distinct 128K byte region of memory) and about one to two megabytes in size.

Although all memory is passed copy-on-write from one process to another, the number of copy-on-write faults is typically small. A typical PASCAL compile/link/load cycle, for example, requires only slightly more than one copy-on-write fault per second. Clearly most of the data passed by copy in Accent is read and not written. The result is that the logical advantages of copy-on-write are obtained with costs similar to that of mapped shared memory [8].

4.3. Programming issues

One of the problems with message based systems has traditionally been the fact that existing programming languages do not support their message semantics. In RIG, a special remote procedure call function was provided called "Call" [14] which took as its arguments a message identifier, a process-port identifier, and operation arguments along with their type information. One of the early decisions in the implementation of Accent was to define all interprocess message interfaces in terms of a high-level specification language. The properties of ports allow them to be viewed as object references. The interprocess specification language is defined in terms of operations on objects. Subsystem specifications in this language are compiled by a program called Matchmaker into remote procedure call stubs for the various programming languages used in the system -- currently C, PASCAL, ADA and Common LISP. The result is that all interprocess interfaces look to the programmer as though they were procedural interfaces in each of these languages. In PASCAL, for example, the interface procedure for writing a string to a window of the screen would look like:

> WriteString(window,string-to-be-written)

All Matchmaker specified calls take as their first argument the port object on which the operation is to be performed. The remote procedure call stub then packages the request in a message, sends it to the port, and waits for a reply message (if necessary). Initial access to server ports is accomplished either through inheritance (by having the parent process send port rights to its children) or by accessing a name server process (a port for which is typically passed to a process by inheritance). A complete description and specification of Matchmaker can be found in [13].

Matchmaker's specification language allows both synchronous and asynchronous calls as well as the specification of timeouts and exception handling behavior. It supports both by-value and by-value-result parameters. It allows types to be defined as well as the specification of their bit packing characteristics in the message. For the server process, Matchmaker produces routines which allow incoming messages to be decoded and server subroutines automatically invoked with the proper arguments.

The support provided by Matchmaker is similar to some of the features which have been introduced in modern languages for managing multiple tasks such as the ADA rendezvous mechanism [2]. Matchmaker, however, supports a number of different programming languages and provides a much greater range of options for synchronous and asynchronous behavior in a distributed environment.

Despite the obvious simplicity of simple "remote procedure call" style interfaces, a suprisingly high percentage of network operating system interfaces take advantage of the asynchronous form of Matchmaker interfaces. Of 225 system interfaces:

- 170 (approximately 77 percent) are synchronous,

- 45 (approximately 19 percent) are asynchronous and

- 10 (approximately 4 percent) represent exceptions.

Runtime statistics show that over 50 percent of messages actually sent during normal system execution are sent as part of asynchronous Matchmaker specified operations -- normally due to the behaviour of I/O subsystems (such as handlers for the PERQ keyboard and display) or basic system servers (such as network protocol servers).

Matchmaker server interfaces account for approximately 10 percent of the total network operating system code -- roughly 75.5k bytes out of 757k bytes. For the Accent kernel itself, the Matchmaker interface is 10280 bytes out of approximately 115k bytes. Runtime costs are considerably less. During a PASCAL compilation, for example, less than 2 percent of CPU time is due to Matchmaker interface overheads.

4.4. Key Statistics[3]

4.4.1. Hardware and basic system performance of Accent

Table 4-2 compares the relative performance of PERQ and VAX-11/780 CPUs. Timings were performed in PASCAL on the PERQ and in C on a VAX running UNIX 4.1bsd.

PASCAL programs written for the PERQ range in overall speed from 1/5 to 1/3 the speed of comparable programs on the VAX 11/780, depending on whether 16-bit or 32-bit operations predominate. In fairness to the PERQ hardware, the underlying microengine is much faster than the PASCAL timings in table 4-2 would indicate. Microcoded operations often run as fast as or faster than equivalent VAX 11/780 assembly language. Note, for example, the relative speeds of the microcoded context switch and kernel trap operations. Moreover, instruction sets better tuned to the PERQ hardware, such as the Accent CommonLisp instruction set, run at speeds closer to 50 percent of the VAX. Nevertheless, for the purpose of gauging the performance of the Accent kernel code, which is written in PASCAL and makes heavy use of 32-bit arithmetic, pointer chasing and packed field accessing, the CPU speed of a PERQ is about 1/5 that of a VAX 11/780.

4.4.2. IPC Costs

Table 4-3 shows the costs of various forms of message passing in Accent. As was previously described, Accent distinguishes between *simple* and *complex* messages to improve performance of common message operations. Simple messages are defined to be those with less than 960 bytes of in-line data that contain no pointers or port references (other than those in the message header). Other messages are considered

---

[3] A more detailed performance analysis of Accent from which this data was taken can be found in [10].

Perq	Vax	Ratio	Operation
2300ns	720ns	.31	Tick (32-bit stack local)
12us	4us	.25	Simple loop (16-bit integer)
20us	3us	.17	Simple loop (32-bit integer)
35us	20us	.57	Null procedure call/return
75us	25us	.33	Procedure call with 2 arguments
80us	400us	5.00	Context switch
132us	264us	2.00	Null kernel trap
30s	9s	.30	Baskett Puzzle Program (16-bit)
50s	10s	.20	Baskett Puzzle Program (32-bit)

Table 4-2: Comparison of Perq and Vax-11/780 operation times

For executing kernel code, a Perq CPU is about 1/5 of a Vax-11/780.

complex. The times for complex messages listed in the table were measured for messages containing one pointer to 1024 bytes of data. The observed ratio of simple to complex messages in Accent is approximately 12-to-1.

Time	IPC Operation
1.15	Simple message send
1.35	Simple message receive
10.	Complex message send (1024 bytes)
10.	Complex message receive (1024 bytes)

Table 4-3: IPC operation times in milliseconds

The average number of messages per second observed during periods of heavy standard version use (e.g., compilation) is less than 30. There were 67378 simple messages and 4279 complex messages sent during one measurement of three hours of editing, network file access, and text formatting, an average of less than eight per second [10].

4.4.3. Accessing file data

One of the reasons for the relatively low message rate of message exchange in Accent is the heavy reliance on virtual memory mapping techniques for transferring large amounts of data in messages. A process making a request for a large file typically receives the entire file in a single message sent back from a file server process. As a result, all file access in Accent is mediated through the memory management system. There are no separate file buffers maintained by the system or special operations required for file access versus access to other forms of process mapped memory. By contrast, in RIG the same operation would have required as many message exchanges between client and server as there were pages in the file.

Table 4-4 shows the costs associated with reading a 56K byte file under UNIX 4.1bsd on a VAX 11/780 with a 30 millisecond average access time Fujitsu disk and under the standard version of Accent with a 30 millisecond average access time MAXSTORE drive.

System	Time	Operation
Accent	66	Request file from server
UNIX 4.1	5-10	Open/close
Accent	5-10	Read a page (512 bytes)
UNIX 4.1	16-18	Read a page (1024 bytes)
UNIX 4.2	16-18	Read a page (4096 bytes)

Table 4-4: File access times in milliseconds

Accent file reading performance is comparable to that of Unix4.1bsd.
File reading performance is dominated by the disk blocking factor.

The measured cost of a file access in Accent as shown in table 4-4 is due, in part, to the cost of a disk write to update the file access time. This disk write is unbuffered in Accent and thus is included in the file request time. The Unix disk write associated with an open is buffered and is excluded from the open/close time.

Accent file access speed is limited by the basic fault time of about four milliseconds (see table 4-5), the average number of consecutive file pages on a disk track and the cost of making new VP entries. Its page size is only 512 bytes, in contrast to 1024 bytes for 4.1bsd and 4096 or 8192 for 4.2bsd.

Once mapped, file access in Accent ranges from somewhat faster than 4.1bsd to slightly slower, depending on the locality of file pages. 4.2bsd file access [16] is considerably faster than either 4.1bsd or Accent. This increase in speed appears to be due almost entirely to the larger (typically 4096 byte) file page size. The actual number of disk I/O operations per second under 4.2 is almost identical to 4.1, about 50-60 per second, and appears to be bounded by the rotational speed of the disk (60 revolutions per second).

4.4.4. Fault handling and copy-on-write

Table 4-5 summarizes the results from test programs that caused 100,000 instances of a variety of memory fault types. It shows the average total times required to handle single faults.

Overall, the costs of copy-on-write memory management are nearly identical to that of by-reference memory mapping. Less than 0.01 percent of the total time associated with an entire rebuilding of the operating system and user programs from source is used to handle copy-on-write faults [10].

Total	Type of fault
0.623	Null fault
3.355	Read fault, zero fill
3.704	Write fault, zero fill
3.760	Read fault, memory fill, small file
4.504	Read fault, memory fill, large file
3.833	Write fault, CopyOnWrite copy

Table 4-5: Fault handling times in milliseconds

5. Mach: Adapting Accent to Multiprocessors

Accent went beyond demonstrating the feasibility of the message passing approach to building a distributed system. Experience with Accent showed that a message based network operating system, properly designed, can compete with more traditional operating system organizations. The advantages of this approach are system extensibility, protection and network transparency.

By the fall of 1984, however, it became apparent that, without a new hardware base, Accent would eventually follow RIG into oblivion. Hastening this process of electronic decay was Accent's inability to completely absorb the ever burgening body of UNIX developed software both at CMU and elsewhere -- despite the existence of a "UNIX compatibility" package.

Mach was conceived as an Accent-like operating system which would provide complete UNIX compatibility. It was also designed to better accommodate the kind of general purpose shared-memory multiprocessors which appear to be on their way to becoming the successors to traditional general purpose uniprocessor workstations and timesharing systems. As of June, 1986 Mach is running on all VAX architecture machines (multiprocessors and uniprocessors) and on the IBM RT PC. Initial performance figures indicate that Mach outperforms Berkeley 4.3bsd in several key measures of virtual memory, IPC and overall system performance [1].

6. Conclusions

The evolution of network operating systems from RIG through Mach was, in a sense, driven by the evolution of distributed computer systems from small networks of minicomputers in the middle 1970s to large networks of personal workstations and mainframes in the early 1980s to networks of uniprocessor and multiprocessor systems today. Not suprisingly, the basic software primitives of Mach -- task, thread, port, message and memory object -- parallel the hardware abstractions which characterize modern distributed systems -- nodes, processors, network channels, packets and primary and secondary memory. Experiences, both good and bad, with RIG and Accent have played an important role in determining the exact definition of the Mach mechanisms and their implementation.

Accent went beyond demonstrating the feasibility of the message passing approach to building a distributed system. Experience with Accent has shown that a message based network operating system, properly designed, can compete with more traditional operating system organizations. The advantages of this approach are system extensibility, protection and network transparency.

7. Acknowledgements

In addition to anything the author may have done, the heroes of the RIG kernel development were Gene Ball and Ilya Gertner. Jerry Feldman was in large part responsible for the initial RIG design and the system's name. The Accent development team included George Robertson and Gene Ball as well as the author. Keith Lantz and Sam Harbison made notable contributions to the design. Mary Shaw contributed the name. Others contributed greatly to Accent's evolution: particularly Doug Philips, Jeff Eppinger, Robert Sansom, Robert Fitzgerald, David Golub, Mike Jones and Mary Thompson. Matchmaker could not have come into existence without the aid of Mary Thompson, Mike Jones, Rob MacLachlin and Keith Wright. Mach was the brainchild of many including Avie Tevanian, Mike Young and Bob Baron. Dario Giuse came up with the name.

References

[1]   Accetta, M., Baron, R., Bolosky, W., Golub, D., Rashid, R., Tevanian, A. and M. Young.
        Mach: A New Kernel Foundation for UNIX Development.
        In *1986 Summer USENIX Technical Conference*. USENIX, June, 1986.

[2]   Department of Defense.
        Preliminary Ada Reference Manual
        1979.

[3]   Almes, G. and G. Robertson.
        An Extensible File System for Hydra.
        In *Proc. 3rd International Conference on Software Engineering*. IEEE, May, 1978.

[4]   Ball, J.E., J.A. Feldman, J.R. Low, R.F. Rashid, and P.D. Rovner.
        RIG, Rochester's Intelligent Gateway: System overview.
        IEEE Transactions on Software Engineering 2(4):321-328, December, 1976.

[5]   Ball, J.E., E. Burke, I. Gertner, K.A. Lantz and R.F. Rashid.
        Perspectives on Message-Based Distributed Computing.
        In *Proc. 1979 Networking Symposium*, pages 46-51. IEEE, December, 1979.

[6]   Baskett, F., J.H. Howard and J.T. Montague.
        Task Communication in DEMOS.
        In *Proc. 6th Symposium on Operating Systems Principles*, pages 23-31. ACM, November, 1977.

[7]   Bisiani, R., Alleva, F., Forin, A. and R. Lerner.
        Agora: A Distributed System Architecture for Speech Recognition.
        In *International Conference on Acoustics, Speech and Signal Processing*. IEEE, April, 1986.

[8]   Bobrow, D.G., Burchfiel, J.D., Murphy, D.L. and Tomlinson, R.S.
        TENEX, a paged time sharing system for the PDP-10.
        Communications of the ACM 15(3):135-143, March, 1972.

[9]   Spice Project.
        Proposal for a joint effort in personal scientific computing.
        Technical Report , Computer Science Department, Carnegie-Mellon University, August, 1979.

[10]   Fitzgerald, R. and R. F. Rashid.
        The integration of Virtual Memory Management and Interprocess Communication in Accent.
        ACM Transactions on Computer Systems 4(2):, May, 1986.

[11]   Hornig, D.A.
        Automatic Partitioning and Scheduling on a Network of Personal Computers.
        PhD thesis, Department of Computer Science, Carnegie-Mellon University, November, 1984.

[12]   Jones, A.K., R.J. Chansler, I.E. Durham, K. Schwans and S. Vegdahl.
        StarOS, a Multiprocessor Operating System for the Support of Task Forces.
        In *Proc. 7th Symposium on Operating Systems Principles*, pages 117-129. ACM, December, 1979.

[13]   Jones, M.B., R.F. Rashid and M. Thompson.
        MatchMaker: An Interprocess Specification Language.
        In *ACM Conference on Principles of Programming Languages*. ACM, January, 1985.

[14]   Lantz, K.A.
        Uniform Interfaces for Distributed Systems.
        PhD thesis, University of Rochester, May, 1980.

[15]   Lantz, K.A., K.D. Gradischnig, J.A. Feldman and R.F. Rashid.
        Rochester's Intelligent Gateway.
        Computer 15(10):54-68, October, 1982.

[16]   McKusick, M.K., W.N. Joy, S.L. Leach and R.S. Fabry.
        A Fast File System for UNIX.
        ACM Transactions on Computer Systems 2(3):181-197 , August, 1984.

[17]   R. Rashid and G. Robertson.
        Accent: A Communication Oriented Network Operating System Kernel.
        In *Proceedings of the 8th Symposium on Operating System Principles*, pages 64-75. December, 1981.

[18]   Spector, A.Z. et al.
        Support for Distributed Transactions in the TABS Prototype.
        In *Proceedings of the Fourth Symposium on Reliability in Distributed Software and Database Systems*,
            pages 186-206. October, 1984.

Network Operating System Kernels for Heterogeneous Environments

Herbert M. Eberle and Hermann Schmutz
IBM ENC Heidelberg
Tiergartenstr. 15

D - 6900 Heidelberg

Abstract

A major function of a Network Operating System Kernel is to free application program developers from the need for designing complex protocols. The Operating System analogy for networks offers a paradigm for inter-program cooperation, as if this cooperation were local.

Instead of forcing programs to send and interpret messages explicitly, the Network Operating System provides services for remote operations on accustomed data structures, whereby the message exchange as such remains invisible to the programmer.

This paper compares two extremes in the range of Network Operating System Kernel interfaces currently under discussion: the "simple" Remote Procedure Call and the more "complex" Remote Service Call. The Remote Service Call is a straightforward generalization of the Remote Procedure Call interface and is designed to encompass the pertinent functions and services which are proven and available in local operating system kernels. By analysis, it will be shown that the "simple" approach, although desirable, does not fulfil all basic functional requirements. Among these are aspects of pluralism, such as protection and scheduling, other system aspects, such as revocability and concurrency e.g. as needed by file servers, aspects of heterogeneity, such as the issue of data presentation, and the support of higher level objects, such as open files. The price to be paid for the complexity in terms of implementation effort, reliability, and portability will be discussed on the basis of a prototype implementation of a Remote Service Call facility in a network of heterogeneous computing systems.

1. Introduction

General resource sharing in networks of heterogeneous operating systems is receiving increased attention with the widespread use of high speed local area networks. Among the many conceivable ways of reaching this goal, there is one which looks particularly attractive and which we term the Network Operating System (NOS) approach. This approach is, among other things, pursued in the Distributed Academic Computing (DAC) project, a joint project between the University of Karlsruhe and the IBM European Networking Center. Similar approaches are described in [16,17] and more recently in [14,15]. Common to these approaches is the idea of providing transparent resource sharing through the extension of heterogeneous Local Operating Systems (LOSs).

In the DAC view we call this extension of the LOSs the NOS. The NOS user remains in the specific environment of his LOS and has access to remote resources as if these resources were local. Users of different types of local systems may thus have different views of the same resources. The NOS is an adjunct to and not a replacement of the local system and has to provide a transparent mapping between differing views of resources in the participating systems.

The subject of this paper is the description of the interface between application programs and NOS system services to the NOS kernel. This interface is of central importance within the NOS. It provides the mechanisms and functions for the cooperation of other NOS components. Figure 1 serves to illustrate the structure of the DAC NOS and the role of the NOS kernel.

Within the DAC NOS we distinguish three layers: Kernel, System Services, and Applications. Existing applications interface with their LOS, whereas new and distributed applications make direct use of the NOS.
The NOS System Services extend the LOS System Services and provide mappings between heterogeneous systems. A typical scenario of remote resource access in the DAC NOS is shown in figure 2 for the Remote File Access (RFA) component. The application program accesses a remote file through the local file system. The RFA service intercepts the local call, transforms it, if necessary, and passes it to the remote file system. Similarly, the results are passed back to the application program in a format determined by its local file system. RFA has to perform two major tasks. The first is the mapping of file systems, which is resource (or service) specific. The second is the transfer of request/response type operations in a network. It is the task of the NOS Kernel to provide the latter in a way which can be used by all System Services and by user-written application programs.

In the DAC NOS, the component responsible for providing this communication mechanism is the Remote Service Call (RSC) component, the external interface of which coincides with the external interface of the NOS kernel. As in Accent, RSC provides the "single powerful ab-

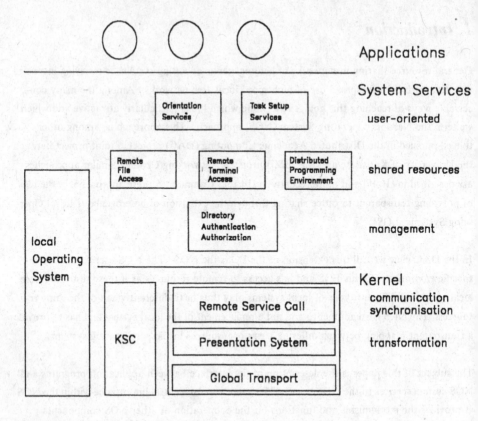

Figure 1. The Components of the DAC Network Operating System

straction" around which the NOS is built [1,2]. The subject of this paper is to describe the motivation which leads to the RSC concepts. To this end we will discuss a well known approach to communication in section 2, and show its limitations by analyzing an example. We will then briefly summarize the requirements of a NOS Kernel interface and describe the concepts on which RSC is based. In section 4 we will present some key features of RSC and analyze their properties. Section 5 summarizes the results.

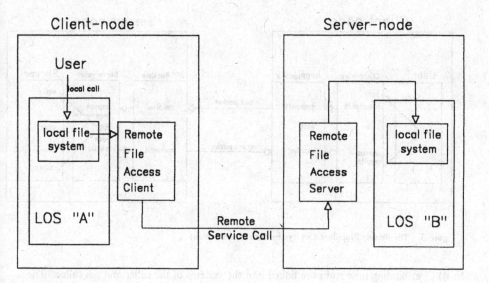

Figure 2. Access to Remote Files in DAC NOS

2. Remote Procedure Call

One of the ideas underlying the design of RSC is the use of local communication as a paradigm for remote communication. This is not new. Perhaps the most popular communication mechanism based on such a paradigm is the Remote Procedure Call (RPC). It has been in discussion at least since 1976 [3,4] and implemented in many systems. It is being used in research prototypes [5] and even in products [6]. Also papers on experiences with RPC have been published [7,8].

RPC is often proposed as a sufficient Interprocess Communication (IPC) mechanism in a distributed system [18]. We will subsequently discuss why we consider the RPC as described in the literature [5] as inadequate for IPC in a network of heterogeneous systems with resource sharing. This discussion will lead to requirements which have to be met by a NOS kernel. The essential elements of RPC are described in Fig. 3.

Client node   Server node

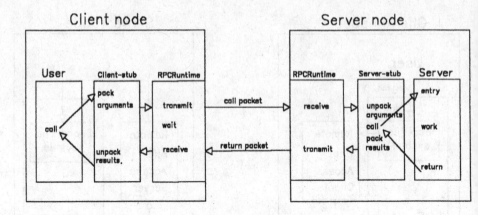

Figure 3. The Remote Procedure Call System (Birrell and Nelson)

In RPC, at binding time stubs are linked into the systems of the caller and the callee. The runtime packages exchange messages and take care of lost messages by means of timeouts, resends, and acknowledgements. The client stub, when actually invoked, packs the arguments and sends them to the server stub, which unpacks the arguments and actually calls the remote procedure. The results are returned in a similar way.

A critical question with RPC is, which types of parameters are supported. In most cases, procedure parameters are not supported as parameters, an exception being [9]. Also linked structures are, in general, not supported. Ultimately, the RPC is restricted by the absence of shared memory, while its paradigm, the local procedure call, relies on shared memory. RPC does not provide a substitute for shared memory. This has consequences for the use of RPC.

Consider the example of a remote file system. Superficially, accesses to a local file system may look like procedure calls [4]. At a closer look, however, an open file is an "abstract object", which regulates the access to the data in the file. In a local system, this "abstract object" is represented by some kind of control block, which is initialized at open time.

Mapping the local situation to the remote situation is not obvious. Semantically, an abstract object like an open file is defined by a set of operations (or set of procedures). Known RPC systems have no means of supporting abstract objects as a data type which can be passed to procedures as a parameter or returned by procedures as a result. Instead, an open file must be identified by data, e.g. a pair *(handle, server)* or the like, where *handle* is some data which permits the *server* to associate an open file with it. To overcome this invonvenience for the client in a sound way, an object package was implemented in Alpine [7] on top of RPC. The object package may then allow clients to refer to files as objects which can be passed as arguments to procedures. With strict RPC, the file server has to maintain and look up a mapping between handles and open files. In addition, the server has to check whether the caller has indeed the right to access the open file. The server has a difficult task in performing this checking. Suppose, a logical node A opened the file and passed it as argument to a procedure at node B. Clearly, node B should have access right to the open file during the call, but how does the file server know?

The straightforward answer to this is: The "object package" also resides at the server node and communication between file clients and file servers is through the object package, which then may take care of whatever is necessary to handle an open file as a single object.

The outlined approach, when implemented on top of RPC, might represent a reasonable solution for a remote file system. But this solution is very unsatisfactory. The file system is just one component of a distributed system. Other system objects, such as device service access points or high level locks, have the same kind of requirements and need a similar "object package". Furthermore, the object package has to handle protection, an issue which is central in any operating system, a fortiori in a NOS, and which should be handled within the kernel consistently for all subsystems. A NOS communication mechanism should satisfy the common needs of all subsystems. The RPC without a generic object package, more precisely without support for "abstract objects", remains in essence a message passing mechanism and is therefore unsatisfactory as the IPC in a NOS.

RPC is a mechanism to provide access to remote services within high level language programs. It looks syntactically like a local procedure call. However, it has restricted semantics and does not provide the mechanisms for transparent support of abstract objects. RSC, which is discussed in the next section is, though guided by a similar paradigm, designed to provide remote IPC in a way which is semantically equivalent to local IPC mechanisms.

3. Remote Service Call

3.1 Introduction

RSC entities are located at individual logical nodes and cooperate to provide a uniform kernel interface for resource sharing in a heterogeneous environment. The set of cooperating RSC entities appears to the user as a single NOS kernel entity. The exchange of messages between logical nodes as result of Service Calls or operations on RSC objects remains hidden to the application program.

RSC combines within it the functions which are commonly needed by applications and handles, transparently for the application program, the kernel services which are usually handled by the local operating system. As such it is similarly motivated as the Common Application Service Elements within the ISO OSI model [13]. Consequently, we would place RSC at the bottom of layer 7 within the ISO OSI Reference Model.

3.2 Objectives of RSC

RSC provides a location transparent mechanism for the invocation of services as needed by the upper layers of a NOS which integrates heterogeneous operating systems. Therefore the following specific requirements have to be met.

Coexistence

The LOS and the local part of the NOS kernel must coexist without placing restrictions on the use of the LOS. This requires a careful design of the interaction between the NOS kernel and the LOS. A separate kernel component (KSC in figure 1) takes care of this interaction.

Presentation Transparency

RSC must support interfaces, which may be used independently of the data representations of the communicating partners. RSC must perform the necessary data conversions transparently for the application programs. The Presentation System of the NOS kernel (see figure 1) assists RSC in actually performing the conversions. However, RSC controls the interfaces to the user and remains responsible for the invocation of the presentation component.

Equivalence to Local Communication

RSC must provide sufficient equivalence to local communication. RSC must be general enough to allow for the mapping of local IPC mechanisms to RSC mechanisms in a way which maintains the functionality of the local mechanisms. LOS mechanisms make use of the fact that the LOS kernel has access to the complete state vector of the caller and callee. Protection, resource usage monitoring, dispatching under fairness constraints, and collection of accounting data can therefore be handled by the LOS kernel completely independent of application programs. Clearly, in the local case a matrix inversion program should not have to care about accounting and maintaining fairness in dispatching. Equivalence then requires that a remote invocation of the same matrix inversion program offers the same potential for accounting, as does the local invocation.

As discussed in section 2, LOS mechanisms make use of the fact that caller and callee share at least part of their memory. Objects are passed via pointers between programs, without changing their protection properties. RSC has to provide mechanisms and objects to which such local objects can be mapped and which can equivalently be passed from program to program in the remote environment.

3.3 Overview of RSC

RSC provides a uniform interface to all RSC processes in a heterogeneous network. RSC processes are tied to a logical node with a virtual address space. The extent to which they share address spaces is at the discretion of the local system.

RSC does not offer higher level objects such as files, procedures, transactions or network processes (i.e. processes not tied to a logical node). Instead, it provides a consistent set of primitives on which higher level objects can be built in a network transparent way.

There are three basic groups of object types in RSC:

Non-Sharable: Process, Event List

Sharable: Port, Window, Account, Lock

Transient: Carrier

Table 1 gives an overview of these objects.

Object	Function	Sharable	Event
Process	issues operations on objects	no	no
Event List	supports wait on a collection of events	no	--
Notice	is used to send a one way message	no	no
Port	is a queue for service requests (Carriers) and Notices	yes	yes
Window	is used to grant access to virtual memory	yes	no
Lock	is used for mutual exclusion of remote Processes	yes	yes
Account	identifies clients to servers in Processes and Carriers and is used to share system information on resource usage and dispatching parameters	yes	no
Carrier	carries a service request and is used to transfer object access rights temporarily to servers	--	yes

Table 1:   The RSC Objects

Sharable objects in RSC are in principle accessible to all RSC processes, but only if the corresponding access rights have been acquired previously.

Figure 4 illustrates a network state with 5 logical nodes. LN1 has the right to wait on and to send to the Port object. LN2, LN3, and LN4 have the right to send to the Port object. LN4 and LN5 share a Lock object. Furthermore the state exhibits the granter/grantee relationship of an access right. For example, LN4 obtained the send right from LN3, which in turn obtained it from LN1.

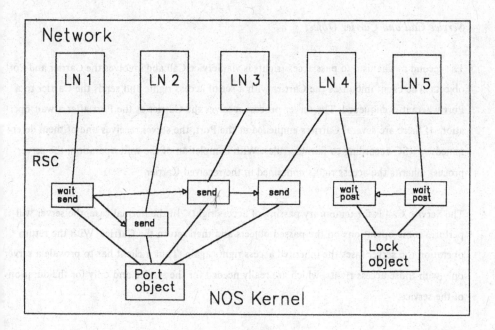

Figure 4. Access Rights in RSC

In general, the creator of an object obtains all access rights to it. The owner of an access right has two mechanisms available for passing the access rights to other logical nodes.

Offer and Share

The first is via explicit offer and share operations, e.g.

LN1 performs:

    port ← create-port (attributes)
            offer (port, 'SERVICE', SEND)

LN2 performs:

    port ← share (LN1, 'SERVICE', SEND)

After these operations LN1 and LN2 share the Port, and LN2 may send Carriers or Notices to this Port. The offering logical node retains autonomous control over the shared object and may retract its offer at any time.

Service Call and Carrier Object

The second mechanism to pass access rights is via Service Call and involves the Carrier and Port objects. The client initializes the Carrier with a set of access rights and sends the Carrier to a Port where it is enqueued. The server process receives the Carrier at the Port after a wait operation. If there are several Carriers enqueued at the Port, the server receives one of them determined by RSC according to fairness rules. With completion of the wait operation, the server process inherits the access rights contained in the received Carrier.

The Service Call is for temporary passing of access rights. In the normal case, the server will perform some operations on the passed objects and then return the Carrier. With the return operation the server loses the inherited access rights again. Thus a client has to provide a server only with those access rights, which are really needed for the service and only for the duration of the service.

In abnormal situations, if the client is for some reason no longer interested in the service, he may retract the Carrier and thus retract the access rights contained in the Carrier. Thus a client retains node autonomy by maintaining full control over the access rights passed to other logical nodes.

The operations supporting the Service Call are described in more detail in [20].

The Account Object

Implicitly every RSC process is associated with an Account object. The Account identifies a logical node uniquely and contains system information (such as priority, resources used, etc.).

When a Carrier is created, it is associated with the Account of the creating Process. A server Process inherits the Account of a received Carrier and, typically, keeps it until the Carrier is returned.

The Account object ensures that RSC is at any time aware of which processes act on whose behalf. It is therefore the basis for selective authorization (e.g. an user may have access to a file system but not to each file), for dispatching decisions and for accounting or other system services, which remain transparent to the application program.

The Port Object

The Port object is the access point for Service Calls. One may view a Port together with its server Process(es) in a conceptual way as a set of operations. A Port is therefore suited to represent files, process groups or any other higher level object. A client sees only the input/output behaviour of (the service behind) the Port independent of its realization.

A Port is associated with a Format Description (FD). The FD enables RSC to perform necessary data transformations in logical nodes transparent to caller and callee. It is specified in a powerful language. The FD associated with a Port specifies the syntax of the service requests to this Port. In general, several types of Service Calls with different sets of parameters are supported.

A Port may be passed as argument or returned as result in Service Calls just like any other sharable object. It is suitable for the realization of higher level abstract objects like procedures, files, or process groups in a distributed, heterogeneous environment.

The Window Object

The Window object is currently the only storage object defined and supported in RSC. The creator of a Window associates a section of his own memory space with the object. Once created, the Window may be passed around to other nodes like any other object. For efficiency reasons, there is no synchronization between processes accessing the shared memory. Thus a process which has passed a section of his virtual memory to some other process with the right to modify the storage contents should avoid accessing the shared section until the Window has been returned or retracted. The rule says, that if only one process modifies and accesses a shared Window, this process will see a consistent memory state. This weak consistency rule permits efficient implementation. In a synchronous Service Call no consistency problem arises. During asynchronous Service Calls the caller must avoid access to the Window and to the section of virtual memory described by the Window for the duration of the Call. These rules are essentially the same as those for shared memory in local systems.

4. Discussion

4.1 Relationship to other work

Besides RSC, which has been discussed in section 2, there are various other approaches to distributed systems, although with different objectives. Perhaps the two approaches which are closest to RSC are Accent and the V system with its predecessor Thoth [1,2,10,11].

Accent is a message based system, but, messages are typed and may carry something like access rights to virtual memory with it. In addition, Accent supports port capabilities in messages. Thus Accent messages are similar to the RSC carrier. Communication in Accent is symmetric, i.e. for a remote operation two ports are necessary, one for each direction, and the Accent kernel cannot be aware, of who is acting on whose behalf. The Accent Communication mechanism does therefore not meet the transparency objectives of RSC.

There is an additional important point which made us exclude communication mechanisms like Accent. The DAC NOS supports a network of autonomous computers. The NOS should not touch the autonomy of nodes. Ownership of a resource remains with the owner of the node at which the resource is located. We did not see a need for the complete transfer of an object from node A to node B, as it is possible in Accent. To be specific, within RSC A may pass its own access rights to B, B may pass any subset of these rights to C. A may then retract the rights from B and C, B only from C. This temporary passing of access rights fits well into a remote operation mechanism such as the RSC Service Call. Accent does not have an equivalent concept. It may, however, be a reasonable approach for a closely coupled network, in which the participating nodes have given up their autonomy and in which there is no need for Accounting or Network Resource Management.

The V interface is based on message passing in a request/respond mode. V messages contain access rights to memory. Thus as in Accent, these messages are something like the RSC Carrier. However, the V kernel does not have objects which are suitable for representing higher level objects like files. It is unclear to us how one could provide this support on top of the V interface in an essentially different way as it would have to be done for RPC (see section 2).

In summary, one may view RSC as the integration of the request/response type communication of the V kernel with the port capability mechanism of Accent into a consistent object sharing system with remote operations in a heterogeneous network of autonomous nodes.

4.2 Object Sharing

The major guideline for the RSC design was object sharing. The notion of a network wide entity
- the NOS Kernel - as counterpart to all users (i.e. logical nodes) was felt as a conceptually
sound basis for achieving network transparency. It is clear that the unprotected sharing of
memory between the LOS and user address spaces in local systems cannot be transferred to a
network of autonomous nodes. In the presence of distrusted nodes in the network the concept
of object sharing among users on different nodes was the closest possible to the memory sharing
of local systems.

Object sharing is an intuitively appealing concept which permits a safe and robust implementation. Object sharing, in particular port sharing, is among known mechanisms the most attractive for the implementation of system objects like files, prodedures, and higher level locks.
This aspect is so important that we would like to illustrate it by examples.

It is straightforward to implement a macro on top of RSC of the following form:

(rc, object list) ← *port* (argument list)

Explanations:

rc   variable to contain the return code
object list   list of RSC objects and data to be returned by the Service Call
port   name of a Port object
argument list   list of RSC objects and data

The macro would expand into initialization of a Carrier with the argument list, have it sent to
the Port, and wait for the returned objects and return code.

Let *f-serv* and *p-serv* be Ports for the file and procedure service, respectively, with the following
interfaces:

Interface to the File Server *f-serv*

(rc, *file*) ← *f-serv* ('open', file-name)
  rc   ← *file* ('read', position, record-variable)
  rc   ← *file* ('write', position, record-variable)
  rc   ← *file* ('status', status-variable)
  rc   ← *file* ('close')

Interface to the Procedure Server *p-serv*

(rc, *proc*) ← *p-serv* ('link', procedure-name)

rc   ← *proc* ('call', argument list)

rc   ← *proc* ('release')

The important property of objects like *file* and *proc* is that they can be passed via calls to other procedures and be used there in the same way as in the calling procedure.

4.3 Triangular Relationships

A service is often implemented by a large set of servers which are coordinated and controlled by a master server. A typical example is a transaction processing system. When a new transaction is initiated by a client the client calls some master server which then determines a server to become responsible for handling the transaction. This may require that the server process has to be started first. After the server has been determined, the master will relay the call to the server and the server will respond to the transaction initiation. All subsequent communication with respect to the transaction takes place between client and server.

A similar scenario applies to time sharing systems, file and archive systems, and possibly other subsystems. The common characteristic is that clients actually communicate with partners (the servers) without knowing the identity of the partner. We call this a triangular relationship between the client, the master and the anonymous server.

A triangular relationship fits nicely into object sharing. The initialization request is a Service Call directed to the master. The master transfers the Service Call to the server which returns a Port to the client. This returned Port is only shared between server and client. It should be noted that it remains completely transparent to the client with whom he actually shares the Port. This is convenient for the client, since he is only interested in the input/output behaviour of the service behind the Port, not in its realization. It is also convenient for the master and server, permits flexible reorganization, and requires only the minimal number of messages. In contrast, handling the triangular relationship on top of binary connections would be clumsy and require additional protocol elements, handle matching, authorization tests, i.e. all kinds of extra work which is not necessary in local systems and from which users should be freed in Network Operating Systems.

4.4 Synchronization

RSC supports asynchronous operations at the client and server side. A client may start Service Calls before all previous calls have completed, and a server may accept service requests before he has processed all previously received service requests to their end. This generality has been subject of controversial discussions.

It would have been easy to establish RSC rules to enforce synchronous operation at any side. But we believe that such a system would be unnecessarily restrictive and of reduced practical use.

An important objective of RSC is equivalence to local mechanisms. RSC was designed by a team with expertise in seven completely different local operating systems, which all support asynchronous communication. In contrast, no really used system was known to us, which is - at the kernel interface level - completely synchronous.

Consider the following scenario. Logical node A starts a transaction involving Service Calls to logical node B and C. If B and C can only complete their respective service requests after having mutually communicated, the synchronous calls from A to B and then B to C would cause deadlock. A situation of this kind has been reported in [19].

In view of such situations it is often said that synchronous calls in combination with light weight processes (which are also supported in RSC) should be used for asynchronous situations. With this circumvention, the outside world has the also an "asynchronous" view of the logical node: the node remains active in the time between a Service Request and the corresponding response.

We consider light weight processes as a means to logically structure independent activities within a logical node. It depends on the situation whether asynchronous calls or separate light weight processes are the proper way of handling concurrency. A scenario which is clearly in favour of the first is the following:

Port A and Port B support independent of each other a simple disk storage. Port C should support a disk storage with very high reliability by storing the same data on A and B transparent to its clients. Both, performing this task within C by means of sequential synchronous calls as well as setting up extra light weight processes for performing these calls in parallel is unattractive, although for different reasons. While it is cumbersome for a user of a strictly synchronous invocation mechanism to have it done asynchronously via light weight processes, it is very easy and straightforward to use the asynchronous RSC client function in a synchronous way.

We thus conclude that asynchronous clients are the basic mechanism on which the preferable synchronous mechanism should be built.

Our discussion was up to now confined to the client side. A simple and convincing example for the advisability of asynchronous servers is the following. Consider a lock, as it is needed in a database system. Within RSC it would be implemented as a Port served by one sequential process. While the lock is being held by some client, the server process collects further lock requests in terms of RSC service requests, which are returned as soon as the requested locking mode becomes compatible with the state of the lock. Any other solution would either be poor programming style (consider the amount of synchronization necessary between independent server processes in accessing and changing the state of the lock) or incur more delay for lock requestors than necessary.

Probably much of the discussion on synchronization is due to a misunderstanding. Should high level languages support asynchronous operations? Maybe, maybe not. But a system programming language definitely should support asynchronous operations. RSC is designed to support both types of languages.

4.5 Bulk Transfer

The need for the convenient and efficient transfer of large amounts of data arises frequently in applications [9]. With object sharing a natural solution arises. There are two situations:

(a) A client expects large amounts of data to be sent to him. In this case he passes a Window of sufficient size to the server in a single Service Call. The server fills the window with data and acknowledges completion by returning the service request. No intermediate acknowledgements are necessary. The Window may be any size. In the DAC NOS this situation shows up when a CMS user wishes to edit a remote file. The CMS editor reads the whole file with one access into virtual memory. In the DAC realization this works as bulk transfer.

(b) If the client wishes to transmit large amounts of data to a server, an efficient way of doing this is by passing a large window with read access. The server then moves the data from the Window to his own memory in segment sizes of his own choice. Each move is confirmed, since read access to a window is a synchronous operation.

Common to both situations is that the receiver of bulk data determines the rate of acknowledgement while the sender determines the actual rate of transfer (in cooperation with the Transport Service, of course).

4.6 System Aspects

Many design decisions in RSC were influenced by evaluating them against the environment in which RSC is expected to work. Resource sharing must not be understood as a situation in which everybody has access to any resource in the network. It should be easy for the owner to grant remote users access to a resource. It should be at least as easy to prevent remote users from using a resource. Similarly, it should be easy to charge for resource usage by remote users. And, it should be easy for the resource owner to provide remote users access to his resources only if there is currently no contention caused by local users or some other preferred user group.

In other words, the NOS must be constructed in a way which maintains the autonomous control of the resource owners on the use of the resources. The necessary mechanisms have to be provided by the NOS as much as possible. We know from local systems that this problem cannot be handled in application programs. Application programs know about their own activity. Control of resource access, allocation of resources, fair sharing of resources must rely on kernel services in local systems. The same applies for the NOS Kernel, however, with drastically increased importance. In a network with resource sharing it is likely that a potentially much larger user group competes for shared resources than in a single, even large, local system.

System aspects need not be totally handled within the kernel. Much of the functions, like authentication, authorization, accounting, and global network resource management can be handled as System Services. These services define the strategy, but RSC has to provide the raw data and the mechanisms for implementing the strategy.

The RSC protection mechanism and the RSC Account object play key roles in this context.

Protection in RSC, as in any local system, is based on trusting parts of the system. However, in a network a user is well advised to distrust other components. The number of trusted entities at any point in time has to be kept at a minimum.

First, a user has to trust his local system and the local part of the NOS. Whenever he accesses remote resources, or when he grants access to remote users, he has, in addition, to trust the Transport System, at least to the extent that messages sent to a remote user are indeed only transmitted unmodified to this user, and messages received carry the identity of the sending user and are received as sent by that user. Otherwise he has to resort to public key encryption, which still is a possibility for rare occasions, but certainly not for the frequent interactions in a local area network.

Finally, a client has to trust a server for the service which he expects of him. But even in this case RSC follows the principle of maximum distrust. As pointed out earlier, a server is only given access to those client resources which are needed for performing the service, and only for the duration of the Service Call.

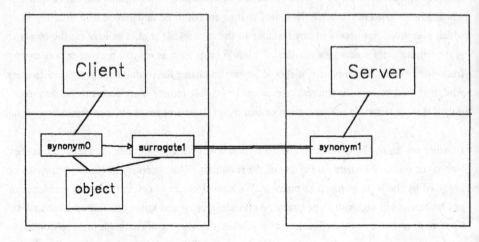

Figure 5. The Synonym-Surrogate Relationship

The RSC protection mechanism is based on access right ownership. Each object has a base location. The base location is the logical node which owns the object. A logical node with shared access to the object maintains a synonym of the object. The object owner has created a surrogate, which corresponds to the synonym (Figure 5). Synonyms and surrogates contain the access rights to an object. If the sharer behaves according to the rules, he will be prevented from performing illegal accesses by his local system due to the synonym mechanism. However, a malicious user of a workstation may modify his system to bypass the synonym mechanism. Therefore all accesses to the object are finally checked at the base location. Without properly corresponding surrogate the access request is rejected.

This strict access control mechanism requires that the base location is informed of any passing of rights via Service Calls. This requires additional protocol elements for the efficient access of subservers to the base objects.

Local OS kernels use the Process Control Block (PCB) to hold state information and system information. The state information contains the data necessary to put the process into a running state. It is essentially determined by the HW architecture. The system information identifies the user and contains data which assist system services such as dispatching, authorization, resource usage monitoring, accounting and system resource management.

RSC processes are bound to a location. Therefore state information remains local. However, system information cannot be confined to a specific logical node. The RSC Account object in connection with the RSC Service Call provides the mechanism for NOS services. As in the LOS, a user need not be aware of this mechanism. Via the Account object, the NOS has complete information about which processes are acting on whose behalf.

Consider, for example, authorization in case of a file system. A client may have access to the file server, but not to each file. The file server keeps a list of authorized user groups with each file. Before accessing a file, the file server has only to check whether the user which is identified by his current Account is a member of one of these groups.

As a second example, consider accounting. At the completion of a Service Call, i.e. with the return of a Carrier, the server RSC writes a log of resources used during the Call. Also the client RSC is informed about the resource usage during the service and may log this information. The logged data are periodically transmitted to the respective Account Servers which then are able to perform accounting services according to the conventions set up in the network.

The Account object mechanism allows for similar solutions to network wide scheduling, a service which determines, at regular intervals, the dispatching parameters.

4.7 Implementation Issues

The RSC design carefully separates interface and implementation. The RSC interface is defined by objects and operations, and the RSC user need not be aware of the internal structures and protocols. The clear separation of externals and internals makes the design extensible and adaptable to future external requirements and also to internal optimizations, such as the selection of optimal data channels for a given user request.

Nevertheless, implementation considerations played a major role in the design of the RSC interface. The Service Call and Window semantics are examples. The Window semantics are such that write operations to the Window do not need an acknowledgement. In fact, a remote server may accumulate several write operations into a local surrogate window and transfer the set of all operations on the window with one message. The absence of acknowledgements supports efficient bulk transfer as described earlier.

The Service Call is defined such that no message transfer needs to take place in a local communication. Only pointers are manipulated. A simple remote Service Call involving only two locations and small windows may be accomplished with two messages. If more than two partners are involved in a Service Call, then acknowledgement is provided to the client at the time of completion of the call. For intermediate steps no acknowledgement is necessary. With these rules there exist scenarios in which no receiver of a message returns an acknowledgement to the sender of the message. Any failure in such a scenario would be detected by the client RSC before request completion. Such optimization is possible on the basis of object sharing and would be too complex to be handled in application programs. However, if implemented in RSC, a thorough testing is possible, including elaborate protocol evaluation [12], to the benefit of many applications that use RSC as if no protocols were involved at all.

Table 2 gives an overview of the RSC protocol elements. By comparison to other approaches, the RSC implementation is complex. A few statistics illustrate this complexity. As defined now, the RSC protocol involves 19 different types of messages, most of which are defined as reliable datagrams (1-way messages) asking the receiver to perform some primitive action on behalf of the sender without acknowledgement to the sender. Protocol elements of type 2-way require two messages representing a request/response pair.

Protocol Element	Function	Type
Connect	get attributes of partner RSC entity	2-way
Sever Connection	inform partner of shutdown	1-way
Share	obtain access rights to an object	2-way
Send Carrier	transmit access rights to a partner	1-way
Return Carrier	return previously received access rights	1-way
Reply Carrier	return previously received access rights and transmit access rights	1-way
Read Window	copy data from Window into own address space	2-way
Write Window	copy data from own address space into a Window	1-way
Make Surrogate	create a surrogate at the RSC entity of object creation	2-way
Set Surrogate	transfer access right to a third RSC entity	1-way
Reset Surrogate	revoke transferred access right	1-way
Delete Surrogate	delete surrogate at the RSC entity of object creation	1-way
Delete Synonym	inform sharer of synonym deletion	1-way
Test Surrogate	retrieve surrogate status	2-way

Table 2: Overview of RSC Protocol Elements

The first implementation of RSC was done by a team of three people during a period of nine months. For adding the missing functions, final testing and performance evaluation, very likely additional six months will be required. RSC version 1 consists of 17.000 lines of C language code, which occupy in the /370 version 80 KBytes of sharable memory.

RSC Version 1 is currently used within the DAC project by three different research groups. Version 1 proves that the concept of object sharing is implementable.

5. Summary and Conclusions

Remote Service Call has been described as an interface to a Network Operating System kernel, which coexists with heterogeneous and autonomous local operating systems in a high speed local area network with resource sharing. RSC provides network transparent Interprocess Communication. It is based on the notions of object sharing and transfer of access rights in a powerful Service Call mechanism.

It may be held against RSC that its implementation is complex. This is not unexpected. RSC was designed to free application program designers from the need to design complex protocols. However, RSC does not provide a collection of unrelated communication mechanisms. It is derived from a consistent abstract model.

With all merits of an approach, which is easy to implement, it is apparent that too simple approaches place the burden on subsystems or applications. In Section 2 we have seen that the Remote Procedure Call requires extension, e.g. by an object package, to provide a satisfactory level of functionality and transparency. The Alpine experience report argues along similar lines. It is clear that any application program interface with strict message passing places a similar burden on the application programmer. Experience shows, that systems with insufficient functions are under pressure for extensions and therefore unstable. Ad hoc extensions to release some of the pressure tend to destroy the originally simple concept and the result is, in general, more complex than necessary.

Desirable is an approach of external simplicity and masterable internal complexity meeting the needs of subsystems and applications, and allowing for efficient implementation. The first version of RSC has been tested and is in use in three independent research groups. RSC has thus been shown to provide a realistic basis for the solution of major unresolved issues in distributed open systems including such important functions as Authorization, Accounting and Network Resource Management.

Acknowledgements

Many colleagues have contributed to the work described in this paper. Michael Salmony and Manfred Seifert contributed to the design of RSC, and Michael Salmony implemented an early prototype. Manfred Seifert developed the basic concepts for the RSC protocol and Kurt Geihs performed a validation of the protocol before implementation. The first version of RSC was implemented by Kurt Geihs, Manfred Seifert, Bernd Schöner and one of the authors.

References

1. R. F. Rashid, G. G. Robertson:
   Accent: A communication oriented network operating system kernel.
   In Proceedings of the Seventh ACM Symposium on Operating System Principles
   (Pacific Grove, California), 64-75 (1981)

2. R. Fitzgerald, R. F. Rashid:
   The Integration of Virtual Memory Management and Interprocess Communication
   in Accent.
   In ACM Transactions on Computer Systems, Vol. 4, 2, 147-177 (1986)

3. J. E. White:
   A high-level framework for network-based resource sharing.
   In Proceedings of the National Computer Conference, AFIPS, June 1976

4. B. J. Nelson:
   Remote Procedure Call.
   Dissertation, Carnegie-Mellon University, Report CMU-CS-81-119 (1981)

5. A. D. Birrell, B. J. Nelson:
   Implementing Remote Procedure Call.
   In ACM Transactions on Computer Systems, Vol. 2, 1, 39-59 (1984)

6. XEROX Corporation:
   Courier: The Remote Procedure Call Protocol.
   XEROX System Integration Standard XSIS038112, December 1981

7. M. R. Brown, K. N. Kolling, E. A. Taft:
   The Alpine File System.
   In ACM Transactions on Computer Systems, Vol. 3, 4, 261-293 (1985)

8. B. E. Carpenter, R. Cailliau:
   Experience with Remote Procedure Calls in Real-Time Control Systems.
   In Software - Practice and Experience Vol. 14, 901-908 (1984)

9. D. K. Gifford, R. W. Baldwin, S. T. Berlin, J. M. Lucassen:
   An Architecture for Large Scale Information Systems.
   In Proceedings of the Tenth ACM Symposium on Operating Systems Principles
   (Orcas Island, Washington), 161-170 (1985)

10. D. R. Cheriton, M. A. Malcolm, L. S. Melm, G. R. Sager:
    Thoth, a Portable Real-Time Operating System.
    In CACM 22, 2, 105-115 (1979)

11. D. R. Cheriton:
    The V Kernel: A Software Base for Distributed Systems.
    In IEEE Software 1, 2, 19-42 (1984)

12. K. Geihs, M. H. Seifert:
    Validation of a Protocol for Application Layer Services.
    In Proceedings of the 9th NTG/GI-Conference on
    Computer Architecture and Operating Systems
    (Stuttgart/W-Germany), NTG/GI Fachberichte Nr. 92, March 1986

13. International Organization for Standardization:
    Information processing - Open Systems Interconnection -
    Specification of protocols for common application service elements.
    ISO/DIS 8650, (1984)

14. S. A. Mamrak, D. W. Leinbaugh, T. S. Berk:
    Software Support for Distributed Resource Sharing.
    In Computer Networks and ISDN Systems 9, 91 -107 (1985)

15. R. F. Gurwitz, M. A. Dean, R. E. Schantz:
    Programming Support in the Cronus Distributed Operating System.
    In Proceedings of the 6th International Conference on
    Distributed Computer Systems, IEEE, May 1986

16. B. C. Goldstein, A. R. Heller, F. H. Moss, I. Wladawsky-Berger:
    Directions in Cooperative Processing between Workstations and Hosts.
    In IBM Systems Journal, Vol. 24, 3, 236-244 (1984)

17. F. N. Parr, J. S. Auerbach, B. C. Goldstein:
    Distributed Processing involving Personal Computers and Mainframe Hosts.
    IBM Research Report RC 10990, February 1985

18. D. Notkin, N. Hutchinson, J. Sanislo, M. Schwartz (editors):
    Report on the ACM SIGOPS Workshop on Accomodating Heterogeneity.
    Technical Report 86-02-01, Department of Computer Science, FR-35,
    University of Washington, Seattle, WA 98195, March 1986

295

19. N. Francez, S. A. Yemini:

Symmetric Intertask Communication.

In ACM Transactions on Programming Languages and Systems,

Vol. 7, 4, 622-636, October 1985

20. M. H. Seifert, H. M. Eberle:

Remote Service Call: A Network Operating System Kernel and its Protocols.

In Proceedings of the 8th International Conference on Computer Communication

(Munich/W-Germany), September 1986

Request-Response and Multicast Interprocess Communication in the V Kernel

David R. Cheriton
Computer Science Department
Stanford University
Stanford, CA 94305, USA

Abstract

Use of local networks to build integrated distributed systems has prompted attention to two classes of protocols, previously largely ignored. Request-response protocols provide a simple transport-level support for page-level file access and remote procedure call. Multicast protocols provide a simple, decentralized and fault-tolerant approach to naming, scheduling, data management and parallel computation. This paper describes the use of these protocols in the V kernel and experiences accumulated over the last 5 years of use.

In addition, we argue that the factors that make these protocols of interest, namely shifts in use of communication as well as changes in the cost and functionality of communication substrate call for a whole new generation of communication systems. The third cornerstone to this new generation of communication system, in addition to request-response protocols and multicast, is a new form of internetworking based on *transport-level gateways*. We also describe some initial work with transport-level gateways.

1 Introduction

Computer communication use has been dominated until recently by remote terminal access and file transfer, basically *inter-system* communication. Transport protocols have been designed with the characteristics of these application in mind, leading to virtual circuit-based protocols. With the development of distributed systems, use is shifting significantly to *intra-system* communication such as remote procedure call, multicast and real-time datagrams. The increasing use of distributed programming leads to greater use of remote procedure call (RPC)[3] and a need to support this form of communication. Page-level network file access, as a specific case of RPC, is performance critical with the increasing use of shared network file servers, especially with diskless workstations. Multicast has been recognized as a powerful facility in distributed systems for implementing decentralized naming[18], distributed scheduling distributed transaction management and replication[24]. Finally, use of clusters of machines for real-time process control, real-time conferencing and data collection is growing. These uses are not well served by current transport protocols.

The common communication substrate has been until recently wide-area networks based on telephone technology and low performance computer switching nodes. Communication bandwidth, node buffer space and node processing power have been the critical resources. A significant shift is taking place to local area networks with multi-megabit data rates, low delay and low error rate. In addition, the future promises high-speed low error rate, wide-area fiber optic channels plus high-performance switching nodes and gateways that take advantage of the low cost of memory, processors and multiprocessor technology. Also, the processing and memory costs for switching nodes, gateways, front-end processors and intelligent network interfaces have fallen significantly, allowing again at least an order of magnitude change in the memory and processing capability of the communication system components. The new *communication economics* introduced by this shift in communication substrate offer a leap in capacity by at least an order of magnitude compared to that previously available. As is well-recognized, an order of magnitude generally constitutes a "qualitative" change, not just a "quantitative" change. This massive change invites a rethinking of communication system design, not just faster communication systems of the current designs.

These two factors prompt the development of a new generation of communication systems, analogous to the generations in other technologies, such as "5th generation" computer systems and 4th generation application languages, to mention but two examples. [1] In the following, we characterize the generations of communication systems and then describe briefly our vision of the 4th generation communication system.

A rough characterization of the generations of (computer) communication systems is as follows.

First Generation: Computers were connected by basic telephone technology, including circuit-switched twisted pair with simple point-to-point connections limited bandwidth and high error rate. Considerable cost and waste arose because of the poor match between PTT assumptions and computer communication characteristics.

Second Generation: Packet-switched wide-area networks were created using basic telephone physical plant augmented with computer (packet) switching nodes, using network protocols for remote terminal access and file transfer, including the NCP, FTP

---

[1]Many technologies are classed into "generations", an era of technological development that represents a certain plateau in progress.

and Telnet suite. Communication use was largely remote terminal access and file transfer.

Third Generation:

Internetworks were created through the interconnection of second generation networks. Internetwork protocols were developed with attention to interoperability in a large-scale and heterogeneous environment. However, the remote terminal access and file transfer modes of communication continued to dominate. In addition, local networks were introduced but largely used as fast, inexpensive second-generation networks, i.e. for remote terminal access and file transfer.

The inadequacies of third generation communication systems are becoming more pronounced. For instance, the DoD Internet is severely strained in trying to deal with real-time communication, mobile hosts, resource management and security. As an example of the latter, the current Internet avoids congestion primarily because of the small window size used by common TCP implementations, and the fact that most uses of the Internet are via standard TCP implementations.

Extensive distributed systems research to date has lead to needs beyond that provided by 3rd generation communication systems. A primary example is the use of multicast communication and support for "group" communication activities in general. In addition, distributed systems have raised issues in resource management, authentication, security and survivability that were given limited attention in the 3rd generation systems. The reduced cost of "bandwidth, cycles and bytes" makes aggressive investigation of these issues attractive now as never before.

What are the key characteristics of a 4th generation communication system? The basic overview is a communication system structured as *communication domains* of common authorization and administration connected by transport-level gateways, with *bridges* or datalink-level gateways connecting networks within a domain, as depicted in Figure 1. Each bridge filters unicast transmissions based on address but otherwise just forwards all packets to the connecting network. Each gateway acts as an agent or communication server to nodes on each local cluster, translating between local protocols and wide-area protocols and controling access to the attached networks.

Using this basic design, a 4th generation communication system can, and must, reflect and support the human organizational structures with less concession to implementation convenience and details of the underlying substrate. For example, communication domains can reflect separate administrative domains. Gateways can isolate one domain from another and require a host to possess access privileges to reach a particular host. Thus, the term *inter-domain communication* may be more appropriate than internetworking, especially since multiple physical networks can be connected transparently within a single domain.

A fourth generation communication systems has three primary novel aspects:

Request-response protocols - request-response pro-

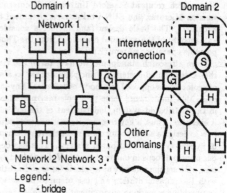

Legend:
B   - bridge
H   - host
G   - transport-level gateway
S   - switching node

Figure 1: A 4th Generation Communication System

tocols are used within a communication domain or local cluster instead of virtual circuit protocols.

Multicast - a full multicast facility provides multi-destination delivery of messages and packets as well as logical addressing.

Transport-level Gateways - The gateways "understand" the transport-level protocol and insulate hosts (to the degree possible) from wide-area characteristics and limitations. For example, local network protocols are used on the local network with translation to wide-area techniques at the gateway. In general, the gateway provides efficient access control, isolation and transformation between communication subdomains defined by human concerns, not network technology concerns.

Finally, communication resource management facilities for real-time, prioritized delivery, and congestion control are provided in a fault-tolerant and secure form.

This vision of the next generation of communication systems is largely based on our experience with communications needs, techniques and hardware in the work on the V distributed system. The remainder of the paper describes our work on V and how it relates to this vision of a new generation of communication system. The next section describes the V distributed system, which has been the software basis for our work. Section 3 describes VMTP, the request-response protocol we have used within a V domain. Section 4 describes our work with multicast at the process, network and internetwork levels. Section 5 describes our work with transport-level gateways as means to internetworking. We close with some conclusions and recommendations for future work.

2   The V Distributed System

The V distributed system[13,20,21] is a workstation-based distributed system developed by the author and his research group at Stanford University. It currently runs on Microvax and SUN workstations connected by Ethernet. The basic design (and in fact a considerable amount of the software) descended from the Thoth portable operating system[11,17].

From a research standpoint, we have taken the view that a distributed system is principally defined as a set of protocols, as opposed to a body of software. Any machine that implements these protocols may participate in the distributed system, independent of the software it executes. Thus, a key research issue is understanding how to design a set of protocols to meet the requirements of various types of distributed systems. This has been our principal focus in our work on the V distributed system, viewing the software as a vehicle for experimenting with the implementation and use of protocols as well as providing a working environment in which we recognize new problems and requirements for new protocols. This paper primarily describes our insights from the experience of developing and using these protocols, and how this experience supports our vision of 4th generation communication systems.

Our work with protocols is roughly divisible into three broad categories corresponding to the ISO Open Systems Interconnection reference model as: (1) network and internetwork layers and "down", including the datalink and physical layers, (2) transport and session layers, and (3) presentation and application layers. Most of our work focuses on the latter two layers. In particular, the system is based on the distributed V kernel, which implements the transport level. Higher-level protocols include the V I/O protocol and the V naming protocol. The following subsections describe the V kernel, the I/O and V naming, primarily from a service interface standpoint.

2.1 V Kernel

The kernel provides an abstraction of processes with operations for interprocess communication (IPC), process management and memory management. As a distributed kernel, the V kernel makes intermachine boundaries largely transparent. Two distinctive aspects of the V kernel abstraction, lightweight processes and highly synchronous message-passing, are borrowed directly from Thoth.

The provision of multiple lightweight processes per address space with dynamic priority-based preemptive scheduling is useful for structuring process-level implementations of network protocols, real-time control programs and parallel programs in general. For example, our implementation of TCP executes as multiple processes sharing an address space with the network reader and writer processes running at higher priority than the connection manager processes. Similarly, the window system uses multiple processes to monitor the mouse and keyboard concurrently with writing to the display and handling client requests.

The highly synchronous message passing is designed, as in Thoth, to support the common case of procedure call invocation of a service, where the application invokes a *stub routine* that formats and sends a message to a server and then waits for a response message in reply,

returning the values in the return message. The basic primitives are Send, Receive and Reply.

```
Send( message, server-pid )
```

sends a message to the process specified by pid and waits for a reply to be returned in the same message buffer.

```
client-pid := Receive( message )
```

returns the next message queued for this process as a result of a Send operation, blocking until a message arrives, it necessary.

```
Reply( message, client-pid )
```

sends a reply message back to the process client-pid. Other primitives are provided for interprocess copy of large parameters.

The V IPC has been extended over that provided in Thoth by allowing a client to send to a *process group*[20] rather than just a single process.

```
Send( message, group-id )
```

sends to all members of the process group and waits for the first reply. Subsequent replies are received using

```
GetReply( message, timeout )
```

if desired. The delivery of the message to all members of the process group and receipt of replies is purely on a best-efforts basis. An additional extension is the so-called *real-time datagram* message, which is sent with datagram reliability and without retransmission or waiting for a reply.

The V IPC primitives are designed to support an efficient remote procedure call facility. They have been used in this form at Stanford in several student projects as well as by a more complete research effort by Almes[1]

2.2 V I/O

All input and output services in V use a common uniform I/O interface[12]. This interface is defined in terms of a set of procedure calls on an abstract object called a uniform I/O object or UIO object. The operations include creating, releasing, reading, writing, querying and modifying these UIO objects. This procedural interface together with a presentation protocol for mapping these procedures onto V messages fully specifies the protocol that V I/O services and their client use on top of the V interkernel protocol that implements the transport level.

The interested reader is referred to the report on this interface for further details.

2.3 V Naming

In V, file names, host names and other high-level character-string names are implemented logically as a global directory providing uniform global interpretation of these names. This global directory is implemented using a *decentralized* approach[18] where each server implements the portion of the global directory naming the objects it implements, as shown in Figure 2. In the figure, one client is able to unicast because it has located the server information in its name cache whereas

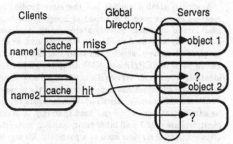

Figure 2: The V Decentralized Naming Directory

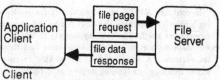

Figure 3: Basic VMTP Message Transaction: Network File Read

the other client is using multicast to locate the named server. Thus, naming is regarded as part of every service that provides objects with high-level names, as opposed to being provided by a distinguished *name server*. For example, the file servers, program managers, window managers and so on all support the *V naming protocol*.

Most naming operations and operations using names can be described as a (remote) procedure call to the server implementing the portion of the name space containing the name. However, the V naming protocol uses multicast to locate the server or servers that implement the correct portion of the name space so that these direct calls can be used. In addition, multicast is used to communicate with multiple servers simultaneously in the case that the specified portion of the name space is handled by more than one server.

The decentralized approach is highly resilient because the system only fails to map a valid name when the server handling the named object is unavailable, or when the network is down (in which case very few facilities can work in any design). This approach is also highly efficient if the client caches the information returned by multicasts since most naming requests, over 99 percent in our measurements, go directly to the correct server, using a standard remote procedure call. In contrast, using a name server introduces an extra remote call on every naming handling operation, one to the name server and one to the server handling the named object.

Further aspects of this approach are described in a technical report[18] and a forthcoming Ph.D. thesis[29]. For the purposes of this paper, the V naming protocol is important as an example of a distributed systems protocol and as a use of the multicast facility provided in V.

The remainder of the paper describes experience with the request-response and multicast protocols in the V distributed system, focusing on the transport-level protocols and their use.

3 VMTP: The V Inter-kernel Request-Response Protocol

The V IPC primitives are implemented using a request-response protocol called the Versatile Message Transac-

tion Protocol (VMTP)[14]. [2] We describe some key aspects to VMTP with reference to both it current performance and experience with previous versions of the protocol.

VMTP is basically a request-response protocol similar to that described elsewhere[3,30] used for client/server and RPC-style interaction. A VMTP session, a *message transaction*, is initiated by a *client* sending a *request message* to a *server* entity and terminated by the server sending back a *response message*. The response acknowledges to the client receipt of the request message. The next request from the client, an explicit acknowledgement or a timeout event acknowledges to the server receipt of the response by the client. The basic VMTP message transaction is illustrated in Figure 3 in an expected common use, namely network file page read. A client can only have one message transaction outstanding at one time although a host may implement multiple VMTP clients. Requests and responses may consist of multiple packets.

A VMTP message transaction takes place between network-visible *entities*: client entities and server entities. An entity may be (for example) a process, port or procedure invocation. In the case of the V implementation, an entity is a process or a process group. Each entity is addressed by an *entity identifier*. A group of entities can be identified by a single entity identifier even if they are distributed across several machines. For example, a single identifier can specify the group of file servers. We do not prohibit an entity from having several different entity identifiers.

The following subsections describe some interesting aspects of this protocol.

3.1 Conversation Support versus Connections

Conversations are required to access many distributed operating system services. By *conversation*, we mean a sequence of user-level related communication actions. Examples include the open-read-write-close operations on a single open file, the terminal read-write actions in a session and the sequence of data transfers that constitute a file transfer. The conventional way to support conversations is to provide a notion of *connection* right in the transport level, such as is done in TP 0-4[28] and TCP[25]. The user level maps its notion of conversation one-to-one onto transport-level connections.

---

[2]This is a refinement of the original protocol as described in previous reports[21,32,13]. At the time of this writing, VMTP is being used in an experimental version of the system and has yet to be included in the "official" version of the V kernel.

In contrast, VMTP provides support for implementing higher-level conversations but it does not implement connections itself. The two key aspects to VMTP conversation support are: stable addressing and message transactions. A *stable address* is one that retains the same "meaning" or binding as long as it remains valid. A *message transaction* is a request-response pair with reliable delivery on both the request and the response messages.

A conversation is typically structured on top of VMTP message transactions as a sequence of message transactions addressed to the same server address and server-specific conversation or connection identifier. Because of stability, the client knows that each request message is delivered to the same server. Using the conversation identifier and client address, the server associates each request message with the same conversation and authorization. Each response is associated with the conversation by its association with a request message that is part of the conversation. In addition, the client and server can check the existence of the other using the other's address, again relying on stability.

Use of these facilities is illustrated by considering open file connections. The client sends an *open* request to the file server. The file server opens the file (assuming file protections are satisfied) and allocates an open file descriptor for file. It records the client address, permissions and other relevant information in this descriptor, allocates and records a local open file identifier and returns this information to the client. Subsequent client requests specify this open file identifier, allowing the server to map to the particular open file descriptor and check the client address against the one previously recorded. Reliable delivery and duplicate suppression for request and response messages are handled by VMTP.

Conversation support in VMTP is preferred over directly implementing virtual circuit connections for several reasons:

- **Minimal Redundancy:** there is one conversation record recording the server-level state, rather than at both the transport and service level.

- **Minimal Transport Level State:** there is one VMTP record per client at the server and client endpoints, not one per conversation as required with a connection-based approach.

- **Flexible higher-level conversations:** The VMTP user can construct a variety of different types of conversations including multicast and real-time message transactions. It is not restricted by a transport-level definition of connection.

We note that all these considerations arise from the new patterns of communication use we have identified with the next generation of communication systems.

3.2 Stable Addressing

Stable addressing is required as a base for constructing user-level conversations. VMTP addresses (or identifiers) are stable in one of two forms: strictly stable or T-stable.

A *strictly stable* identifier has the same binding or meaning forever once it is bound although it may appear invalid at times. Strictly stable entity identifiers are typically only used to identify a service such as the group of file servers, serving the same function as *well-known ports* in TCP/IP and XNS. In this example, the identifier refers to file service if any of the file servers are operational and otherwise appears invalid in the sense of "unbound". Generally, strictly stable identifiers should be administratively bound and used sparingly. T-stable identifiers are used for all other entity naming, including instantiations of services such as a particular file server, processes to execute application programs, and groups of entities such as jobs and parallel programs that have a transient existence.

A *T-stable* identifier has at most one valid binding or meaning over a time interval T, but unlike a strictly stable identifier, can be rebound over time. A T-stable entity identifier is guaranteed not to be reused for at least T seconds after it becomes invalid. Thus, if the entity identifier is checked for validity more frequently than every T seconds, a change to the binding of the identifier will be detected. (VMTP provides a facility to probe the validity of an entity identifier.)

The use of T-stable identifiers provides an infinite supply of identifiers over time with T-stability limiting the confusion that reusing these identifiers can produce. T-stable identifiers are also useful in duplicate detection, as described in the next section. In contrast, the use of only strictly stable identifiers would require a very large identifier space since (for example) every process creation would permanently consume an identifier. VMTP uses a relatively large entity identifier, however, the structuring of this identifier for efficient mapping to network addresses reduces the effective number of identifiers.

Implementing T-stable identifiers is much easier if the size of the number of participants in the "address space" is relatively small and can be reached as a group fairly easily. These restrictions allows the addresses to be fairly small, such as 32-bits each. They also allow an individual node to negotiate with the other participating nodes directly for an allocation of one or more unique addresses. With a large domain, the identifiers have to be large and reliable negotiation is very expensive in elapsed time and processing..

Thus, the basic requirement of stable addressing argues for use of local domains of naming for local request-response protocols with more static techniques used in the wide-area case. Despite its cost, stable addressing appears to be essential in implementing request-response protocols without underlying (expensive) connection support. Stable addressing also aids with the solution of several other problems, including duplicate suppression, as described below.

3.3 Duplicate Suppression

Duplicate suppression is an important part of transport protocols, especially in highly variable internetwork environments where it is difficult to prevent long delayed duplicate packets. Conventional duplicate suppression uses the so-called 3-way handshake to catch duplicates on connection setup, and then a large sequence number and the connection itself to avoid duplicates while the connection is being used. However, connections are neither available in VMTP nor appropriate for the 4th generation communication environment. Instead, duplicates are handled as follows.

A VMTP request is identified by its source address, which is a T-stable identifier, and a transaction identifier. A client can guarantee that the transaction identifiers associated with a particular client identifier are *T-monotonic*, i.e. strictly increasing modulo 2^{32} within any time window of less than T seconds. This is easy to do if the client remembers the last used transaction identifier for each valid client identifier. Invalid client identifiers cannot be reused in less than T seconds so the property holds independent of the transaction identifier selected.

The server end keeps a record of the last transaction identifier used by each client from which it received a message in the last T seconds. A request that arrives with a lower transaction identifier than what it has stored is discarded. A request with a transaction identifier that is higher or for a client for whom no current record is maintained is accepted as a new request.

Duplicate response packets are handled similarly. That is, it is discarded if it does not match the transaction identifier of the client to which it is addressed, or the client address is invalid.

As we have argued elsewhere[14], a long maximum packet lifetime and a long propagation delay significantly increases the time frames for discarding client packets at the server and for reusing VMTP identifiers. Again, a localized domain of nodes with stricter control of packet lifetimes and propagation delay, as argued for the 4th generation communication systems, considerably simplifies the implementation of VMTP.

3.4 Selective Retransmission

Earlier communication systems assumed that retransmissions were required primarily because of errors in transmission. However, communication hardware has evolved to provide lower error rates and higher speeds so retransmissions are now more frequently due to overruns, either in the destination host or intermediate gateways and packet switches. The overrun problem is accentuated by several factors. First, a server such as a file server can suffer from overrun because it can be the focus of several clients' transmissions at the same time. The frequency with which this occurs increases with increasing load on the server. Second, a client can suffer packet overrun in receiving a multi-packet transmission from a server because the server have faster hardware than the client, i.e. the shared resource is made more powerful because it must serve multiple clients. Third, bridges and gateways interconnecting high-speed networks usually cannot handle simultaneous traffic bursts from their connecting networks, causing packet drop or overrun at these intermediate hops. Finally, many hosts find it easier and more efficient in processing cost to send a large amount of data as a single blast of packets, transmitted as fast as the network allows, rather than pacing the flow.

When packets are lost due to overruns, systematic errors are more likely. For instance, the receiver may drop every 4th packet. If a sequence of packets are retransmitted at the same speed, the same packets will be dropped again.

These characteristics make selective retransmission attractive. For instance, if a group of N packets are transmitted at a speed that slightly overruns the receiver or some intermediate node, every kth packet may be dropped. Selectively retransmitting these dropped packets, possibly with a slightly longer inter-packet gap is significantly more efficient than retransmitting from the first dropped packet. A selective retransmission mechanism also gives a greater indication to the sender of whether overrunning is the problem and how severe the problem is. For instance, if the selective acknowledgement indicates that every kth packet was dropped, overrunning is the likely cause, with the value of k indicating the degree of speed mismatch.

VMTP supports selective retransmission using a bitmask scheme to indicate the missing blocks. Packets are grouped into *packet groups* of at most 32 blocks, one per bit of the bitmask. The bitmask indicates the packets in a packet group that were received.

The bitmask provides a simple, fixed length way of specifying which packets were received as well as indicating the position of a packet within the packet stream. The combination of packet groups and a (re)transmission mask provide a flexible selective retransmission facility with no variable length header or acknowledge information. Also, the sender can efficiently recover from slightly overruning the receiver or intermediate network nodes and is provided with information that aids in adjusting the transmission rate to reduce loss.

With the higher performance of local domain communication expected for 4th generation communication systems, overruns become a greater problem. Yet, conventional protocols such as TCP and TP do not support selective retransmission.

3.5 Packet Group-Based Flow Control

VMTP uses packet group-based flow control in combination with *rate-based flow control*, similar to that proposed for NetBlt[23]. The transmitter sends groups of packets of at most 32 packets and adjusts the inter-packet transmission time within a packet group, and the inter-packet group transmission time, to avoid overrunning the receiver. Selective acknowledgement and retransmission within a packet group provides a means of dynamically detecting when the transmission rate is too high. An implementation should increase the inter-packet delay until packets are not being dropped due to overruns. Appropriate inter-packet and inter-packet group intervals can be estimated to minimize packet loss due to overruns, again similar to that suggested by Clark et al.[23].

Packet group and rate-based flow control have several advantages. First, the transmitter of a packet group can transmit a complete packet group as one "blast" (with appropriate inter-packet gaps), knowing that the receiver will accept the entire packet group. For example, a file server can reply to a sender with a packet group, knowing that the reply will complete in the time to transmit these packets. In contrast, with sliding window flow control, the file server may be flow controlled and blocked by the client indefinitely.

Second, the packet group with proper inter-packet gap times allows the transmitter to send packets near the maximum rate acceptable to the client, without the overhead of packets to advance a flow control window. This is particularly important in a 4th generation communication environment where the transmitter and receiver are often roughly balanced and delay is low, so the transmitter is always waiting at the end of the flow control window, incurring maximal flow control packet overhead or else poor network performance. The worst case of this problem can occur with TCP where the window size can become as small as 1 byte, known as the *silly window syndrome*[22]. The solution to this problem is to force the protocol to simulate rate-based and packet group-based flow control by having the transmitter wait until it has a significant amount of data to send and then sending it at a reasonable rate. We note that all sliding window schemes must either go to single-packet windows or else use rate control to reduce overruns when this problem arises.

Finally, the packet group model allows a transmitter to send a group of packets in one scheduling operation either at the process or network module level, thus reducing the cost of scheduling packets for transmission. For example, in the V kernel, a packet group is queued for transmission as a unit with round-robin scheduling on the network transmission queue introducing acceptable inter-packet gaps when the server is loaded.

In general, overrunning can only be efficiently solved by rate control. The selective retransmission mechanism gives, as a side-effect, information that allows adjustment of packets rates. The packet group concept allows a simple selective retransmission mechanism using a bitmask.

Packet group-based flow control is made feasible by the reduced cost of memory. It now is feasible for a file server to be prepared to accept up to 16 kilobytes of data off the network most of the time. Similarly, it is feasible for a sending client to transmit 16 kilobytes as one blast as well as receive 16 kilobytes as a response. In addition, the 4th generation uses of communication systems generate worst-case behavior with the conventional sliding window techniques, making new techniques of this nature essential.

3.6 Idempotency

A client retransmits the request message periodically until it receives a response. When retransmission of the request arrives after the response has been sent, the response must be retransmitted. Normally, VMTP saves a copy of the response message until the transaction record times out in case it needs to be retransmitted. However, for efficiency, the server can specify that the transaction is *idempotent*[3]. In this case, a retransmission of the re-

[3]By *idempotent* transaction, we mean that processing multiple copies of the associated request received consecutively from the same client has the same effect as processing a single request,

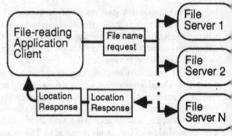

Figure 4: VMTP Group Message Transaction: File Name Query

quest after the response has been sent is passed on to the server as a new request, allowing it to provide the response message anew. (Note that a duplicate of the last request is the only event that causes the server to retransmit the response.)

Specifying a transaction as idempotent eliminates the overhead of the VMTP transport module making a copy of the response for possible retransmission. Eliminating this copy can be a significant saving on a file server, where read requests are the dominant operation and the file server buffer pool makes the server-level regeneration of the response fairly inexpensive, especially given the infrequency with which they should arise. Making a copy of a large response can double the processing cost of sending the response judging by our experience with the V kernel.

Idempotency is not provided in conventional protocols yet its use has significant benefit for 4th generation communication systems. Moreover, it is not clear how to define idempotency in the context of a conventional connection-oriented protocol. This points out another advantage of the request-response framework, namely flexibility, as discussed in the next section.

3.7 Variants on the Basic Message Transaction

VMTP provides three variants on this basic VMTP session to widen its applicability and efficiency.

In the first variant, the *group message transaction* the client sends or *multicasts* to a *group* of server entities and may receive multiple responses. The message transaction is terminated by the client initiating a new message transaction. Normally, a client only retransmits until the first response is received. A group message transaction session is illustrated in Figure 4 by a client multicasting to a group of file servers to locate a file. In this figure, two file servers respond, indicating they both have a copy of the named file.

In the second variant, the client sends the request as a *datagram* with an indication that no response is expected. The request may be sent reliably or not (acknowledged or unacknowledged). In the former case, the client may not issue another transaction until the datagram is acknowledged. In the latter case, the datagram may be discarded by VMTP if the client issues a new

both on the state of the server and the response that is returned

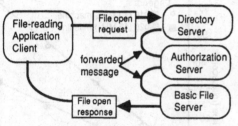

Figure 5: Forwarded Message Session: Authorized File Open

message transaction before the datagram is received.

In the final variant, the request message may be *forwarded* to another server in which case the other server responds directly to the client, as illustrated in Figure 5. In this figure, the client sends an open request to a directory server which, after locating the specific file, forwards the request to the authorization server, which after authorizing access, forwards the request to the basic file server. The latter server can refuse to accept open file requests unless they are forwarded to it by the authorization server. Viewed as a series of remote procedure calls, the sequence logically behaves as though the directory service called the authentication service which called the basic file service. However, the return from the basic file server is optimized by going directly back to the client.

These variants can be combined in various forms, e.g. datagram group message transaction. Most of these forms have been used extensively in the V distributed operating system[13,20,19]. The interested reader is referred to the cited references for further example uses.

There are several advantages to the basic session structure and its variants.

Higher Level Conversation Support: The basic VMTP session provides an efficient and reliable base for implementing multiple types of higher level conversations, one example being open file connections. It also avoids redundancy, overhead and restrictiveness of using connections provided directly as part of the transport level.

Minimal Packet Exchange:
The basic message transaction provides the minimal 2-packet exchange for simple connectionless interactions such as file query, getting the time and simple remote procedure call. In contrast, a virtual circuit protocol such as TCP would require 8 packets for most implementations (although 6 is adequate in theory) unless the message transaction uses an existing connection.

Ease of Use: Providing the multicast, datagram and forwarding variants as part of the same protocol means that they are accessible using the same service interface. Thus, clients and servers that use multiple types of message transactions can share considerable common application code for the handling the different cases. For instance, it is not un-

usual for a server in the V distributed system to receive basic message transactions, group message transactions and datagram message transactions. Since they are all supported by one protocol, the server uses a single, simple "receive" operation to wait for the next message transaction, which may be any of the variants.

Shared Communication Code: The variants are all simple modifications of the basic message transaction and therefore have considerable code in common. Thus, a VMTP implementation can support reliable request-response transactions, , multicast and datagrams at a lower space cost than providing each as a separate protocol. For instance, a datagram message transaction uses the normal VMTP transmission mechanism; it just does not wait for a response. Similarly, delivery of datagram requests is handled by the same mechanism as other types of requests. In general, these variants are simple and efficient to implement as extensions of the basic VMTP implementation.

A common alternative is to realize RPC support using a virtual circuit protocol. This approach provides limited advantages and imposes significant costs and restrictions. In theory, a virtual circuit provides: (1) low-cost addressing once a circuit is established, (2) synchronization and streaming flow control, and (3) authentication on circuit setup that is amortized over use of the circuit. However, the benefit of small addresses is limited and seldom realized because it requires state in the network or internetwork that "understands" these smaller addresses. TCP, for example, sends the full Internet host addresses plus port identifiers of the source and destination in every packet. Also, VMTP achieves efficient flow control, synchronization and authentication without virtual circuit support.

In addition to the variants described above, we are also exploring the provision of *streams* of message transactions using techniques similar to those used in the Δt protocol of the Fletcher and Watson[27]. A *stream identifier* associates a set of consecutive message transactions. The start and end flags plus the consecutive transaction identifiers within the stream are used to detect dropped message transactions. Rate-based flow control is used between transactions.

3.8 VMTP Packet Format

All VMTP packets use the same fixed format. This format, shown in Figure 6, is logically structured as 4 portions:

- **entity and transaction identification** - including authentication identifier, domain, source, destination, forwarder and transaction identifier.

- **packet group control** - including checksum, control flags, stream identifier, function code and delivery mask.

- **user message control block** - system flags, user data and segment size.

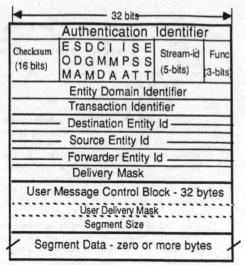

Figure 6: VMTP Packet Format

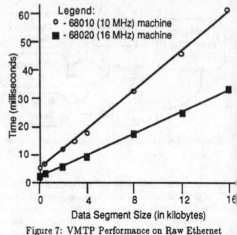

Figure 7: VMTP Performance on Raw Ethernet

- segment data - a maximum of 16 kilobytes of segment data, possibly further limited by the maximum packet size.

The *authentication identifier* field is used in one of two modes depending on whether the message is encrypted or not. If not encrypted, the authentication identifier indicates the authentication of the sending entity, such as a user account within the entity domain. If encrypted, the authentication identifier is used to route the encrypted message to the right decryption key within the receiving node, the same as the *conversation key* used by Birrell[4]. In fact, we propose following Birrell's design for secure VMTP communication. As pointed out elsewhere[20], this design extends to handle multicast.

The use of several other of these fields should be evident from the previous sections. The interested reader is referred to other reports[14,15] for a detailed explanation.

The V kernel implementation of VMTP consists of about 1700 lines of C code resulting in 7.8 kilobytes of code and initialized data on a Motorola 680X0. (This figure does not count the basic Ethernet driver or IP support.) Figure 7, taken from [14], gives the performance of VMTP running between two Motorola 68020-based machines[4] and two Motorola 68010-based machines [5] using raw Ethernet interconnection. Figure 7 shows that, as expected, the time for transmission grows linearly with size of data segment. With 16 kilobyte requests, the data rate is roughly 461 kilobytes per second or 3.69 megabits per second, a reasonable performance level compared to most TCP implementations. Packet loss and network load would of course degrade this performance, although these factors have not been an issue in our experience to date. With the 32-byte user data area in the packet header, exchanging small amounts of

---

[4]SUN 3's with 16.6 MHz clock rate.
[5]SUN 2's with 10 MHz clock rate.

data is very efficient, 2.5 milliseconds round trip on the 68020 machines.

Our network environment is ideal for VMTP, providing low delay, high data rate and low error rate, as is typical of local networks. Using VMTP on top of an internetwork architecture and spanning several different networks transparently at the network level means that these properties cannot be assumed. In particular, the amount of delay and its variance becomes significant The use instead of transport-level gateways with VMTP as described in Section 5, provides a means of internetworking which provides better control of delay and does not incur a cost on local network VMTP performance.

4 Multicast

Imagine a world in which each person could only communicate with a single other person at a time. Meetings and group discussions would be infeasible as would classroom lecturing and other group activities. This hypothesized world points out how dependent we are on broadcast or multicast communication. Yet this restricted world is precisely what is provided by current computer communication systems.

In our work with V, we have provided a process-level multicast facility as well as developed a model and prototype implementation for internetwork multicast support. This section describes key aspects of the process-level facility including its use and implementation. We also describe the lower levels.

Multicast is the transmission of a packet, message or remote procedure call (depending on level) to a set of zero or more communication endpoints, i.e. hosts, ports or modules. Multicast is distinguished from *broadcast* which is the transmission to *all* communication endpoints. That is, multicast restricts the communication to a subset of the reachable endpoints, but supports multiple endpoints. Broadcast is not a generally useful facility since there are few reasons to communication with all endpoints. In fact, we argue that the broadcast facility in a broadcast network is best viewed as an "accident of the technology" in the same way as self-modifying programs are for stored program computers: just because the technology provides the facility as a side-effect of its implementation or design does not make it proper engineering practice to use it.

A proper multicast facility allows efficient transmission to multiple endpoints while avoiding the unnecessary loading of communication facilities and endpoints that arises with broadcast. The following subsections describe the V multicast abstraction and implementation.

4.1 Groups: The V Multicast Abstraction

The multicast facility in V is based on the principle: *For each entity and operation on that entity, the same operation should be applicable on groups of these entities.* This principle arises in law as well, where (for example) a group of individuals can be treated as an individual "body" by becoming a corporation. Thus, V IPC multicast arises from applying the Send operation to a group of processes, rather than a single process. We apply this principle to more than just communication primitives. For instance, one can also destroy a group of processes and suspend a group of processes. However, these other group operations are all implemented in terms of the group communication primitives.

We have applied the group principle to hosts as well, leading to the concept of *host groups*[16] as a model for internetwork multicast. Similarly, *network groups* are a useful concept in the implementation of host groups. In particular, a process group defines a corresponding host group, namely the hosts that contain processes in that group. A host group defines a corresponding network group, namely the networks that include hosts in that host group. Delivery to a process group can be structured as delivery to hosts in the corresponding host group which is structured as delivery to the networks in the corresponding network group. At each of these levels, primitives are provided for creating, joining, leaving and querying groups.

4.2 Logical Addressing

The *group* concept we associate with multicast introduces a level of indirection in the naming of individuals, in addition to providing multi-destination delivery. For example, one can have an identifier for the group of hosts supporting name service, in addition to the names for individual hosts. The names of these groups provides a form of so-called *logical addressing*, where a communication endpoint can be addressed by function, as opposed to (and in addition to) by location or randomly allocated identifier. We have made extensive use of this facility in V, assigning static *well-known* group identifiers to logical groups of services, such as time, storage, display and processing. These functional groups support the implementation of operations on groups at a higher level, such as querying the disk space available across the group of storage servers. They are also used as part of the decentralized naming[18], as described below and in Section 2.3.

4.3 Use of Multicast

The two generic uses of multicast are multi-destination delivery and logical addressing (as defined above). We list some specific uses of multicast in the V distributed system.

Name Lookup: Multicast is used to query a group of servers to determine which one implements the portion of the name space specified in the name[18]. Multicast makes this facility very flexible and resilient compared to using a dedicated name server. Since clients cache this information, the multicast is infrequently used so the burden on the servers and network is minimal.

Scheduling: Multicast is used to locate an unloaded or lightly loaded machine to remotely execute[31] a program. The parallel query and response provided by multicast makes this scheduling fast, especially when the client caches scheduling information.

Time Synchronization: Time synchronization algorithms require communication among a group of time servers to periodically ensure consistency of the clocks. Multicast is used in V to send out synchronization messages between time servers[9], reducing the cost of synchronization over the use of unicast.

Distributed Games: Multicast is used by a version of our multiplayer distributed game, Amaze[2], to maintain a real-time database describing the state of the game across multiple machines. Update messages are multicast to the group of game managers as real-time datagrams, with each new update from a particular game manager subsuming its previous message. This real-time distributed program can run concurrently with the full multi-programming environment provided by V.

Atomic Transaction Management: Multicast is used for communicating commit, abort and prepare-to-commit operations across a set of data servers participating in a distributed atomic transaction[7]. The group mechanism is also used to identify the transaction and the set of servers involved in the transaction.

Distributed Parallel Programs: Multicast is used for maintaining shared state in parallel programs[19]. For example, a checkers playing program runs multiple move evaluators assistants in parallel, implementing parallel $\alpha - \beta$ search. Multicast is used to exchange search information, thereby reducing the and focusing the search effort. Similarly, multicast is used in a parallel program solving the traveling salesman problem to communicate the cost of the best current path, reducing the time wasted exploring inferior routes.

In addition to these examples, we see clear benefit in the use of multicast for updates of replicated files and databases[6], replicate procedure call[24], and numerous other uses, although we have yet to explore these uses ourselves.

In many of these examples, multicast is being used to maintain *shared state*. Shared state is most easily maintained by shared memory, which of course is not available in the usual sense. We have recognized that many of these examples are effectively implementing a *problem-oriented shared memory*[10] in support of state sharing. The memory is problem-oriented since the consistency constraints and "read" and "write" operations are optimized for the problem or application in each case, thereby requiring significantly lower communication overhead than conventional shared memory.

Finally, multicast as the basis for addressing in the VMTP implementation. In this implementation, unicast is viewed as an optimization of multicast. In general, one transmits to a group with some indication of which one is intended, as in the naming protocol. However, when the sender happens to have the information necessary to address the specified individual, multicast is refined to unicast. This is precisely the approach we take in mapping VMTP entity or process identifiers. When the host that implements a particular process is not known, the message is transmitted to all hosts, with each host filtering on the process identifier. With the request-response behavior of VMTP, a reasonably sized host address cache and the communication patterns we observe, very few packets are multicast.

In many uses of multicast, aspects arise that are distinctive to a *group protocol*, a protocol for communicating with a group, as opposed to with an individual. For example, we have made extensive use of a group convention that models human communication, namely: When a server receives a query message about something it has no knowledge, if it is a group message, it simply ignores the message. However, if it is sent directly to this server, the server responds with an error indication. In V, we provide support in the transport level to cause replies to be discarded in this case, thereby avoiding a flood of "I don't know" messages on such queries. There are several other techniques of this nature that we are just beginning to explore.

4.4 Transport Level Implementation Issues

The ideal transport level multicast service is a natural extension of the unicast operations, supports a wide range of applications, and operates efficiently with minimal complexity in the transport module and protocol. In particular, it should take advantage of network-level broadcast and multicast facilities whenever available. The first point is basically accomplished by using the same operations for multicast and unicast with the "address" of the destination determining which is being used. That is, multicast is specified by the destination address being that of a group; otherwise, a unicast. This design in V allows some programs to be obvious as to whether they are communicating with a group or which an individual process. The second two points must be addressed with an understanding of the issues of the reliability model, performance and complexity.

With unicast communication, the reliability model is fairly clear: either the data is delivered with positive acknowledgement or the sender is signaled that a failure occurred. With multicast, there are a wider range of possibilities. We define a multicast facility to be *k*-

reliable if either at least *k* members of the destination group received the message or the sender receives an failure indication. It is *all-reliable* if k is the size of the group.

In V, the transport level (the kernel) provides 1-reliable and 0-reliable multicast. 0-reliable multicast is a real-time datagram multicast (where there is no reply). This approach leads to a simple, efficient implementation at the kernel level. In particular, it is compatible with the basic request-response model since the message primitives and their implementation support receiving one or zero replies to each unicast message. In addition this model together with the ability to receive multiple replies allows one to implement efficient all-reliable communication outside of the kernel. All-reliable communication is implemented in one of two ways. First the sender may collect replies from each member of the group, explicitly retransmitting to those from which it did not receive replies. Second, it may send each message to a globally available message log which assigns a sequence number to the message and then forwards the message to the group. A member detects that it has missed a message by detecting a gap in the sequence numbers of messages it receives. It then retrieves the missing message from the message log. This is basically the scheme described by Chang and Maxemchuk[5]. In general, all-reliable multicast is expensive, either in messages or in worst-case delay in message delivery.

Implementing the all-reliable model in the kernel would require a kernel-level message log or the kernel maintaining exact information on the membership of the multicast group. Besides the all-reliable model being expensive, it is not required in many applications and often can be achieved with significantly less cost using application-specific techniques. For example, request for scheduling information need only return sufficient data to allow choosing a good candidate host for remote execution, not information from all hosts.

In general, we argue that the basic transport-level multicast facility should define reliable multicast as reliable communication with at least one member of the multicast group. The request-response model with the provision of multiple responses provides a good base to build different notions of reliable group communication without complicating the basic transport service.

4.5 Scope

Transmission of a multicast message is controlled by two parameters of scope, one being the destination address and the other being the "distance" to the members in the group. In particular,

```
Send ( message, group, distance )
```

transmits the specified message to members of the group within the specified *distance* of the sender. Distance may be measured in several ways, including number of network hops, time to deliver and what might be called administrative distance. Administrative distance refers to the distance between the administrations of two different networks. For example, in a company the networks of the research group and advanced development group might be considered quite close to each other, networks of the corporate management more distant, and networks of other companies as much more distant. One may wish to restrict a query to members within one's own administrative domain because servers outside that domain may not be trusted. Similarly, error reporting outside of an administrative domain may not be productive and may in fact be confusing.

Besides limiting the scope of transmission, the distance parameter can be used to control the scope of multicast as a binding mechanism and to implement an expanding scope of search for a desired service. For instance, to locate a name server familiar with a given name, one might check with nearby name servers and expand the distance (by incrementing the distance on retransmission) to include more distant name servers until the name is found. To reach all members of a group, a sender specifies the maximum value for the distance parameter.

The distance parameter can be viewed as an extension of the time-to-live or hop count parameters that are used in several internetwork architectures to prevent infinite routing cycles. In those cases, the distance parameter basically ensures that the delivery mechanism only expends a finite amount of work in delivery and therefore discards a packet caught in a routing loop. The distance parameter in our multicast facility refines this finite bound into further gradations.

The specific semantics of the distance parameter should be well-defined. However, there is a particular need for well-known boundaries values that coincide with administrative domains. For instance, there is a need for a distance value that corresponds to "not outside this local network". These logical scopes are analogous in some ways to logical addresses.

4.6 Network and Internetwork Level Implementation Issues

Given the predominance of broadcast local-area networks and the locality of communication to individual networks, the delivery mechanism must be able to exploit the hardware's capability for efficient multicast within a single local-area network. In addition, the delivery mechanism must scale in sophistication to efficient delivery across the internetwork as internetworks acquire high-speed wide-area communication links and high performance gateways, as expected in the next generation of communication systems. Finally, the delivery mechanism must avoid "systematic errors" in delivery to members of the group. That is, a small number of repeated transmissions must result in delivery to all group members within the specified distance, unless a member is disconnected or has failed. We refer to this property as *coverage*. In general, most reliable protocols make this basic assumption for unicast delivery. It is important to guarantee this assumption for multicast as well or else applications using multicast may fail in unexpected ways when coverage is not provided. For efficiency, the multicast delivery mechanism should also avoid delivering multiple copies of a packet to individual hosts.

Failure notification is not viewed as an essential requirement given the datagram semantics of delivery.

However, a host group extension of internetwork architectures such as IP and XNS should provide "hint"-level failure notification as the natural extension of their failure notification for unicast.

The remainder of the section describes aspects of implementing multicast in a conventional internetwork, since that subsumes the considerations for local networks and store-and-forward networks.

Implementation of network or internetwork multicast involves implementing a binding mechanism (binding internetwork addresses to zero or more hosts) and a packet delivery mechanism (delivering a packet to each host to which its destination address binds). With a broadcast network technology such as Ethernet, packet delivery is automatically to all hosts. One need only provide filtering based on addresses with each host able to select a large number of addresses, normally its unicast address plus several multicast addresses.

Without broadcast coverage, this functionality fits most naturally into the gateways of the internetwork and the switching nodes of constituent point-to-point networks (as opposed to separate machines) because multicast binding and delivery is a natural extension of the unicast binding and delivery (i.e. routing plus store-and-forward). That is, a multicast packet is routed and transmitted to multiple destinations, rather than to a single destination as with a unicast packet.

A gateway in a host group internetwork is thus viewed as a "communication server", providing multicast delivery and host group management. The multicast delivery service is invoked implicitly by sending packets addressed to host groups, with unicast delivery as a special case. The group management service is invoked explicitly using a request-response transaction protocol between the client hosts and the server gateways. In addition to the operations for managing transient host groups the gateway supports operations for administrative allocation of permanent group addresses, including static, single-host group addresses (i.e. unicast addresses).

A mentioned earlier, a host group defines a *network group*, which is the set of networks containing current members of the host group. When a packet is sent to a host group, a copy is delivered to each network in the corresponding network group. Then, within each network, a copy is delivered to each host belonging to the group.

To support such multicast delivery, every gateway maintains the following data structures:

routing table : conventional internetwork routing information, including the distance and direction to the nearest gateway on every network.

network membership table : A set of records, one for every currently existing host group. The *network membership record* for a group lists the network group, i.e. the networks that contain members in the group.

local host membership table : A set of records, one for each host group that has members on directly attached networks. Each *local host membership record* indicates the local hosts that are members of the associated host group. For networks that support multicast or broadcast, the record may contain only the local *network-specific multicast address* used by the group plus a count of local members. Otherwise, local group members may be identified by a list of unicast addresses to be used in the software implementation of multicast within the network.

A host invokes the multicast delivery service by sending an internetwork datagram to an immediate neighbor gateway (i.e. a gateway that is directly attached to the same network as the sending host). The gateway forwards the packet to each network that has members in this group. Member networks that are beyond the datagram's distance constraint are ignored. The network membership records and the network-specific multicast structures are updated in response to group management requests from hosts. If the host is the first on its network to join a group, or if the host is the last on its network to leave a group, the group's network membership record is updated in all gateways. The updates need not be performed atomically at all gateways, due to the datagram delivery semantics; hosts can tolerate misrouted and lost packets caused by temporary gateway inconsistencies, as long as the inconsistencies are resolved within normal host retransmission periods. In this respect, the network membership data is similar to the network reachability data maintained by conventional routing algorithms, and can be handled by similar mechanisms.

Multicast routing among the internetwork gateways is similar to store-and-forward routing in a point-to-point network. The main difference is that the links between the nodes (gateways) can be a mixture of broadcast and unicast-type networks with widely different throughput and delay characteristics. In addition, packets are addressed to networks rather than hosts (at the gateway level).

This basic implementation strategy meets the delivery requirements stated above. However, it is far from optimal, in terms of either delivery efficiency or group management overhead.

A simple optimization within a network is to have the sender use the local multicast address of a host group for its initial transmission. This allows the local host group members to receive the transmission immediately along with the gateways (which must now "eavesdrop" on all multicast transmissions). A gateway only forwards the datagram if the destination host group includes members on other networks. This scheme reduces the cost to reach local group members to one packet transmission from two required in the basic implementation [6] so transmission to local members is basically as efficient as the local multicast support provided by the network.

A similar opportunity for reducing packet traffic arises when a datagram must traverse a network to get from one gateway to another, and that network also holds members of the group. Again, use of a network-specific multicast address which includes member hosts plus gateways can achieve the desired effect. However, in this case, hosts must be prepared to accept datagrams

that include an inter-gateway header or, alternatively, every datagram must include a spare field in its header for use by gateways in lieu of an additional inter-gateway header.

A refinement to host group membership maintenance is to store the host group membership record for a group *only* in those gateways that are directly connected to member networks. Information about other groups is cached in the gateway only if it is being used by hosts supported by that gateway. When a gateway receives a datagram to be forwarded to a group for which it has no network membership record (which can only happen if the gateway is not directly connected to a member network), it takes the following action. The gateway assumes temporarily that the destination group has members on *every* network in the internetwork, except those directly attached to the sending gateway, and routes the datagram accordingly. In the inter-gateway header of the outgoing packet, the gateway sets a bit indicating that it wishes to receive a copy of the network membership record for the destination host group. When such a datagram reaches a gateway on a member network, that gateway sends a copy of the membership record back to the requesting gateway and clears the copy request bit in the datagram.

Copies of network membership records sent to gateways outside of a group's member networks are cached for use in subsequent transmissions by those gateways. That raises the danger of a stale cache entry leading to systematic delivery failures. To counter that problem, the inter-gateway header contains a field which is a hash value or checksum on the network membership record used to route the datagram. Gateways on member networks compare the checksum on incoming datagrams with their up-to-date records. If the checksums do not match, an up-to-date copy of the record is returned to the gateway with the bad record.

This caching strategy minimizes intergateway traffic for groups that are only used within one network or within the set of networks on which members reside, the expected common cases. Partial replication with caching also reduces the overhead for network traffic to disseminate updates and keep all copies consistent. Finally, it also reduces the space cost for data in large internetworks with large numbers of multiple host groups.

We have not addressed here the problem of maintaining up-to-date, consistent network membership records within the set of gateways connected to members of a group. This can be viewed as a distributed database problem which has been well studied in other contexts. The loose consistency requirements on network membership records suggest that the techniques used in Grapevine[26] could be useful for this application.

Further details are described elsewhere[20,16] as well as some forthcoming Internet RFC's describing the DoD Internet implementation.

4.7 Security

Encrypted multicast messages can be provides using an encryption and key distribution scheme similar to that described by Birrell[4], assuming that a multicast group can be regarded as representing one *principal*. In this

---

[6] One unicast transmission from sender to gateway and one multicast transmission from gateway to local group members.

case, there is an encryption key for each group, potentially in addition to a key for each individual in the group.

In addition, there are three main security issues that arise with multicast:

- controlling who may send to a group.
- controlling access to membership information.
- controlling the membership of the group.

We have placed no restrictions on who may send to a group in V since there are no restrictions on who may send to a single process. Most communication systems only control the right to send by allowing the receiver to refuse or discard the communication, such as in refusing a circuit setup request. In general, the control on who can send to a multicast group should match the control on who can send to an individual process or port.

Access to membership information can be restricted in the query operations. In addition, with the V primitives, we allow a group member to reply anonymously, with the group identifier used as the source address for the response message.

Finally, controlling who may join the group is important since that determines who receives a given message and, in a request-response model, who may respond to the message. In distributed applications, we have found the most useful model to be as follows. The creator of the group is allowed to add other entities to the group. As part of adding a group member, it may confer the right to the new member to add members as well. We have also used *unrestricted* groups, for which there are no restrictions on joining or leaving. The original approach we used[20], based on user accounts, did not work satisfactorily since many groups crossed authorization boundaries.

5 Internetworking and Transport-level Gateways

Most of our work in V has used raw Ethernet as the network and datalink level, thereby providing us with a simple implementation, high performance and ready availability of efficient network-level multicast. In looking at the ways to carry the same functionality to handle multiple networks, there appears to be three options to connecting networks:

Bridges - devices that connect two or more networks into a single logical network, effectively link-level gateways.

Internetwork Gateways - conventional packet forwarding gateways that move internetwork datagrams between networks. This approach adds a layer like IP or XNS on top of the Ethernet layer.

Transport-level Gateways - gateways that understand the local domain transport protocol, VMTP in our case and provide access to communication endpoints in other domains by mapping local communication actions onto remote actions and vice

versa. This may involve state, protocol translation and name or address translation.

The bridge approach works well when the multiple networks have the same host address and packet format (i.e. the same link-level protocol) with non-overlapping host address spaces. It also requires acyclic interconnection of networks since each bridge forwards all multicast packets to the connecting network. This approach has been used extensively at Stanford and elsewhere to connect multiple Ethernets into a single logical network[8]. It is appropriate when one logical network is desired but physical length limitations and load capabilities of the physical network preclude using a single physical network. However, the physical networks must be compatible, as described above, and the collection of networks should be under the same administrative control and authorization. That is, there are no user-level reasons to isolate these networks from each other.

The use of a current internetwork architecture poses a number of problems. First, a datagram-based internetwork architecture requires that each VMTP packet be embedded in an internetwork datagram packet. For IP, the header adds at least 20 bytes to each V packet, over half of which are less than 80 bytes long according to our network measurements, leading to a 25 percent overhead. It also requires an additional level of demultiplexing as well as code for handling packet reassembly. Secondly, the VMTP implementation must deal with a far greater complexity in setting good timeout values as well as inter-packet times in an internetwork environment to achieve good performance. When the typical client in our environment is a diskless workstation accessing a local set of file servers, these overheads are significant. Finally, none of these architectures currently supports multicast so that would have to be added to the architecture and its implementation to make it usable at all with our current protocols.

In spite of these overheads, internetwork architectures do nothing to control access to a network, such as is desired for reliability and security. They also do nothing to control delay. An internetwork gateway is just a transparent packet forwarder. These issues caused us to examine the use of transport-level gateways for internetworking[8] and initiate work on a prototype of a transport-level gateway for VMTP.

With a transport-level gateway, the gateway "understands" the transport protocol and can cause different transport-level behavior to take place between two gateways than between an endpoint and a gateway. In particular, not every packet sent to the gateway results in a packet send between the gateway and the gateway(s) required to reach the destination network. [7]

The V transport-level gateway provides inter-gateway transport, naming and access control. In order for a local process to communicate outside of its local network, there must exist a local *alias process* for the destination process. The alias is a representative in the local name space, accessible using the local domain protocol, for the

---

[7]Note that fragmentation may cause an IP gateway to generate several packets in response to a single IP packet it received. However, it does not reduce the number of packets transmitted relative to those it receives.

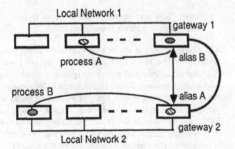

Figure 8: Internetworking using Transport-level Aliases

remote process. Similarly, before a remote process B can communicate with a local process A, the gateway must create a local alias B' for the remote process B and allow the remote process to use this alias, as depicted in Figure 8. The local and remote gateways are responsible for translating between the two name spaces, and transporting data between the networks. Except for performance, the internetwork communication should appear the same as local network communication. Because the gateway never acknowledges a packet transmission until the other endpoint has acknowledged it, the transport-level gateway does not violate the end-to-end semantics of the transport protocol.

This design has several advantages. First, the gateway can strictly control access to and from the network based on the creation of these alias processes. The criteria for providing these alias processes can include authentication of authorization, accounting and origination of the request. However, the exact policy is outside our discussion. The alias process is a mechanism for a variety of different access control policies. This control carries over to multicast access as well. A gateway can prevent unauthorized multicast communication from entering the network.

Second, this design provides internetworking without any cost on the local network communication. All packets are send to, and received from, alias processes the same as any other local process. In particular, there is no extra level of packet header or code to handle this header information and no knowledge of internetwork timeout periods, error rates and round-trip delays. Except for the gateways, the nodes of the network only need to know about the network to which they are attached, and thus can be optimized accordingly.

Third, internetwork communication is almost identical to local network communication, just as with the internetwork architectures.

Finally, the gateways can perform several optimizations on the communication between the gateways, including:

Retransmission Damping - local retransmissions can be "damped" to a rate consistent with the wide-area round-trip delay and underlying communication properties. In particular, a local node may retransmit every K milliseconds yet the gateway may filter out these retransmissions, reducing it to every 4K milliseconds (for example) between the gateways.

Packet Merging - multiple local packets can be merged into a single internetwork transmission unit for more efficient handling. Packet merging can span multiple unrelated communications if the costs of wide-area communication warrant it.

Data Compression - data packets can be compressed using standard data compression techniques to reduce further the amount of data between exchanged.

In general, the gateway can optimize the wide-area communication independent of the local network protocol, providing that it does not violate end-to-end semantics. For the retransmission damping, this appears to require the ability to respond to a retransmission with "encouragement" to keep retransmitting periodically without actually acknowledging the packet, as provided in VMTP.

There is a flaw in this transparency that arises because the process identifier name space is local to a logical local network, so process identifiers cannot be passed and used between networks transparently. An identifier on network 1 must be translated to an alias process identifier on network 2 before being passed to a process on network 2. Otherwise, the process identifier would be interpreted incorrectly on network 2.

Several considerations reduce the problems arising from this flaw. First, process identifiers are passed around fairly infrequently. Second, the use of process identifiers in messages is primarily in server modules and these servers modules can be properly programmed to handle the alias translation. Finally, for security, accounting and even scheduling, servers should be aware when dealing with a remote process. Thus, handling process identifiers in the messages differently is not an extension of the information they need to know. (In V, testing an identifier to see if it designates an alias process is efficient since it is indicated by a bit in the process identifier.)

This transparency flaw can be removed by using a internetwork-wide global identifier space. However, that requires larger identifiers than required for a single network, imposing a cost on all local hosts.

The major disadvantage of this approach is the state (such as the alias process) that transport-level gateways must maintain with its potential disruption of communication when a gateway crashes. We have been looking at techniques for recovering the state as well as replicating the state to solve this problem.

To date, we have built a prototype gateway to explore these ideas. The gateway is currently structured to communicate with another gateway (running the same software) using a TCP[25] connection. This gateway implements the design as described above with a number of the transmission optimizations. Work remains to be done before this prototype can be put into service. However, the experience developing and using this prototype have convinced us of the basic soundness of this design as well as, more generally, the merits of a cluster and transport-level gateway approach to internetworking.

6 Conclusions

The experience with a request-response or transaction transport protocol and multicast in the V distributed system has been extremely favorable. In particular, the resulting communication facility provides high-performance for remote procedure call, multidestination delivery, logical addressing and real-time communication. Building a transport-level request-response and multicast facility on top of a network-level multicast facility leads to a simple, efficient and elegant implementation.

Our experimental work in this area have been facilitated by using the "raw" Ethernet packet service as the network level. This approach avoids developing software to handle the internetwork datagram layer. It also provides direct access to the multicast facility of the Ethernet. However, it excludes the use of standard internetwork architectures and gateways. Instead, we can implement a limited form of internetworking using datalink- or physical-level gateways, called(*bridges* and *repeaters*) respectively, providing: (1) the networks are compatible, (2) the network host address spaces do not overlap, and (3) the interconnection of the networks is acyclic. In particular, a large number of Ethernets can be interconnected by bridges to form a single logical local network, across which our protocols can function with no change.

Two problems arise with the logical network approach. First, these low-level gateways do not provide sufficient access control (for reliability and security) between the networks, especially if some of the networks are under a different administration. Second, the packet delay across multiple networks becomes unpredictable relative to that of a single or small number of networks. Our simple timeout mechanisms interact poorly with this variable delay. One might hypothesize a third problem, namely that multicast packets are transmitted to all networks, excessively loading the networks. Unicast packets also have this behavior with physical level gateways or repeaters. However, we have minimized the use of repeaters. Also, the number of multicast packets per second, despite our reliance on multicast is insignificant, averaging 4 per second according to our network measurements. In our experience, multicast packets only become a problem under host or application failure, thus relating back to the access control problem for reliability. [8]

Conventional internetwork architectures provide no assistance in dealing with the problems of access control and delay yet impose a significant packet processing and code complexity overhead. In addition, they do not define a multicast facility.

These considerations have prompted us to explore the use of transport-level gateways. We have designed and implemented a prototype transport-level gateway that deals with the access control and packet delay problems, reducing the need for individual hosts to handle these considerations. Our limited experience with this approach to date suggests that this type of gateway is

the key to the next generation of communication systems and the requirements we have identified for them.

In general, we have identified significant changes in communication use and substrate that we conjecture will lead to a new generation of communication systems, the 4th generation in our taxonomy. The approaches, protocols and techniques described in this paper are but one modest step in the direction of realizing this new generation. We expect to report on further development of these ideas as part of our ongoing research in the future, particularly the transport-level gateway.

7 Acknowledgements

This work was sponsored in part by the Defense Advanced Research Projects Agency under contract N00039-83-K-0431, by Digital Equipment Corporation, by the National Science Foundation under Grant DCR-83-52048, and by ATT Information Systems, Bell-Northern Research and NCR.

Various members of the Distributed Systems Group at Stanford have contributed to the work described here including Tim Mann, Lance Berc, Steve Deering, Ross Finlayson, Paul Wang, Willy Zwaenepoel, Bruce Hitson, Domingo Lampaya and Michael Stumm. VMTP is currently being reviewed as a candidate for the Internet request-response protocol by the Internet End-to-end Task Force, chaired by Bod Braden. The design of VMTP has significantly benefited from discussions as part of this review. I am also grateful to Dave Clark of MIT who has been a major participant in these discussions.

References

[1] G. Almes.
The impact of language and system on remote procedure call design.
In *Proc. 6th Int. Conf. on Distributed Computer Sys.*, IEEE Computer Society, May 19-23 1986.

[2] E. Berglund and D.R. Cheriton.
Amaze: a multiplayer computer game.
IEEE Software, 2(3):30–39, May 1985.

[3] A. Birrell and B. Nelson.
Implementing remote procedure calls.
ACM Trans. on Computer Systems, 2(1), February 1984.

[4] A.D. Birrell.
Secure communication using remote procedure calls.
ACM. Trans. on Computer Systems, 3(1), February 1985.

[5] J.M. Chang and N.F. Maxemchuck.
Reliable broadcast protocols.
ACM Trans. on Computer Systems, 2(3), August 1984.

[6] Jo-Mei Chang.
Simplifying distributed database systems design by using a broadcast network.
In *SIGMOD 84*, ACM SIGMOD, 1984.

---

[8] For example, an application can fail sending out real-time multicast messages in a tight loop, generating a significant system load.

[7] D.R. Cheriton.
Fault-tolerant transaction management in a
workstation cluster.
1986.
To appear.

[8] D.R. Cheriton.
Local networking and internetworking in the
V-system.
In *8th Data Communication Symposium*,
IEEE/ACM, 1983.

[9] D.R. Cheriton.
Multicast-based clock synchronization.
1986.
Paper in progress.

[10] D.R. Cheriton.
Problem-oriented shared memory: a decentralized
approach to distributed systems design.
In *6th Int. Conf. on Distributed Computer
Systems*, IEEE Computer Society, May 1986.
Boston, MA.

[11] D.R. Cheriton.
*The Thoth System: Multi-process Structuring and
Portability.*
American Elsevier, 1982.

[12] D.R. Cheriton.
UIO: a uniform I/O interface for distributed
systems.
ACM Trans. on Computer Sys., 1986.
to appear.

[13] D.R. Cheriton.
The V kernel: a software base for distributed
systems.
IEEE Software, 1(2), April 1984.

[14] D.R. Cheriton.
VMTP: a transport protocol for the next
generation of communication systems.
In *Proceedings of SIGCOMM'86*, ACM, Aug 5-7
1986.

[15] D.R. Cheriton.
VMTP: Versatile Message Transaction Protocol.
Technical Report RFC ??, Defense Advanced
Research Projects Agency, 1986.
To appear.

[16] D.R. Cheriton and S.E. Deering.
Host groups: a multicast extension for datagram
internetworks.
In *9th Data Communication Symposium*, IEEE
Computer Society and ACM SIGCOMM,
September 1985.

[17] D.R. Cheriton, M.A. Malcolm, L.S. Melen, and
G.R. Sager.
Thoth, a portable real-time operating system.
Communications of the ACM, 22(2):105–115,
February 1979.

[18] D.R. Cheriton and T. Mann.
A Decentralized Naming Facility.
Technical Report STAN-CS-86-1098, Computer
Science Department, Stanford University, April
1986.
Also available as CSL-TR-86-298.

[19] D.R. Cheriton and M. Stumm.
Multi-satellite star: structuring parallel
computations for a workstation cluster.
Distributed Computing, 1986.
To appear.

[20] D.R. Cheriton and W. Zwaenepoel.
Distributed process groups in the V kernel.
ACM Trans. on Computer Systems, 3(2), May
1985.

[21] D.R. Cheriton and W. Zwaenepoel.
The distributed V kernel and its performance for
diskless workstations.
In *Proceedings of the 9th Symposium on Operating
System Principles*, ACM, 1983.

[22] D.D. Clark.
Window and Acknowledgement Strategy in TCP.
Technical Report RFC 813, Defense Advanced
Research Projects Agency, 1982.

[23] D.D. Clark, M. Lambert, and L. Zhang.
NETBLT: A Bulk Data Transfer Protocol.
Technical Report RFC 969, Defense Advanced
Research Projects Agency, 1985.

[24] E. Cooper.
Replicated procedure call.
In *10th Symp. on Operating Systems Principles*,
pages 63–78, December 1985.
Also published as Operating Systems Review
19(5),1985.

[25] DARPA.
DOD Standard Transmission Control Protocol.
Technical Report IEN-129, Defense Advanced
Research Projects Agency, January 1980.

[26] Birrell et al.
Grapevine: an exercised in distributed computing.
Communications of the ACM, 25(4), 1982.

[27] J.G. Fletcher and R.W. Watson.
Mechanism for a reliable timer-based protocol.
Computer Networks, 2:271–290, 1978.

[28] *Connection Oriented Transport Protocol.*
International Standards Organization, 1983.
DP 8073.

[29] T.P. Mann.
A Decentralized Naming Facility.
PhD thesis, Computer Science Department,
Stanford University, 1986.

[30] A. Spector.
Performing remote operations efficiently on a local
computer network.
Communications of the ACM, 25(4):246–260,
April 1982.

[31] M.M. Theimer, K.A. Lantz, and D.R. Cheriton.
Preemptable remote execution facilities in the
V-system.
In *10th Symp. on Operating System Principles*,
ACM SIGOPS, 1985.
also published in Operating Systems Review.

[32] W. Zwaenepoel.
Message Passing on Local Network.
PhD thesis, Computer Systems Lab., Stanford
University, October 1984.
Also available as Technical Report
STAN-CS-85-1083.

Communication Support in Operating Systems for Distributed Transactions[1]

Alfred Z. Spector
Department of Computer Science
Carnegie-Mellon University
Pittsburgh, PA 15213

21 November 1986

Abstract

This paper describes the communication functions required for distributed transaction processing. The paper begins with a discussion of models that illustrate how a communication subsystem fits into a proposed system architecture. Then, it describes the system and user activities that depend on the communication subsystem. Finally, it uses these activities to motivate the facilities that should be provided by a communication subsystem that supports transaction processing.

1. Introduction

Communication subsystems permit their client programs to invoke operations on remote sites and to perform auxiliary control tasks. For example, an application-level client may send a SQL Select operation and associated data to a remote database, which then returns the data that matches the selection criteria. A system control facility client may piggyback time information on application-level messages to maintain a consistent global notion of time. The exact nature of the applications and system control facilities using a communication subsystem substantially influence the functions it must implement.

The functions may be complex if they must support *distributed transaction processing*, that is, the execution of transactions on data stored in multiple partitioned and replicated databases on various network nodes. Distributed transaction processing provides applications with access to shared data that are stored with high data integrity and availability. Such applications require the usual communication functions such as datagram, file transfer, data streaming, network virtual terminal, and RPC. They may also require more unusual facilities to support support data replication, atomic commitment, a coherent system-wide notion of time, and authenticated protected data access. In addition, the volume and frequency of communication performed by distributed applications may require high bandwidth, low-latency communication if users' response time and throughput requirements are to be met.

Because of such complexity, this paper takes the view that a communication subsystem for supporting distributed transactions must be designed to fit within the complete framework of processing that will occur. With this viewpoint as a basis, this paper first describes generic system, computation and architecture models. Only then does it describe the communication facilities that are required. While this approach has the disadvantage of limiting the discussion of communication to that required for one particular set of models,

---

[1] This work was supported by IBM and the Defense Advanced Research Projects Agency, ARPA Order No. 4976, monitored by the Air Force Avionics Laboratory under Contract F33615-84-K-1520. The views and conclusions contained in this document are those of the author and should not be interpreted as representing the official policies, either expressed or implied, of any of the sponsoring agencies or the US government.

it shows more clearly how a communication subsystem is *integrated* into the greater system — the major point of the document. Furthermore, the system, computation, and architectural models are general enough to apply to many real systems.

Following the discussion of models, Section 3 discusses the system activities that require communication support. Section 4 uses these activities as a basis to describe important communication primitives and implementation strategies. These primitives are motivated by the Camelot distributed transaction processing system, developed at the Carnegie Mellon Computer Science Department [Spector et al 86]. The paper concludes with Section 5, which is a brief summary.

2. Three Models

There is substantial agreement on the underlying *system model* for distributed processing. The model has processing nodes and communication networks, as illustrated in Figure 2-1. Processing nodes are fail-fast and may be either uniprocessors or shared memory multiprocessors. In general, there are many types of processing nodes on the same distributed system. Processing nodes are assumed to have independent failure modes.

Storage on processing nodes comprises volatile storage — where portions of objects reside when they are being accessed, non-volatile storage — where objects reside when they have not been accessed recently, and stable storage — memory that is assumed to retain information despite failures. The contents of volatile storage are lost after a system crash, and the contents of non-volatile storage are lost with lower frequency, but always in a detectable way. Stable storage can be implemented using two non-volatile storage units on a node, or using a network service [Daniels et al. 86].

The system model's communication network provides datagram-oriented, internetworked OSI Level 3 functions [Zimmermann 82] such as the Arpanet IP protocol [Postel 82]. In other words, the network comprises both long-haul and local components and permits processes to send datagrams having a fixed maximum size. Some local area networks may specially support multicast or broadcast, and the network protocols are assumed to support these features for reasons of efficiency. Because applications using the system may need high availability, communication networks should have sufficient redundancy to render network partitions unlikely. However, network partitions can nonetheless occur, so higher levels of the system must take measures to protect themselves against the erroneous computations or inconsistencies that could result.

The *computation model* comprises applications that perform processing by executing transactions performing operations on data objects. Data objects are distributed across the network and are encapsulated within protection domains that (1) export only operations that make-up the defined interface and (2) guarantee that the invoker has sufficient access rights. Data objects may be nested. This model applies to many systems, including R*, Argus, TABS, and Camelot [Lindsay et al. 84, Liskov and Scheifler 83, Spector et al. 85, Spector et al 86].

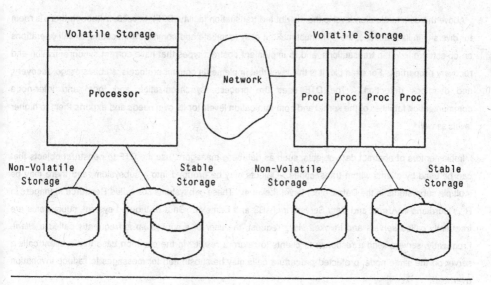

Figure 2-1: Hardware Model

This figure shows the components of the hardware model. Stable storage and non-volatile storage do not necessarily have to be implemented on disks.

The model further defines transactions as encapsulation units that provide three properties [Gray 80]: *Synchronization* properties, such as serializability, guarantee that concurrent readers and writers of data do not interfere with each other. *Failure atomicity* simplifies the maintenance of invariants on data by ensuring that updates are not partially done. For example, failure atomicity guarantees that a transaction that updates two distributed copies of a replicated file will either succeed and modify both, or fail and modify neither. *Permanence* ensures that only catastrophic failures in stable storage will corrupt or erase previously made updates. Transactions can be nested to reduce the likelihood that they need to abort completely and to provide protection from concurrently executing processes within a transaction.

The *architectural model* describes how processing on a node is organized; that is, it describes how to realize the computation model on the system model. It is structured in five logical levels, as shown in Figure 2-2. As one might hope, few calls proceed from lower levels to upper levels. (The levels referred to in this model are distinct from the OSI levels, and subsume functions in OSI levels 4 to 7.)

At the base in Level 1 is the operating system kernel that implements processes, local synchronization, and local communication. Examples are the V, Accent, and Mach kernels [Cheriton 84, Rashid and Robertson

81, Accetta et al. 86], though V and Mach also include inter-node communication facilities which this model includes in Level 2, the communication level. This level is the subject of this paper and the following sections analyze it in more detail.

Above the communication level is the distributed transaction facility (DTF), Level 3. Although there is room for diversity in its functions, the DTF must make it easy to initiate and commit transactions, to call operations on objects from within transactions, and to implement abstract types that have correct synchronization and recovery properties. For example, it is this level that implements commit protocols, stable storage, recovery, and deadlock detection. The DTF uses the process, synchronization and local and inter-node communication facilities of the kernel and communication levels for its own needs and exports them to higher levels as well.

Implementors of abstract data objects, such as database managers, use the DTF to construct objects that can be used by clients within transactions. Objects may be grouped into a subsystem, and there may be multiple subsystems in the Data Object Level (Level 4). These subsystems are called Resource Managers in R^*, Guardians in Argus, and Data Servers in TABS and Camelot. On a distributed system, subsystems are frequently called *servers* and invoked via a request; the user of a subsystem is frequently called a *client*. Frequently, servers send a response to clients to return a result. In the common case that a client calls a server on the same node, protected procedure calls may be substituted for messages to reduce invocation overhead.

In Level 5, applications use the DTF to begin, commit, and abort transactions and to execute operations on objects. Example applications include a banking terminal system and an interactive interface to a database manager.

This architecture provides two benefits over traditional architectures that blur the distinction between Levels 3, 4, and 5: First, because many of the components that support transactions are standardized and moved lower into the system hierarchy, there is the potential to implement them more efficiently. Second, the architecture provides a common notion of transaction and data object for all objects and applications in the system, and permits more uniform access to data. This permits an application, for example, to update transactionally a relational database containing indexing information, a file containing image data, and a hierarchical database containing performance records. All the system components also use standardized facilities for performing remote accesses, for transaction commitment, etc.

Having characterized the computational activities required for distributed transaction processing, it is now possible to examine the activities that use the communication subsystem. We can then turn our attention to requirements of the communication subsystem and how to meet them.

```
┌─────────────────────────────────────────────────────────────┐
│                                                             │
│   Level 5: Applications                                     │
│                                                             │
│   Level 4: Data Objects                                     │
│                                                             │
│   Level 3: Distributed Transaction Facility                 │
│                                                             │
│   Level 2: Communication                                    │
│                                                             │
│   Level 1: Operating System Kernel                          │
│                                                             │
└─────────────────────────────────────────────────────────────┘
```

Figure 2-2: Five Level Architecture Model

This figure illustrates the five system levels. The kernel level provides processes and inter-process communication. The communication level provides inter-node communication. The distributed transaction facility provides complete support for transaction processing on distributed objects. Data objects are maintained in Level 4. The applications that use them are in Level 5.

3. Activities Requiring Communication Support

Distributed transaction facilities require communication for data objects and applications in Levels 4 and 5 and system activities in Levels 1 through 3. This section lists those activities and then describes a set of communication subsystem functions that will support them.

3.1. Communication-related Activities of Data Objects and Applications

Before data objects and applications can begin to access other data objects, they must first establish a communication path to them. A *name service* locates servers that encapsulate objects and returns lower-level names that can be used to establish communication. To support distributed replication algorithms, a single object name may be associated with a several copies of objects, each stored on a different node. Typically, replication techniques specify the number of copies of an object to which they require access.

The name service must manage the name space to prevent unintended name duplication and to ensure appropriate authorization for name insertion and deletion operations. This is clearly a distributed data management problem, which is best solved by a collection of trusted Level 4 objects that can use the DTF. Thus, while the name service is logically related to communication services, it need not be implemented within the communication subsystem. However, the communication subsystem must provide the name service with well-known connections through which name servers can communicate with each other.

Questions arise concerning the permanence of name mappings, the granularity of objects that are named, and the management of the name space. There are many feasible answers to these questions, but here are some reasonable ones:

- Name mappings are relatively useless for objects that are inaccessible; when an object is unusable, it usually does not help to know its location. Hence, the motivation for replicating name mappings on another node is to reduce communication, not to provide availability.

- Objects registered in the system-wide name service should be coarsely grained; e.g., a database name rather than the names of all its relational tables. More detailed name resolution can be performed in an object-specific fashion. This decision is almost a necessity, both to reduce the number of communication paths to a server and to obviate the need for a uniform name space for each data object in an entire distributed system.

- Name mappings should survive node crashes to reduce the amount of work required to restart a node.

Once the name service has located an object, the lower-level object name can be passed to the communication subsystem and a session created. A session is required between a client and server for many reasons:

- **Authentication and protection**. If clients are to be certain they are accessing a particular server and servers are to check the access rights of a caller, then the communication session must be authenticated in some way, possibly using encryption techniques [Needham and Schroeder 78]. Prevention of active and passive attacks on the communication channel is also a desirable service for many applications [Sansom et al. 86, Birrell and Nelson 84].

- **Flow-control and pipelining**. Large amounts of data may be passed to and from objects and require flow-control. For example, a request to a remote object could result in a response containing megabytes of data. Pipelining may be useful on networks having long latencies. Even on local area networks, the increasing use of networks interconnected by bridges or gateways tends to increase delays and the consequent need for pipelining.

- **Crash detection**. There must be a mechanism for determining if a server has crashed after its first use and prior to commit. Timing out while awaiting a response from an object is one crash detection technique, but a session failure provides more uniform and timely information for most errors. For example, sessions can detect most crashes even when a client is not calling its server.

Communication on a session usually takes the form of a (synchronous) remote procedure call having *at-most-once* semantics. While there is room for diversity in the definition of these semantics, all definitions guarantee that an operation on a server will be performed at most one time, despite network failures and retransmissions. Providing higher service levels in the communication subsystem (atomicity, or exactly-once semantics) is unneeded because the DTF can completely abort arbitrary units of work, which may then be retried. As mentioned above, requests and responses may have unlimited lengths so intra-message flow control may be needed.

Sometimes, more general forms of remote procedure call may be useful [Spector 82]. Asynchronous RPC's permit a client to continue processing and to receive a signal when a response is returned. Multicast RPC's issue a request to multiple servers. A multicast RPC primitive may await all responses before returning or it may signal the client as each response arrives. The latter organization is useful when a client invokes an operation on multiple servers but does not need all responses before continuing work. Multicast RPC's may be implemented on multiple sessions, or may use a single session having multiple destinations. The latter is required if low-level network multicast primitives are to be used.

3.2. Communication-related Activities of System Levels

The data objects and applications of Levels 4 and 5 require communication primitives that provide very general functions: support for arbitrarily long messages, authentication, and the like. In Levels 1, 2, and 3, communication is more constrained and there is more a-priori knowledge of message contents. For example, knowledge that a message usually fits within a network packet permits a simpler transmission protocol to be used. Similarly, some data can be piggy-backed on messages sent by Levels 4 and 5.

Communication services required by Levels 1, 2, and 3 must support at least five different functions: time services, commit processing, deadlock detection, certain higher-level communication services themselves, and failure inducement for testing purposes.

Logical time services, as provided by a Lamport distributed clock [Lamport 78], order events in a distributed system. All observable dependencies between events are reflected in time values provided by the clock; that is, if Server 1 observes the time as A and it sends a message to Server 2, and then Server 2 receives the message, Server 2 will then observe the time as B, with B > A. Such a mechanism is useful for various types of synchronization, for example, for supporting hybrid atomicity [Herlihy 85]. The underlying algorithm makes use of a counter on each node and a field included in each inter-node message that may update the counter.

A distributed real time service that is synchronized across nodes supports synchronization algorithms and performance measurement techniques. Many implementations of such mechanisms require periodic exchange of time information, which is done by appending information to existing message traffic and sending short messages during idle periods.

To perform atomic commit processing, the DTF must send control messages such as **Prepare-to-Commit**, **Prepare-Ack**, **Commit** and **Commit-Ack** [Lindsay et al. 79]. Some of these messages are typically sent to one or more of the nodes involved in a transaction. Regardless of protocol, the communication subsystem should maintain appropriate information on the nodes involved in the transaction, and control messages should be sent with low-overhead. Usually, they can be sent as network datagrams because messages are short and reliable transmission is not needed; the transaction manager must deal with node crashes anyway. Even though data encryption and authentication may be needed for Level 4 and 5 communication, control messages are difficult to forge and they contain so little data that is valuable to outsiders that there may be no reason to encrypt them. However, certain commit protocols can benefit from transmission to a multicast address that is incrementally developed as the transaction executes.

A DTF that supports nested transactions also requires a lock-resolution protocol in addition to the commit protocols. This protocol is invoked to determine if a nested transaction can inherit a lock from a relative in the tree of transactions. Depending upon the frequency of lock inheritance, this protocol may be invoked often and require high performance.

Distributed deadlock detection algorithms typically require piecing together enough of the distributed "wait-for" graph to break cycles. This requires the periodic transmission or the piggy-backing of information on other messages.

The communication facilities themselves require communication in addition to the usual demands for session establishment and the transmission and acknowledgment of user-supplied data. For example, control messages are sent by authentication servers as part of session creation. "Are you there messages" may be periodically sent on sessions to rapidly detect server or node crashes.

Finally, to aid in *reliability* testing, the communication subsystem should enable users to test the system under conditions of communication failures: lost, duplicate, and corrupted packets; partitions; and delays. Being able to simulate these conditions is an important feature. Also, facilities for monitoring the performance of the communication subsystem are useful. Methodical, empirical testing is needed to develop robust systems.

4. Communication Subsystem Functions and Implementation

This section lists a plausible set of functions that a communication subsystem should provide, given the requirements described in the previous section. It also describes the broad outlines of an implementation strategy for them. This ideas are loosely based on our design of TABS and Camelot with additions from other systems where needed.

4.1. Name Service

The name service should provide primitives to associate a name with one or more servers that implement the named object. It may also associate a lower-level name used by a server to distinguish between the multiple objects it implements. The name service also provides primitives to lookup and delete names. The lookup primitive should permit the caller to specify how many servers should be returned and to set a timeout interval after which control will be returned. While the name service does not need to be part of the communication subsystem, it is closely related and worthwhile to include in this section.

One implementation strategy is to have multiple name servers on the network that communicate with each other. Because of the desirability of storing name bindings permanently (so as to not have to register objects after a crash), the name service should be implemented as transactional (Level 4) servers, which can utilize stable storage. The DTF's services also simplify the consistency management of the name space. For example, new names can be added within a transaction. Locally storing recently used name bindings (hints) reduces the amount of inter-node communication, provided that applications are willing to detect and handle potentially out-of-date information.

4.2. Session-based Communication

Sessions should be the basis for communication between Level 4 and 5 entities. They are also appropriate for some of the communication between system-level entities. At minimum, sessions should support an efficient implementation of RPC with "are you there" messages to detect crashes. However, a session's required functions and implementation (including the amount of state that must be maintained) varies with the requirements of the DTF and the structure of the underlying system. For example, sessions supporting multicast RPC should have multiple recipients to take advantage of low-level multicast facilities. Differing needs for asynchronous RPC's, protection, authentication, conversion of heterogeneous data, and arbitrary internetworking also influence the functions and implementation of sessions.

There are at least two facilities for supporting a DTF that a communication subsystem can perform: It can record the participants in a transaction by watching the messages and the transaction identifiers contained in them. This information is needed at commit time. Additionally, the communication subsystem can incrementally distribute a network multicast address to all the sites within a transaction so that network multicast can be used during the two-phase commit protocol. This multicast address can be related to the global transaction identifier and be piggy-backed on request messages. Cheriton describes a design for this in the V System [Cheriton 86].

4.3. Datagram-based Communication

The raison d'etre of datagram-based communication is to reduce transmission latency and CPU overhead. In order to keep datagram-based communication sufficiently lightweight, it is inevitably restricted in function: limited datagram sizes, lack of protection or authentication, etc. New functions that slow datagram transmission should be avoided.

Certainly, datagram communication should support unreliable point-to-point transmissions; also, it should support multicast, because many networks provide necessary hardware support. Both of these two services require little protocol layering. Possibly, there should be some datagram support that is tailored to operation on a single local area network recognizing that there are services that would not be used over a long-haul network. For example, a stable storage server (log) would almost certainly be on the same local area network as its client nodes [Daniels et al. 86].

4.4. Miscellaneous Features

There are a collection of miscellaneous features that a communication subsystem should support: a distributed (logical and/or real) time service, the parameterized insertion of errors or creation of network partitions, and a communication performance monitor. Other features may be needed for real-time applications or some high-availability architectures.

5. Summary

After examining the uses of a communication subsystem in a distributed transaction processing environment, it is not surprising to find that sessions and datagrams are the two most important facilities. However, in this environment where there is closely-coupled distributed processing, atomic commitment, replication, and a strong emphasis on reliable, highly available operation, there are some additional features that a communication subsystem should support. These include multicast, logical time, real time, performance evaluation, and fault insertion services. Higher level protocols not part of the communication subsystem but closely related to it are needed for commitment, nested transaction lock resolution, deadlock detection, and name resolution.

All these additional facilities necessarily require standardized interfaces and protocols to support open systems. In some instances, these facilities are being considered by standardization committees. In others, they represent new protocols not yet under consideration. Further work on prototypes (e.g., Argus [Liskov 84], Camelot, ANSA [ANSA 86]) is needed to develop the necessary experience with them.

Acknowledgment

Thanks to Jeffrey Eppinger and who read and commented on this paper.

References

[Accetta et al. 86] Mike Accetta, Robert Baron, William Bolosky, David Golub, Richard Rashid, Avadis Tevanian, Michael Young. Mach: A New Kernel Foundation for UNIX Development. In *Proceedings of Summer Usenix*. July, 1986.

[ANSA 86] *Functional Specification Manual (Release 1)* 1986.

[Birrell and Nelson 84] Andrew D. Birrell, Bruce J. Nelson. Implementing Remote Procedure Calls. *ACM Transactions on Computer Systems* 2(1):39-59, February, 1984.

[Cheriton 84] David R. Cheriton. The V Kernel: A Software Base for Distributed Systems. *IEEE Software* 1(2):186-213, April, 1984.

[Cheriton 86] David R. Cheriton. Fault-tolerant Transaction Management in a Workstation Cluster. 1986.Stanford University.

[Daniels et al. 86] Dean S. Daniels, Alfred Z. Spector, Dean Thompson. *Distributed Logging for Transaction Processing*. Technical Report CMU-CS-86-106, Carnegie-Mellon University, June, 1986.

[Gray 80] James N. Gray. *A Transaction Model*. Technical Report RJ2895, IBM Research Laboratory, San Jose, California, August, 1980.

[Herlihy 85] Maurice P. Herlihy. *Availability vs. atomicity: concurrency control for replicated data*. Technical Report CMU-CS-85-108, Carnegie-Mellon University, February, 1985.

[Lamport 78] Leslie Lamport. Time, Clocks, and the Ordering of Events in a Distributed System. *Communications of the ACM* 21(7):558-565, July, 1978.

[Lindsay et al. 79] Bruce G. Lindsay, et al. *Notes on Distributed Databases*. Technical Report RJ2571, IBM Research Laboratory, San Jose, California, July, 1979. Also appears in Droffen and Poole (editors), *Distributed Databases*, Cambridge University Press, 1980.

[Lindsay et al. 84] Bruce G. Lindsay, Laura M. Haas, C. Mohan, Paul F. Wilms, Robert A. Yost. Computation and Communication in R*: A Distributed Database Manager. *ACM Transactions on Computer Systems* 2(1):24-38, February, 1984.

[Liskov 84] Barbara Liskov. *Overview of the Argus Language and System*. Programming Methodology Group Memo 40, Massachusetts Institute of Technology Laboratory for Computer Science, February, 1984.

[Liskov and Scheifler 83] Barbara H. Liskov, Robert W. Scheifler. Guardians and Actions: Linguistic Support for Robust, Distributed Programs. *ACM Transactions on Programming Languages and Systems* 5(3):381-404, July, 1983.

[Needham and Schroeder 78] Roger M. Needham, Michael D. Schroeder. Using Encryption for Authentication in Large Networks of Computers. *Communications of the ACM* 21(12):993-999, December, 1978. Also Xerox Research Report, CSL-78-4, Xerox Research Center, Palo Alto, CA.

[Postel 82] Jonathan B. Postel. Internetwork Protocol Approaches. In Paul E. Green, Jr. (editor), *Computer Network Architectures and Protocols*, chapter 18, pages 511-526.Plenum Press, 1982.

[Rashid and Robertson 81] Richard Rashid, George Robertson. Accent: A Communication Oriented Network Operating System Kernel. In *Proceedings of the Eighth Symposium on Operating System Principles*, pages 64-75. ACM, December, 1981.

[Sansom et al. 86] Robert D. Sansom, Daniel P. Julin and Richard F. Rashid. *Extending a Capability Based System into a Network Environment*. Technical Report CMU-CS-86-115, Carnegie Mellon, April, 1986. To appear in *SIGCOMM '86: Futures in Communications*, August 1986.

[Spector 82] Alfred Z. Spector. Performing Remote Operations Efficiently on a Local Computer Network. *Communications of the ACM* 25(4):246-260, April, 1982.

324

[Spector et al 86]   Alfred Z. Spector, Dan Duchamp, Jeffrey L. Eppinger, Sherri G. Menees, Dean
        S. Thompson. The Camelot Interface Specification. September, 1986.Camelot Working Memo 2.

[Spector et al. 85]   Alfred Z. Spector, Dean S. Daniels, Daniel J. Duchamp, Jeffrey L. Eppinger, Randy
        Pausch. Distributed Transactions for Reliable Systems. In *Proceedings of the Tenth Symposium on
        Operating System Principles*, pages 127-146. ACM, December, 1985. Also available in *Concurrency
        Control and Reliability in Distributed Systems*, Van Nostrand Reinhold Company, New York, and as
        Technical Report CMU-CS-85-117, Carnegie-Mellon University, September 1985.

[Zimmermann 82]   Hubert Zimmermann. A Standard Network Model. In Paul E. Green, Jr. (editor), *Computer
        Network Architectures and Protocols*, chapter 2, pages 33-54.Plenum Press, 1982.

Replicated Distributed Processing

by

S.K. Shrivastava
Computing Laboratory, The University, Newcastle
upon Tyne, England

Abstract: Replicated processing with voting provides a powerful means of constructing highly reliable computing systems. We will consider a functionally distributed computing system intended for real-time applications, where each functional module - a node - has been configured in an NMR (N-modular redundant) fashion. Such a system receives processing requests from 'actuators' (the entities that demand services) that require distributed processing at various nodes. The paper will discuss various approaches to scheduling computations to ensure that each processor of an NMR node processes input messages in an identical order. The concept of exception handling for voters will be developed to detect failures in the system.

Keywords: replicated processing, voting, exception handling, fault-tolerance, distributed systems, message passing, Byzantine agreement.

1. Introduction

One of the great challenges facing computer scientists is to design and build computer systems that provide *guaranteed services* in the presence of a finite number of failures. We will assume - and this hardly needs any justification - that many applications require computer systems which 'closely' approximate ideal systems that never fail (since physical systems will eventually fail, we can only approximate the ideal). In order to be able to provide any kind of guarantee of service, one must precisely specify what kinds of, and how many, component failures a system is supposed to tolerate.

Suppose our system is constructed out of 'n' components (where a component can either be a hardware or a software module); then its reliability specification could be along the lines that *provided* there are 'm' or less component failures (where m, m < n, characterises the redundancy in the system) *and* each failure is of an assumed type, *then* the system will function as specified. Note that in such a specification, failure mode assumptions for components need to be stated explicitly, since if a component failure occurs that is outside the failure mode assumptions made, then no guarantee of normal services can be given. For many components (e.g. a microprocessor, an operating system) it is often hard to predict all possible failure modes that could occur. Then there is no alternative but to make *minimal* (and if possible *no*) assumptions about failure modes of such components and to design a system under the assumption that a failed component can behave in an *arbitrary* manner. In subsequent sections we will examine the consequences of this assumption on system design, but it is hoped that the reader will appreciate that design and construction of such systems is a remarkably difficult task and a substantial amount of research work is required. The purpose of this paper, which is tutorial in nature, is to acquaint the reader with some current research work in the area of highly reliable systems for real time processing.

2. On Redundancy and Byzantine Agreement

Let us first consider a fundamental reliability problem for interacting processes. Assume that a process A transmits a value to some other processes (B,C and D) and we wish to ensure that these processes do indeed receive the same value. Now, it is possible for A to fail in such a manner that - assuming that the value is binary - 'Yes' is sent to say B and C and 'No' to D (since we assume that a failed component can behave arbitrarily, we cannot discount this possibility). It is therefore necessary for processes B,C and D to take part in an *agreement protocol* to ensure that all of them decide on some common value. What if one or more of the processes taking part in the protocol also fail in the aforementioned manner? In a classic paper [1], Pease *et al* proved that, to tolerate m component failures, a consensus among non-faulty components can only be reached if the total number of components, n, is greater than 3m (n > 3m). Details of such protocols, popularly known as Byzantine agreement protocols [2] need not concern us here, it is enough to assume the existence of such protocols, and to note that they tend to be quite expensive in terms of message requirements.

Since the only way of providing fault tolerance is through the introduction of redundancy, we are often faced with the sort of problem discussed above. For example, B, C and D could be three processors that redundantly carry out the same task, in which case we must ensure that they receive the same input data before processing begins. A system intended for real time applications must provide prompt responses to service requests, so excessive use of agreement protocols must be avoided as far as possible, giving us a *design rule* for highly reliable real time systems:

> *A real time system should be structured so as to minimise the requirements for Byzantine agreement.*

Let us now examine how redundancy can be introduced into a system. To be specific, let us consider how we can construct an "ideal" processor out of a number (say n) of ordinary processors. There can be two radically different approaches:

(1)   **Fail Stop Processors (FSPs):** All of the n processors *interactively* (by making use of Byzantine protocols) maintain the following abstraction: any processor failure causes all of the non-faulty processors to stop (hence the name). We thus obtain an (almost) ideal processor - the FSP - that does not possess arbitrary failure modes, rather, it either works or simply stops [3]. FSPs can thus be used as building blocks for the construction of reliable systems.

(2)   **N Modular Redundant (NMR) Processors:** All the n processors *work in isolation* and use voting to *mask* outputs from faulty processors (see fig.1, where n = 3, giving us the well known triple modular redundant - TMR - system). It is assumed that a processor of an NMR node performs both voting and task processing functions (V and P) and that results from each P are sent to all the Vs of the next node where further processing will take place after voting.

In the system shown in fig. 1, the failure of a single V-P combination in node N_i can be masked by non-faulty voters of the subsequent node N_j provided that the non-faulty processors of Ni produced identical results which arrive at N_j uncorrupted. Assuming that the only communication paths are those implied by the figure, no assumption about the behaviour of failed processors need be made.

Out of the above two techniques, the NMR based approach appears more suitable, based on the following two observations: (i) FSPs do not mask failures, so application programs need recovery facilities which is not the case with the NMR approach ; and (ii) FSPs require extensive use of Byzantine agreement protocols and require a rather complex internal structure [4].

In the rest of this paper we will concentrate on distributed systems composed out of NMR nodes. An NMR node must satisfy a synchronization

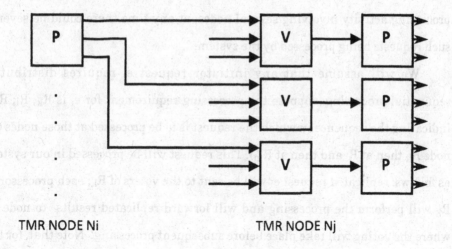

TMR NODE Ni   **TMR NODE Nj**

FIG. 1

requirement, which is that all of its n processors be 'roughly in step' with each other. If this synchronization is to be achieved by the processors communicating with each other then we have to introduce extra communication paths in the system (not shown in the figure) and the need for Byzantine agreement surfaces once more. In the subsequent sections we will examine design requirements for distributed NMR systems processing replicated computations.

3. Distributed NMR System Architecture

We will assume a functionally distributed system architecture consisting of a number of NMR nodes fully connected by an N-redundant communication system. Each NMR node R_i, manages (or represents) some resource (e.g. sensors). The environment of the system consists of a set of *initiators* (entities that demand services from the system at arbitrary times) and a set of sensors (entities that monitor parameters of the environment, such as temperature, air pressure and so on). A service request from an initiator can give rise to a processing activity involving several nodes; at any time there could be several such requests being processed by the system.

We will assume that any initiator request e_i requires distributed sequential processing: suppose the processing requirement for e_i is R_k; R_l; R_m, indicating the sequence in which the request is to be processed at those nodes (at node R_k then at R_l and then at R_m). This request will be processed in our system as follows: replicated request e_i will be sent to the voters of R_k; each processor of R_k will perform the processing and will forward replicated results to node R_l where the voting will take place before subsequent processing. Note that, for the sake of simplicity, we are restricting the processing of a request to be sequential; in particular this means that there are no synchronization requirements for the processing of any two requests e_i and e_j - other than the fact that each resource is to be used exclusively.

In general, a processor of an NMR node can receive several requests for processing , from other NMR nodes and initiators, which suggests the software architecture within a processor to be as depicted in fig. 2. A non-faulty processor maintains a pool of buffers for storing incoming messages. A voter performs voting as soon as it can form a majority on a given set of messages received from some node or an initiator. The voted messages are stored in a voted message pool. Some task scheduling policy (to be discussed shortly) is employed for selecting voted messages from the pool and queueing them in the voted message queue (VMQ) for processing. The messages in the VMQ are processed on a FCFS basis by the task process (P_i in fig.2).

We assume that Pi maintains some state which affects the execution of a task, and further that the execution of a task can modify the state. Assuming that all the non-faulty processes of a node have identical initial states before task processing begins, we require the following *sequencing condition* for an NMR node:

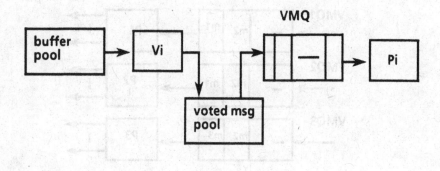

FIG. 2

SEQ: all non-faulty task processes of an NMR node process voted messages in an *identical order*.

It will be the function of the task scheduler of a processor to satisfy the sequencing condition. Application level requirements may dictate some further constraints on selecting voted messages for processing - for example some messages could have a higher priority over others for processing (e.g. 'alarm' messages). Task scheduling is discussed at length elsewhere [5]; we will briefly address some approaches to meeting just the sequencing condition in the next section.

A violation of the sequencing condition in an NMR node will be termed a *sequencing failure*. It should be clear that sequencing failures reduce the failure masking capability of a node. Consider for example the situation depicted in fig. 3, where the third processor of the TMR node has a VMQ in a state different from the other two. Assume that all the processors are non-faulty. It is quite likely that the results produced by processes P_1 and P_2 for message m_2 are different from those produced by P_3. Processor P_3 can thus appear to behave like a faulty processor.

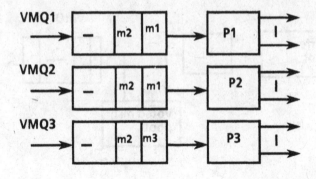

FIG. 3

4. Task Scheduling Approaches

In a concurrent processing environment, the sequencing condition can be particularly hard to meet. As an example, consider an NMR node N_k which can receive requests from two different NMR nodes N_i and N_j (fig. 4). Suppose N_i and N_j send their (replicated) results to N_k at about the same time and that messages can experience variable transmission delays. It is thus possible for N_i and N_j messages to be voted at N_k voters in a different order: this can cause - unless some preventive measures are taken - the messages to be queued in a different order in the VMQs of node N_k.

There can be several ways of meeting the sequencing condition, some of which are briefly discussed here.

(1) **Use of atomic broadcasts:** An atomic broadcast message sending facility exhibits the following three properties [6]: (i) it delivers every message broadcast by a non-faulty sender to all non-faulty receivers within some known time bound (*termination*); (ii) it ensures that every message whose broadcast is initiated by a non-faulty sender is either delivered to all correct receivers or to none of them (*atomicity*); and (iii) it guarantees that

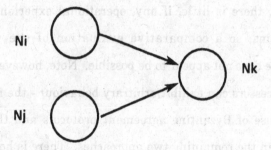

FIG. 4

all messages delivered from all non-faulty senders are delivered in the same order at all non-faulty receivers (*order*). It can be seen that by employing atomic broadcasts, it can be ensured that all non-faulty voters of a node receive messages in identical order, which in turn makes the task of meeting the sequencing condition straightforward. The remaining approaches do not rely on atomic broadcasts.

(2) **Use of an agreement protocol:** In this approach [7], non-faulty processors of an NMR node periodically take part in an agreement protocol to ensure that their respective VMQs are identical to each other, thereby guaranteeing that the sequencing condition is met.

(3) **Identical message selection:** This approach [5] requires that all of the non-faulty task schedulers of a node pick up messages from their voted message pools in an identical order for queueing in their respective VMQs.

(4) **Exception detection with recovery:** In contrast to the previous three approaches that prevent the occurrence of sequencing failures, this approach employs the philosophy of fault tolerance: sequencing failures are permitted, but are detected by voters as exceptions; specific exception handlers can then be employed for recovering from such failures [8].

As far as we know, there is little, if any, operational experience with distributed NMR systems, so a comparative evaluation of the various approaches discussed here does not appear to be possible. Note, however, that - assuming that failed processors can exhibit arbitrary behaviour - the first two approaches require the use of Byzantine agreement protocols and therefore appear less attractive than the remaining two approaches. There is however a 'hidden' synchronization overhead in all of the approaches presented here; this is the requirement that the clocks of all of the non-faulty processors be kept synchronized to some accuracy. If we assume that a faulty clock can exhibit arbitrary behaviour, then any clock synchronization algorithm must rely on a Byzantine agreement protocol [see 9 for example]. We thus see that the need for agreement protocols can not be ruled out entirely in replicated systems with voting.

5. Failure Detection

The possibility of detecting sequencing failures as exceptions during voting was mentioned earlier. We can go a step further and use voting for detecting processor failures and node failures (a node will fail when a majority of its processors have failed). Detecting the latter two types of failure is extremely important in systems that require on-line reconfiguration. The details given here are, of necessity, rather sketchy; for a detailed exposition the reader is referred to a previous paper [8].

We require the facility of *authenticating* messages [10] before voters can be used for detecting the above mentioned failures. We assume that an actuator request is a triple $<s(t_i), t_i, r_i>$, where r_i is the data part containing an encoding of the processing requirements, t_i is a unique timestamp, and $s(t_i)$ is a unique signature derived from t_i. During the processing of a request, only the data part will be modified (as the message travels from one node to the other). The objectives of message authentication are to ensure that any corruption of a message or a forgery of the signature of a message can be detected by a non-faulty receiver . As a result, a non-faulty voter can maintain a pool of only authenticated messages for voting. Assuming that a voter votes as soon as a majority of messages from a given node are available, it can detect the following situations:

(i)   All the messages being voted upon are identical. This represents the normal situation.

(ii)   The timestamps of the messages being voted upon are not identical. This can only mean that the node supplying the messages has suffered a sequencing failure.

(iii)   The timestamps of the messages being voted upon are identical, there is a disagreement in the data part of the messages, but a majority vote is possible. This can only mean that those processors supplying the disagreeing messages failed during the processing of the messages.

(iv)   Same as (iii) except no majority vote is possible. This indicates a failure of the node supplying the messages during processing.

We thus see that, in addition to masking failures, a voter can also perform the important task of detecting them.

6. Concluding Remarks

Functionally distributed real-time systems are no longer a thing of the future - they are here and in regular use. For example, the on-board computing system for the F/A-18 aircraft contains *fourteen* computers (twelve for performing sensor oriented computations and two for performing mission oriented computations) connected by a bus [11]. This system employs a number of *ad hoc* techniques for achieving fault tolerance. Modular redundancy in the form of replication of processing modules with majority voting provides a systematic and powerful means of introducing fault tolerance in systems. We have discussed an architecture for replicated processing of computations in distributed NMR systems, highlighting the various task scheduling approaches to meet the sequencing condition. In addition, the use of majority voters for detecting failures in the system was also briefly discussed. We believe that the concepts presented here form a sound basis for constructing highly reliable distributed systems. We conclude this paper by observing that the architecture presented here can be adapted to run *diverse software* (e.g. N-version programs [12]) to obtain a measure of tolerance against design faults in application programs.

References

[1] M. Pease, R. Shostak and L. Lamport, '*Reaching agreement in the presence of faults*', Journal of ACM, April 1980, pp.228-234.

[2] L. Lamport, R. Shostak and M. Pease, '*The Byzantine Generals problem*', ACM TOPLAS, July 1982, pp.382-401.

[3] R.D. Schlichting and F.B. Schneider, '*Fail-Stop processors: an approach to designing fault tolerant computing systems*', ACM TOCS, August 1983, pp. 222-238.

[4] F.B. Schneider, '*Byzantine generals in action: implementing fail-stop processors*' ACM TOCS, May 1984.

[5] P.D. Ezhilchelvan and S.K. Shrivastava, '*Task scheduling for replicated processing in distributed real time systems*', Tech. Report, Computing Laboratory, University of Newcastle upon Tyne (to appear).

[6] F. Cristian, M. Aghili, R. Strong and D. Dolev, '*Atomic broadcast: from simple message diffusion to Byzantine agreement*', Digest of papers, FTCS-15, Ann Arbor, June 1985, pp.200-206.

[7] L. Mancini, '*Modular redundancy in a message passing system*', IEEE Trans. on Software Eng., Jan. 1986, pp.79-86.

FORMAL SPECIFICATION IN OSI

Chris A. Vissers, Giuseppe Scollo (*)
Twente University of Technology
Dept. Informatics
7500 AE, Enschede, NL

(* on leave from University of Catania,
Istituto di Informatica e Telecomunicazioni)

ABSTRACT

Formal Description Techniques (FDTs) that should be capable to express
the OSI Protocols and Services are confronted with unprecedented
requirements in terms of the abstraction level at which the OSI
architectural concepts need to be expressed as well as the high
complexity of the OSI standards.

The development of the ISO FDTs Estelle and LOTOS for the formal
specification of OSI standards has been accompanied and guided by a
long period of extensive trial specifications. These exercises proved
highly necessary to introduce and justify several FDT concepts and to
test the appropriateness and expressive power of these FDTs for the
intended application area.

The OSI requirements for FDT can be roughly divided into a category
related to implementation independent specification and a category
related to the structuring of complex specifications to achieve
conciseness and readability.

Requirements that fall into the first category are related to
architectural concepts like service access points, service primitives,
connectionidentification mechanisms, implementation options, and non-
deterministic operations of the protocol. It appears that the
exercises in formal description did not only lead to better FDT
concepts, but at the same time led to better insights into basic
architectural concepts that underly OSI protocols and services.

The complexity of specifications can be coped with by the introduction
of suitable structuring facilities such as composition operators,
parameterization and recursion.

The language features of the FDTs must be based on a consistent
semantical model. The FDTs Estelle and LOTOS have exploited quite
different but complementary semantical models to tackle the
architectural problems out of the application area.

This paper presents a few small, but significant, examples of
fundamental architectural specification problems and discusses the
different ways in which one may deal with them in LOTOS and in
Estelle.

It is concluded that a few differences in elementary choices for the
semantical model of the FDT, reflecting differences in interpretation
and representation of basic architectural concepts, lead to vast
differences in appearance of formal specifications. Some implications
on current standardization work are outlined.

[8]   L. Mancini and S.K. Shrivastava, *'Exception handling in replicated systems with voting'*, Digest of papers, FTCS-16, Vienna, July 1986, pp.384-389.

[9]   L. Lamport and P.M. Melliar-Smith, *'Synchronizing clocks in the presence of faults'*, Journal of ACM, Jan. 1985, pp.52-78.

[10] R. Rivest, A. Shamir and L. Adleman, *'A method for obtaining digital signatures and public-key cryptosystems'*, Comm. ACM, Feb. 1978, pp.120-126.

[11] T.V. McTigue, *'F/A-18 Software development - a case study'*, Proc. of AGARD Conf. on software for avionics, Sept. 1982 (AGARD - CPP - 330).

[12] A. Avizienis, *'The N-version approach to fault-tolerant software'*, IEEE Trans. on Software Eng., Dec. 1985, pp.1491-1501.

1. INTRODUCTION

Formal specification of OSI in ISO is currently focussed around the development of the FDTs Estelle [DP9074rev] and LOTOS [DP8807rev] and their application to OSI protocols and services. Within the CCITT a corresponding activity is focussed around the development and application of SDL [SDL, SARACCO]. Since these developments all concentrate on fundamental architectural and specification aspects of (open) systems, notably their unambiguous interpretation and representation, it seems appropriate to start our discussion with a demand for unambiguous definitions of these basic concepts.

This demand seems more than ever necessary because, even while we are in several years of OSI development [desJARDINS], there are major differences of opinions on extremely basic architectural concepts, such as the service and the protocol concept: what are they 'tout court'. Let's mention a few, at least debatable opinions, which are still a matter of concern:
- a service can do more than a protocol;
- service primitives should not be shown in protocol definitions;
- a connection-endpoint-identifier is not a parameter of the service primitive;
- one can design a protocol by designing one protocol machine and deriving its peer entity from it (see e.g. [CHOI]);
- to test a protocol one needs only to test the exchange of protocol data units;
- a service need not be standardized and we need no conformance clauses for service definitions [VISSERS];
- the notion of interface should not play any role in standards for open systems;
- etc.
Differences in the understanding of basic OSI concepts evidently affect all aspects of OSI, notably (formal) specification and its immediate derivative, (formal) testing; if one does not precisely know what to specify, how should one specify, and what and how to test [BRINKSMA2].

The aim of this paper, therefore, is to relate formal specification concepts to basic architectural concepts by showing how seemingly small differences in the understanding of basic architectural concepts may affect the development of FDTs, resulting in major differences in the final appearance of formal specifications. The evidence is given in section 5 on basis of a few small but significant examples taken from OSI. Sections 2, 3, and 4 are leading up to section 5 by clarifying in section 2 some basic notions of formal specification, by clarifying in section 3 some basic OSI architectural concepts, and showing in section 4 how the Estelle and LOTOS groups have interpreted these concepts leading to different choices for their basic semantical model. In section 6 we assess the current position of these FDTs. Section 7 gives some conclusions.

2. WHAT IS FORMAL SPECIFICATION

A formal specification is the definition of an object, or a class of objects, by way of a Formal Description Technique (FDT).

What is description?

Description is a symbolic representation of a certain object in a given language. The language may use various kinds of symbols such as graphical, vocal, or gesture symbols.

What is formal description?

The term formal description has a very specific meaning viz. that the description language uses strict rules for the construction of language expressions, the "formal syntax", and strict rules for the interpretation of, or the assignment of meaning to, well formed language expressions, the "formal semantics".

The assignment of meaning to a well formed expression is defined on basis of the mathematical concepts of a semantical model that is defined for the formal description language. These mathematically defined concepts are unambiguously related to basic architectural concepts out of the application area from which the objects of description are chosen. In our case the application area is OSI, and the objects are OSI services and protocols.

What is specification?

When we talk about (formal) specification a specific object is to be expressed in a (formal) description technique. To express an object formally requires to relate the semantics of the FDT to the realm of the real world object, and thus the necessity to precisely understand the basic concepts or construction elements from which the real world object is composed. In our case this demands for a precise understanding of basic OSI concepts, or construction elements, such as services, protocols and derived concepts like service primitives, service access points, protocol data units, etc., so that the FDT semantics can be unambiguously related to its intended "architectural semantics", to be founded on the basic notions of OSI. As mentioned in the introduction, these basic OSI concepts seem to be not always clear.

Obviously the formal semantics is the dominant factor in formal specification as it tells us how a string of symbols, which could in principle express anything (including nothing at all), defines the construction of a (complex) object, such as a service or protocol, in terms of more basic architectural concepts such as service access points, service primitives, protocol data units, connections, in sequence integrity of data, multiplexing, concatenation, etc., to which it gives a mathematically precise meaning.

The use of the term "formal" has also another important connotation viz. that the object of description is the "functional behaviour" or "architecture" of something that can be implemented in many different ways (see also next section). A formal specification thus describes a class of implementations where each member of the class satisfies the same functional requirements. The specification abstracts from details in which implementations may differ and therefore is an abstraction of all valid implementations.

We also want to use formal description for concrete implementations of OSI objects, implying the description of "real interfaces". This allows us to use the FDT throughout the complete protocol development cycle, from abstract specification to concrete implementation. In principle this objective does not present major difficulties insofar as the implementation need not be described at a very low level of detail (e.g. program language instructions, transmission line drivers, etc.).

There is also a desire to express "very high level" objects such as the OSI Reference Model itself. This presents really unprecedented difficulties, as can be anticipated from the theme of this paper, because it seems not clear at all what should be expressed when one wants to describe for example that a protocol should be "connection oriented", where this could be a network, transport, session or presentation protocol. In this respect the Reference Model often gives us a warm feeling, but little grip.

3. WHAT ARE BASIC OSI CONCEPTS

Although it is not our aim to present a tutorial on architecture, nor is it our aim to shake OSI on its foundations, it seems necessary for the aim of this paper to provide at least some clarity for the confusion about basic architectural concepts mentioned in the introduction. This provision is necessarily elementary and far from comprehensive.

Basic requirements of OSI

To achieve some clarity about basic OSI concepts it seems sensible to analyse basic OSI objectives and requirements.

In principle we can view an OSI system as an arbitrary configuration of data processing equipment interconnected by data communication equipment, whose main objective is to allow a defined (i.e. standardized) cooperation of a number of user defined (i.e. not standardized) distributed application processes (figure 1).

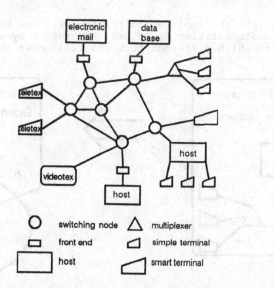

Figure 1: Configuration of an arbitrary OSI system.

Since the configuration of the equipment can be rather arbitrary (as determined by different implementation approaches), and since cooperation implies the obedience of all cooperating application processes to the same rules, we need a both <u>configuration independent</u> and <u>equipment independent</u> definition of this cooperation.

To this end, we introduce an "implementation independent", i.e. "abstract", definition of the cooperation between the user defined application processes, by way of a shared process which we provocatively will call "the application service" (figure 2).

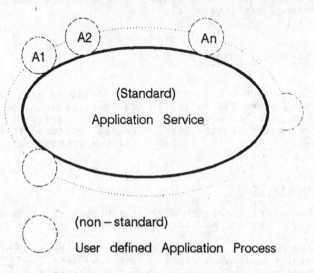

Figure 2: Abstract definition of the cooperation between user defined distributed application processes

Since we know that the application service will not be implemented by a monolithic implementation structure, but rather by a configuration of equipment of which different portions fall under the responsibility

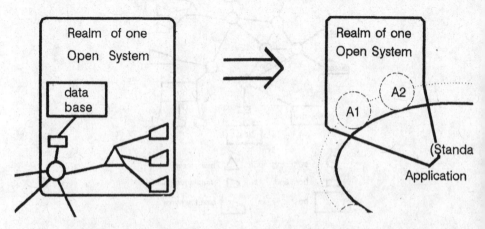

Figure 3: Realm of one Open System: the domain of an implementation authority

of different implementation authorities, we have first to delimit the
realm of the implementation authorities, and second to decompose the
application service, and assign portions of this service to individual
implementation authorities.

We delimit the domain of the implementation authority by introducing
the concept of Open System (figure 3), so that one or more application
processes are incorporated in an Open System.

The decomposition of the application service yields portions of
behaviour, let's call them "sub-systems", such that a sub-system can
be assigned to an Open System as an implementation responsibility and
as the contribution of the Open System in the implementation of the
application service. Naturally sub-systems need to be interconnected
by communication media. In this way we get a system of Interconnected
Open Systems (figure 4). Given the implementation independent
specification of a sub-system, each implementation authority can now
go ahead and implement the sub-system and associated end-user
application processes in any desired way as long as the specification
of the sub-system is observed.

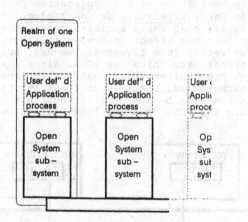

Figure 4: A system of Interconnected Open Systems

Complexity

It would be quite profitable if we could actually proceed as described
above: it would result in the most simple definition of the subsystem. The technical problem that we usually have to face, however,
is complexity: in practice (and certainly in OSI) the problem is too
complex to derive in a single design step the specification of the
sub-system from the application service. Therefore the method of
stepwise problem solving, also called "separation of concerns", is
applied. This method leads us in a first design step to a partly
decomposition of the application service in a "layer" of "application
layer sub-systems" and a "presentation service" (figure 5). Each
application layer sub-system consists of one or more functions called
"application protocol entities". In this decomposition process we cope
with a limited number of technical problems in the layer of
application protocol entities, and defer remaining problems to the
decomposition of the presentation service.

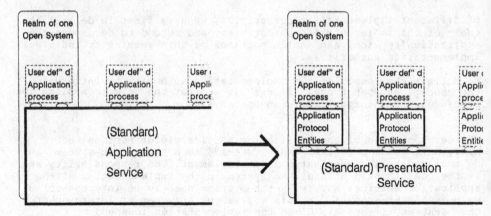

Figure 5: First decomposition of the Application Service

Separation of concerns in the above case is achieved when the
definition of the activities of the application protocol entities is
done, at least in part, separately or independent of the presentation
service. This implies that these entities must be able to exchange
information that does not affect, nor does it alter, the presentation
service (and vice versa) (see figure 6). We call such units of
information "Protocol Data Units" or PDUs. Since these PDUs are
exchanged between protocol entities that are implemented by different
implementation authorities, their unambiguous interpretation must be
guaranteed. This can only be achieved if their bit coding is defined.
This is the reason why PDUs need complete coding schemes.

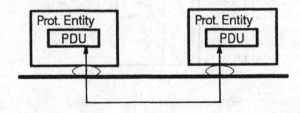

Figure 6: Separation of concerns requires transparent transfer of PDUs

Having done this exercise for the application service, we can now
apply the same decomposition method to the presentation service,
yielding presentation protocol entities and a session service. And we
can repeat this for each lower level service. It should be noted,
however, that the decomposition of a service is dictated by technical
criteria: if these demand for a different decomposition strategy,
which for example is the case at network interconnection level, then a
different sub-structure of a service results.

What do we learn

From the above reasoning we can derive a number of conclusions that
have a determinant impact on the development of FDTs, as well as on
the understanding of the OSI architecture.

Important conclusions for formal description are:

- The starting point for the development of formal description(technique)s is the service boundary. This applies to the description of the service as well as of the protocol (see section 5). It implies that the architectural concepts that are used to define services should be clearly defined. Whereas the service boundary is a boundary within Open Systems, it should be described completely implementation independent, i.e. at the highest possible level of abstraction, so as not to constrain unduly the freedom of the implementation authority. From this view we can start the development of derived concepts, such as service access points, service primitives, etc.

- A protocol is derived from a service through decomposition of this service, and the better definition for the concept of (N)-protocol would be a composition, or cooperation, of (N)-layer protocol entities and the (N-1)-service, such that the composed behaviour is equivalent with a predefined (N)-service. This makes the definition of the (N)-protocol subordinate to the definition of the (N)-service. It also follows that the exchange of PDUs between protocol entities is a derived mechanism, and not the everything dominating factor as many people tend to believe.

Architecturally we can conclude the following:

- From the users point of view, the service is the dominating factor, whereas for the implementer the protocol is the dominating factor.

- A service can do nothing in addition to a protocol, rather a protocol, as it is defined at a lower level of abstraction, defines more detail of how a service is rendered.

- We indeed can (and should!) define an application service. The Reference Model has submerged the unknown part of the application process and the defined application protocol entity in one application layer which, in our opinion, is an error of thinking and probably a source of confusion that hampers the development of application layer protocols.

- The number of layers that result from the stepwise decomposition of an (application) service is determined by the amount of problems that can be coped with in each decomposition step (and is not necessarily seven). Effective design demands for as few layers as possible. This applies also to ISDN systems.

- A medium should be considered a service. This service is not further decomposed because it is considered to be simple enough and can be obtained directly from some manufacturer or carrier. Its sub-structure, therefore, falls outside the scope of the OSI standardization activity. The technology to implement a medium service, however, can still be quite complex and the manufacturer or the carrier may find it convenient to further decompose it.

For readers familiar with the (development of the) Reference Model it will be evident that the latter was not developed with these views in mind but rather based on a sound intuitive notion that the layering principle comprises a good structuring discipline. Therefore the authors do not expect that all views presented here will find immediate support.

4. THE ISO FORMAL DESCRIPTION TECHNIQUES

Role of semantical model

The semantical model chosen for an FDT binds the semantics of the FDT to the real world architectural concepts and is therefore dominant in the use of the FDT. We briefly discuss the semantical models of Estelle and LOTOS. We appeal to a general understanding that an object, or more precisely its abstract representation in the semantical model, can be defined in two steps: first the definition of the means through which the object communicates with its environment, the inputs and outputs, and next the relationship between inputs and outputs in terms of the order in which inputs and outputs may occur and their value dependencies. We assume that the environment is also describable in terms of the semantical model chosen.

Estelle

In Estelle the abstract mechanism defined for inputs is value passing, i.e. the environment creates a value of a certain type which is rendered to the object. The passing is defined as an atomic event, i.e. no other activity can take place during the passing of the value. The mechanism for output is defined correspondingly. In this choice Estelle follows conventional views on inputs and outputs as the passing of information.

To model the dependency of outputs from inputs Estelle uses a finite state transition model (figure 7). Starting from a defined initial state the object, in Estelle called "module", makes transitions from one state to a next state etc. A transition is normally conditioned by an input. A transition must be made when the transition condition is enabled, whereas an output may be produced during the transition if so defined. In this respect the machine obeys the rules of a synchronous state machine. Refinements of the model allow for spontaneous transitions, i.e. a transition with no condition, for non-deterministic choice among alternative transitions, and timing constraints on transitions. The interested reader will find a tutorial on Estelle in [LINN]. Figure 7 shows an example of application of the model illustrated by a directed graph.

The model demands for some extensions to be useful in practice: First the number of states that an object of practical interest, like a protocol entity, may exert is generally very high, giving rise to the so-called "state explosion" problem. The problem can be countered by decomposing the state space by way of a number of state variables whose values together span the state space. In Estelle one of these variables is called STATE, the "major state variable", while the others are sometimes called "minor state variables", or "context variables". Variables, inputs and outputs are typed according to standard PASCAL facilities, allowing to use conventional PASCAL operations on these types.

Next the complexity of practical applications demands for additional structure, e.g. objects may exert behaviour which demands for the dynamic creation and extinction of state machines. An example is the setting-up, maintaining and closing of a connection. In Estelle this is solved by the introduction of the "parent-child" concept, where a parent can create a child, which may live a life of its own. Siblings may communicate either via channels by sharing common variables.

```
specification EXAMPLE;

type
    input    = (* PASCAL definition*);
    output   = (* PASCAL definition*);
channel INPUT_OUTPUT (user, provider);
    by user:
        In1, In2, In3 : input;
    by provider:
        Out1, Out2, Out3 : output;

module P (GATE : INPUT_OUTPUT);
    trans
        from Start
        to Next
        when GATE.In1
            begin
                GATE.Out1
            end
        from Next
        to Rep
        to Wait
        when GATE.In2
            begin
                GATE.Out2
            end
        from Rep
        to Start
            begin
                GATE.Out3;
            end
        from Wait
        to Start
        when GATE.In3
end;
```

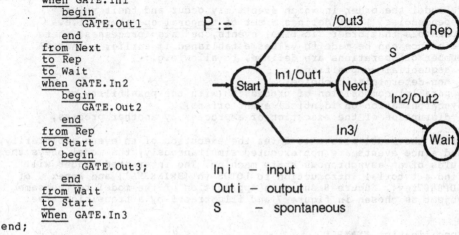

In i — input
Out i — output
S   — spontaneous

Figure 7: Example in (simplified) Estelle, illustrated by a state transition graph

Most crucial in our opinion, however, is the fact that Estelle needs to define FIFO buffers between finite state machines to buffer sequences of imputs/outputs. This need accrues from the assumption of atomicity of transitions: a synchronous state machine model where an output is produced synchronously with the input during the transition. Since the environment may be temporarily unable to accept the output as its input (because it may not be in the appropriate state or may choose to process another input) it may be necessary to temporarily buffer the output. Estelle uses FIFO queues, called "channels", for this purpose to preserve sequencing between different outputs. Since it is undetermined when the environment will be in the position to accept output from the FIFO while an arbitrary number of outputs may be generated, the FIFO in principle must be infinite.

LOTOS

In LOTOS the abstract mechanism defined for communication with the environment makes no distinction between inputs and outputs, but is based on the interaction concept. I.e. during an interaction with the environment the object, in LOTOS called "process", and the environment together establish a value of a certain sort to which both the process

and the environment can refer. Like in the Estelle model interactions
are atomic events. Process and environment may impose constraints on
the value that may be established during the interaction. Depending on
the conjunction of these constraints five possibilities can occur:
- no interaction,
- pure synchronization, or "value matching",
- "value passing" from environment to process (i.e. the traditional
input),
- "value passing" from process to environment (i.e. the traditional
output),
- non-deterministic "value generation".
Corresponding with the concept of spontaneous transition in Estelle
the LOTOS model defines the internal event (written "i"). This event
may result, in fact, from parallel composition with internal
synchronization.

To model the order in which events may occur and their value
dependencies, LOTOS defines a set of temporal operators, i.e.
operators that order (in time) events, but also processes, where
reference can be made to values established in earlier events. A
number of operations are defined, to allow e.g.:
- sequential composition,
- non-deterministic choice,
- parallel composition of processes (with the possibility of
synchronization on identical event offers),
- disruption of the execution of a process by another process,
etc.
Since the model allows to defer the execution of an event arbitrarily,
and since events are not executed simultaneously, the model obeys the
rules of an asynchronous state machine. The interested reader will
find a tutorial introduction to LOTOS in [BRINKSMA1] and Annex C of
[DP8807rev]. Fugure 8 shows an application of the model to the same
object as chosen in figure 7 and illustrated by a transition tree.

specification EXAMPLE

type input is (* ACT ONE definition *) endtype
type output is (* ACT ONE definition *) endtype

process P[GATE]():noexit :=
   GATE?In1:InputSort;
   GATE!Ou11;
   (i;
   GATE!Out3;
   P [GATE] ()
   []
   GATE?In2:InputSort;
   GATE!Out2;
   GATE?In3:InputSort;
   P [GATE] ()
)
endproc

endspec

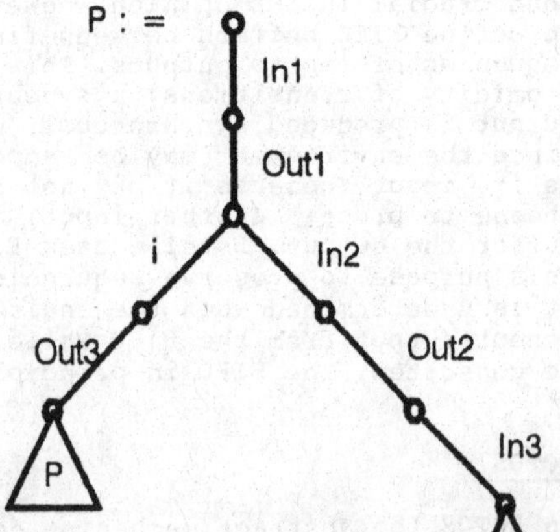

Ini, Outi, i — labels
i   — internal event

Figure 8: Example of figure 7 in (simplified) LOTOS,
illustrated by a transition tree

To be useful in practice, the language demands for a number of
abstraction facilities. These can be categorized as catering for
process abstraction and data abstraction.

Process abstraction allows to define behaviour identifiers, possibly
paramete- rized, which may occur, possibly recursively, in behaviour
expressions. Both parameterization and recursion extend the model to
transition systems, where the number of states may be (countably)
infinite.

Data abstraction allows to constructively define the data value
domains by way of, and together with, the operations on them. The
Abstract Data Type (ADT) language ACT ONE [EHRIG] is employed as a
sub-language of LOTOS. In addition, a standard data type library is
provided (in Annex A of [DP8807rev]) to relieve the user from
everywhere occurring definitions like Boolean, Natural Number, Set,
etc. ACT ONE features powerful combinators of data type definitions
such as enrichment, renaming, actualization of parameterized types by
actual types, etc. A subset of ACT ONE has also been taken as the
kernel of the revised ADT part of SDL, Recommendation Z.104
[Z.104rev]. LOTOS and SDL, therefore, have a common kernel for the
semantics of ADT specifications.

Estelle-LOTOS correspondence and differences

Estelle and LOTOS are both based on the general model of transition
system, which is formally defined by a 4-tuple

$$\langle S, T, A, S0 \rangle$$

where S is the state space, T is the set of labelled transitions, A is
the alphabet of labels for transitions, and S0 (element of S) is the
initial state.
In Estelle the states are explicitly indicated by the values of the
state variables, in LOTOS the states are implicitly indicated by the
nodes in the process tree. Some further relationships are suggested in
[AHOOJA].

The major differences between Estelle and LOTOS are in the way
transitions are handled. Estelle is based on the synchronous state
machine concept: a transition is made and the output is produced as
soon as the transition is enabled. Asynchrony between different
machines is achieved through the introduction of the queues in the
channels. LOTOS is based on the asynchronous state machine concept:
when all involved processes make an event offer then the event will
eventually occur, whereas an (output) event resulting from an (input)
event is treated in LOTOS as a sequence of two transitions (possibly
interleaved with other transitions).

5. APPLICATION TO OSI

In the following we show some corresponding and different ways in
which Estelle and LOTOS deal with the specification of OSI standards.
The examples that we take from OSI all refer to the Transport Service
and Transport Protocol but generally have a broader scope. First we
show the global approach towards the formal specification of a service
and protocol. Next we show some details of formal specification of a
few specific architectural problems derived from the global approach.

In doing so we necessarily have to remain on a very superficial level: a full treatment of the subject would require the reader's knowledge of the FDTs, of the OSI standards and of their formal specification in the respective FDTs; a requirement nobody in the world can (yet) fulfil. Furthermore, by taking examples out of their original context and simplifying them, we necessarily do injustice to the designers of the specification. For this we offer our apologies. We nevertheless hope that the reader will find the discussion illustrative and basic in nature.

Example 1: Service

According to the analysis in section 3, an (N)-service should be specified as an abstract machine, usually called the "(N)-service provider", that is accessible via (N)-service access points for several (N)-service users (see figure 9). The abstract machine should define the composite behaviour of all layers below the (N)-service boundary as observable by the (N)-service users by an as simple as possible specification (rather than "the description of the capability of the (N)-layer and the layers beneath it" as a reader might expect in analogy to the definition of the service concept in [IS7498]).

The specification of the (N)-service provider should be done in terms of (N)-service primitives (representing inputs and outputs) defined at peer (N)-service access points (SAPs), the local ordering of these primitives per service access point, the global ordering of these primitives between service access points, and the dependencies of parameter values of these primitives.

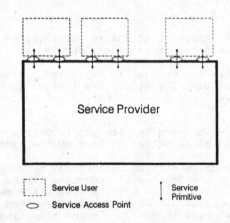

Figure 9: Abstract view of a Service

For the Transport Service it appears that both the Estelle editing group (see [SC6/WG4/N53]) and the LOTOS editing group (see [SC6/WG4/N116]) have followed this approach. Both groups model a connection in terms of an initiator-responder interaction and multiple connections by defining that there may be an arbitrary configuration of initiators and responders at service access point (addresses). The major difference is that Estelle uses channels to communicate service primitives between user and provider, whereas LOTOS uses direct interaction.

Example 2: Protocol

According to the analysis in section 3, where an (N)-protocol is considered as a decomposition of an (N)-service, an (N)-protocol should be specified as a composition of an arbitrary number of abstract (N)-protocol entities and a (N-1)-service, where each (N)-protocol entity and the (N-1)-service are described by an abstract machine. The composite behaviour of these abstract machines should exhibit the same behaviour as the (N)-service of which it is a decomposition (except possibly for a decomposition of (N)-service primitives).

In practice, however, it is sufficient to describe only one (N)-protocol entity (see figure 10). This is due to the fact that (up till now) OSI protocols are symmetric, i.e. at specification (not necessarily at implementation!) level all (N)-protocol entities comprise the same functionality, whereas the (N)-service is specified by the editing group of the (N-1)-layer. This pragmatism however does not change the principle that the formal specification of the (N-1)-service should be available together with the formal description of the (N)-protocol entity in order to be capable to verify the (N)-protocol.

The specification of the (N)-protocol entity should be done in terms of (N)-service primitives and (N-1)-service primitives (representing inputs and outputs) defined at the (N)- and (N-1)-service access points of the protocol entity, the local ordering of these primitives per service access point, the global ordering of these primitives between service access points, and the dependencies of parameter values of these primitives.
(N)-protocol data units should be treated as values of variables internal to the description of the entity. Their encoded representation should be visible at the boundary of the entity as values of the "user data" parameters in (N-1)-service primitives. It could be argued whether the specification of the entity should be given on basis of the encoded or abstract representation of the PDUs. The latter, for reasons inherent to our objective of abstract specification, is compulsory.

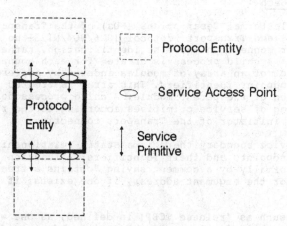

Figure 10: Abstract view of a Protocol Entity

For the Transport Protocol it appears that both the Estelle editing group (see [SC6/WG4/N123]) and the LOTOS editing group (see

[SC6/WG4/N117]) have followed this approach. Both groups have made an effort to make the protocol and service description consistent with each other by copying elements of the service description in the protocol description insofar as applicable. The structure of the formal descriptions of the protocol entities, however, is quite different although common approaches to the grouping of protocol classes, and the handling of Transport Connections separately from Network Connections, may be recognized.
Again, Estelle uses channels to communicate service primitives across the layer boundaries, whereas LOTOS uses direct interaction.

In the above we have seen that the editing groups apparently have the same understanding of the service and protocol concepts as evidenced by the same global approach towards structuring the formal descriptions. In all these approaches behaviour observable at the service boundary is the starting point for description. Next we show a few examples related to the establishment of this behaviour, where the approaches are (or may be we have to say: must be) quite different, as imposed by basic choices of the respective semantical models.

Example 3: Connection endpoint identification

In the OSI standards it is formulated that a connection endpoint identification mechanism should be provided to distinguish among concurrent connections accessing the same service access point. It is explicitly stated that such an identifier is not a parameter of the service primitive whereas the definition of this mechanism is left undetermined.
For the formal specification this implies that although the means by which such a function is implemented are a local matter (local to each SAP) it is necessary to specify the requirement that such a function is available at the service and protocol boundary (no matter how it is made available, e.g. whether it is provided by the service provider, or by the service user, or by the local cooperation of both).

Estelle approach

In the Estelle Formal Descriptions (FDs) of the Transport Service [SC6/WG4/N53] and Transport Protocol [SC6/WG4/N123] no requirement is specified for connection endpoint identification, rather a mechanism is described: a child process is created for each connection endpoint identifier out of an array of modules indexed by addresses and connection endpoint identifiers. This array is created at initialization of the parent module. A child process describes the local ordering of service primitives according to the role of either responder or initiator of the Transport connection.

Across a service boundary there is a static relationship between connection endpoints and their identifiers. A function 'get_TCEP' is declared informally by a comment saying "obtains a free endpoint identifier for the argumant address, if one exists; if none exists, returns a 0".

No function such as 'release_TCEP' is defined, as one would probably expect, that would deallocate a connection endpoint identifier at connection termination for later reuse. This is probably due to, and consistent with, the fact that connection termination is not a synchronized occurrence for user and provider because of the delay of the disconnect service primitive in the channel (see next example). Therefore the involved processes (user and provider, or protocol entities) have no common view as to when the identifier is free again. A child process is killed at connection termination.

LOTOS approach

.In the LOTOS FD's of the Transport Service [SC6/WG4/N116] and the Transport Protocol [SC6/WG4/N117] the requirement on connection endpoint identification is formalized by putting constraints on the execution of Transport Service primitives, i.e. a connection endpoint identifier is treated as if it were a parameter of the service primitive:
- if the primitive is a connect request or indication, it can only be executed if the identifier is not in use,
- if the primitive is not a connect request or indication, it can only be executed if the identifier is in use,
Immediate administration of identifiers in use is done by appropriate process parameterization. The same mechanism is applied at connection termination to make the identifier again available for later reuse.
The constraints described and the related administration are specified within the scope of a service access point address by way of a process that is defined for that purpose only. This approach is an aspect of separation of concerns that is applicable because of the possibility that any number of processes can participate in the execution of an event [TOCHER].

Example 4: Connection termination

In this example we only want to address the local view of connection termination which is invoked by the disconnect primitives. Since service primitives are local interactions between processes (between service user and service provider or between two protocol entities) and because the termination is unconditional, the execution of a disconnect primitive is a sufficient condition to terminate the connection.

The global view of connection termination shows different forms of terminations, including collision cases. We found the complexity of its specification really a testcase for the FDT bent for clarity and conciseness. However, its falls beyond the scope of this paper. The interested reader is further invited to examine and compare the Estelle and LOTOS specifications.

Estelle approach:

In Estelle a disconnect primitive produced by one state machine generally does not immediately reach the receiving state machine because of the queuing of primitives between these processes by the channel interconnecting them. This implies for example that data primitives preceding the disconnect primitive in the queue are processed by the receiving process (which may affect the execution of derived primitives at other service boundaries) before the disconnect primitive terminates this processing. This necessarily implies processing overhead for data that might otherwise have been discarded by the disconnect. Also the receiving machine might have previously issued a disconnect itself, in which case it is obliged to discard data in the input queue.

LOTOS approach:

Since in LOTOS the event concept allows to specify that both user and provider are simultaneously aware of the execution of a disconnect primitive by describing this primitive by an event, local connection termination is as simple as the occurrence of the disconnect primitive. The possibility of a local collision of a disconnect request or disconnect indication is excluded by the fact that events are atomic, excluding the simultaneous occurrence of these primitives, and the termination of the connection after occurrence of either primitive in the specification of both user and provider.

The example above illustrates a remarkable difference between the two FDT's under consideration. In LOTOS the occurrence of an interaction has an immediate effect on all processes involved, including the environment. In Estelle the input/output asymmetry caused by the queues is reflected in the different temporal scopes that an event finds in the specified process and in its environment, viz. a process' output affects immediately the process itself, whereas it affects the environment later on (if ever) when the event has reached the environment via the queue and when the environment is ready to accept the corresponding input.

Example 5: Backpressure

The term 'backpressure' is used in the context of flow control descriptions. Consider a directed flow of data from a sending entity to a receiving entity through some "medium", e.g. a service provider. The receiver may exert back- pressure flow control on the sender's ability to transmit data by refusing to remove data from the medium. Under the assumption of a finite capacity of the medium, the flow control exerted by the receiver is "transferred back" to the sender, where it is effectuated by the provider's inability to accept data from the sender. From the local point of view, to which we again will limit our attention, backpressure requires that a consumer of data can prevent a producer from producing data.

Estelle approach:

In Estelle the receiver and provider of data are described by finite state machines interconnected by an infinite queue. This implies that the receiver cannot prevent the provider from putting data in the queue. Similarly the provider cannot prevent the sender from putting data in the queue. Consequently a formal description of backpressure in Estelle is not possible [SC6/WG4/N123 page 20]. For the same reason it will not be possible to describe the transfer of backpressure exerted on the service access point of the receiver to the service access point of the sender when anywhere in the chain between these service access points there is an infinite queue.

LOTOS approach

In LOTOS an event will only occur if all involved processes are prepared to engage in the event by making the appropriate event offer. Consequently the execution of a data primitive can be prevented by any process by specifying in the appropriate process state that the event offer is not made. This means that the local view on backpressure is expressed in a natural way. The transfer of backpressure is dealt with by the language facilities for non-determinism.

It is an architectural question, however, which assumptions have to be made about the conditions under which backpressure should not be transferred. More precisely, given a non-zero capacity assumption about the provider it is a question whether such an assumption applies to each connection independently, or to the whole set of connections. In the latter case availability of resources for some connections may depend on removal of resources from other concurrent connections (thereby introducing a subtle distinction between "concurrency" and "absence of (explicitly specified) causality", which we submit to the attention of General Net theorists). We can conclude that the specification of backpressure in LOTOS, like many other such requirements, is possible, but the precise formulation depends on the specifier's understanding of the technical meaning of the requirement.

The last three examples show that elementary choices for the semantical model of an FDT may impose drastic constraints in the way basic architectural concepts of a service or protocol can be expressed. If we extrapolate this to the description of more complex architectural concepts of the service and protocol we may expect vast differences in the way their formal description will be structured. This effect is amplified by the introduction of different language features of the FDTs and the different specification styles of the applicators of the FDTs.

It can be concluded that formal descriptions based on different FDTs will have complete different appearances if their semantical models are not based on the same understanding of basic architectural concepts and if the specifiers do not have the same understanding of the architecture and choose the same structure for the description.

This effect will have a large impact on implementations since it will be extremely hard, if not impossible, for an implementer to interpret a specification only in terms of external observable behaviour, detaching completely from the structure of a description. Rather the latter will form the basis of the implementation. Therefore there is a high responsibility for a specifier, whatever (in)formal language is used, to design a minimal, but effective, structure for a specification [SCOLLO].

6. EVALUATION

The development of the FDTs Estelle and LOTOS within ISO started roughly in February 1981 with the installation of the respective WG1/FDT subgroups B and C. At this moment the status of these FDTs is second Draft Proposal for an ISO standard. Standards are anticipated for February 1988.

It can be observed that certainly outside ISO these developments have triggered a growing interest in the application of the FDTs, the development of tools that support the FDTs [ANSART, BRIAND, EIJK], and further research in the development of verification techniques [SEDOS1, SEDOS2], and extensions and refinements of the FDTs.

Within ISO official application of Estelle started in October 1984 with the installation of special editing groups for Transport and Session [SC21/N1707]. Later such editing groups were formed for LOTOS. These editing groups have already produced quite complete specifications [SC6/WG4/N53, N123, N116, N117, SC21/WG6/N187, N188]. Applications to other layers are coming along [SC6/N3938].

A consensus "to go formal" now that the FDTs are available and stable
does not seem to emerge (yet) despite of the pioneer work of the
special editing groups.
Rather the persistent reluctance of protocol experts to get involved
with FDTs, because their nature is too abstract and not well
understood, seems still prevailing (see e.g. [DIS8824] for which a
semantical definition is lacking).

The fact, however, that FDTs seem abstract and difficult to understand
is not because their designers want them to be so, but because the
application area, viz. OSI, is abstract and difficult to understand.
The previous sections have tried to exemplify this. It is the main
reason why the FDT developments have taken several years. The
application of the FDTs to OSI standards proves the real testcase.
Here it can be observed that the specifiers (who currently seem rather
FDT experts than technical experts) have a very hard, though
intriguing, job in precisely understanding the verbally described
standards and to develop complete, unambiguous, clear and concise
formal descriptions. The fact that many defect reports on the verbal
description have been produced by the editing groups shows how
necessary formal description is. Moreover it shows that the
development of these standards should be done together with their
formal descriptions. Given the stable status that the FDTs have now
achieved, this goal is now within reach.

There seems to emerge yet another sophism that formal description is
not (longer) necessary because of the activities on conformance
testing: "if one can test an implementation, than a correct
implementation can be obtained by testing it and improving it until
errors are no longer discovered; so why should we need a formal
description to produce a correct implementation"?
This reasoning can be simply withstood by arguing that only exhaustive
tests can provide this security whereas exhaustive testing of
functions, as complex as a protocol entity, is beyond our capacity.
Undetected errors, therefore, are always possible. Moreover the
quality of tests that are not based on a formal test theory, that on
its turn is based on a formal description method, seems quite
disputable [BRINKSMA2].

7. CONCLUSIONS

It has been shown that a small number of differences in elementary
choices for the semantical model of an FDT have far reaching
consequences for the way basic architectural concepts of OSI can be
formally described. This, on its turn, has a high impact on
specification style and leads consequently to vast differences in
appearance of formal descriptions. On its turn this will have a yet
unknown impact on implementation.
Where formal description is subordinate to its (OSI) application area
and not the other way around [SC16/N1408], and where the roots of
these differences lay in the apparent unclarity of the semantics of
(basic) OSI architectural concepts, it can be concluded that ISO, in
particular its SC21/WG1/FDT Subgroup A [SC6/WG4/N123 page 20]
[SC21/WG1/N79], should provide this clarity.
The difficulties in the widespread acceptance of FDT show that more
attention is to be given to the education of technical experts to
encourage them in a threefold direction: 1) to apply FDT during a
technical development rather then as an effort afterwards, 2) to
understand the (basic) architectural problems and to improve their
formalization, 3) to relate specification to testing, considering that

testing cannot replace (formal) specification, rather (good testing) is dependent on it [BRINKSMA2, MEER, FAVREAU].

FDT, its development and application, is still in an early state of the art. It will depend on the well-understood interests, and insights, of many whether it will stay there or make progress.

8. ACKNOWLEDGEMENT

We like to acknowledge the experts of ISO/TC97/SC21/WG1 who have through their patience and understanding provided the framework in which the ISO FDTs Estelle and LOTOS could develop towards standards. We like to acknowledge in particular the FDT experts of this group who have had a hard job in making this happen. We like to acknowledge our colleagues from the CCITT for their cooperation.

9. REFERENCES

[AHOOJA] R. Ahooja, B. Sarikaya: 'Comparing Normal Forms Obtained from Estelle and LOTOS Specifications', proc. 6th IFIP WG6.1 Workshop on Protocol Specification, Testing, and Verification, Montreal, June 1986, in print.

[ANSART]   J.P. Ansart et al: 'Software tools for Estelle', Proc. 6th IFIP WG6.1 Workshop on Protocol Specification, Testing, and Verification, Montreal, June 1986, in print.

[BRIAND] J. P. Briand, et al: 'Executing LOTOS Specifications', proc. 6th IFIP WG6.1 Workshop on Protocol Specification, Testing, and Verification, Montreal, June 1986, in print.

[BRINKSMA1] E. Brinksma: 'A Tutorial on LOTOS', Proc. 5th IFIP WG6.1 Workshop on Protocol Specification, Testing and Verification, Toulouse-Moissac, June 1985, North-Holland, Amsterdam, 1986, pp171-194 (also as Annex C of DP8807rev).

[BRINKSMA2] E. Brinksma, G. Scollo, C. Steenbergen: 'LOTOS specifications, their implementations, and their tests', Proc. 6th IFIP WG6.1 Workshop on Protocol Specification, Testing, and Verification, Montreal, June 1986, in print.

[CHOI] T. Y. Choi: 'A Sequence Method for Protocol Construction', proc. 6th IFIP WG6.1 Workshop on Protocol Specification, Testing, and Verification, Montreal, June 1986, in print.

[EHRIG] H. Ehrig, B. Mahr: 'Fundamentals of Algebraic Specification 1', Springer Verlag, Berlin 1985.

[EIJK] P. van Eijk: 'A Comparison of Behavioral Language Simulators', proc. 6th IFIP WG6.1 Workshop on Protocol Specification, Testing, and Verification, Montreal, June 1986, in print.

[FAVREAU]   J.P. Favreau, R.J. Linn: 'Automatic generation of test scenarios from protocol specifications written in Estelle', Proc. 6th IFIP WG6.1 Workshop on Protocol Specification, Testing, and Verification, Montreal, June 1986, in print.

[desJARDINS] desJardins R., Foley S.F.: 'Open Systems Interconnection, A Review and Status Report', Journal of Telecommunications Networks, 1985, 194-209.

[LINN]   R.J. Linn: 'The features and Facilities of Estelle, a Formal Description Technique based upon an Extended Finite State Machine Model', Proc. 5th IFIP WG6.1 Workshop on Protocol Specification, Testing and Verification, Toulouse-Moissac, June 1985, North-Holland, Amsterdam, 1986, pp271-296 (also as 'Provisional Estelle tutorial', ISO/TC97/SC21/N937, December 1985).

[MEER]   J. de Meer: 'Derivation and validation of test scenarios based on the formal specification language LOTOS', Proc. 6th IFIP WG6.1 Workshop on Protocol Specification, Testing, and Verification, Montreal, June 1986, in print.

[RAYNER]   D. Rayner: 'Towards Standardized OSI Conformance Tests', Proc. 5th IFIP WG6.1 Workshop on Protocol Specification, Testing and Verification, Toulouse-Moissac, June 1985, North-Holland, Amsterdam, 1986, pp441-460.

[SARACCO] Saracco, R: 'Response to Standardization of Formal Description Techniques for Communication Protocols', proc. IFIP 86, 10th World Congress, in print.

[SCOLLO] G. Scollo, C.A. Vissers, A. di Stefano, 'LOTOS in Practice', Proc. IFIP 86, 10th World Congress, in print.

[TOCHER] A. J. Tocher: 'OSI Transport service: A Constraint-Oriented Specification in LOTOS', ESPRIT/SEDOS/C1/WP/Z1/IK, July 1986.

[VISSERS]   C.A. Vissers, L. Logrippo: 'The importance of the service concept in the design of data communication protocols', Proc. 5th IFIP WG6.1 Workshop on Protocol Specification, Testing and Verification, Toulouse-Moissac, June 1985, North-Holland, Amsterdam, 1986, pp3-17.

[DIS8824] ISO/TC97/SC21/WG5: 'Specification of Abstract Syntax Notation One', ISO Draft International Standard DIS8824, June 1985.

[DP8807rev] ISO/TC97/SC21/WG1/FDT/C: 'LOTOS, A Formal Description Technique based on the Temporal Ordering of Observational Behaviour', ISO Second Draft Proposal DP8807, July 1986.

[DP9074rev] ISO/TC97/SC21/WG1/FDT/B: 'Estelle, A Formal Description Technique based on an Extended State Transition Model', ISO Second Draft Proposal DP9074, August 1986.

[IS7498] ISO/TC97/SC16/WG1: 'Reference Model for Open Systems Interconnection', ISO Standard IS 7498, 1984 (also CCITT recommendation X.200).

[SC6/N3938] ISO/TC97/SC6/WG2: 'Protocol for providing the Connection-less Mode Network Service, Addendum 2: Formal Description of ISO 8473, ISO/TC97/SC6/N3938, January 1986.

[SC6/WG4/N53] ISO/TC97/SC6/WG4/Ad Hoc Group: 'Formal Description of the Transport Service in Estelle', ISO/TC97/SC6/WG4/N53, November 1985.

[SC6/WG4/N116] ISO/TC97/SC6/WG4/Ad Hoc Group: 'Formal Specification of IS 8072 (Transport Service) in LOTOS', ISO/TC97/SC6/WG4/N116 February 1986.

[SC6/WG4/N117] ISO/TC97/SC6/WG4/Ad Hoc Group: 'Formal Specification of IS 8073 (Transport Protocol) in LOTOS', ISO/TC97/SC6/WG4/N117, February 1985.

[SC6/WG4/N123] ISO/TC97/SC6/WG4/Ad Hoc Group: 'Formal Description of ISO 8073 (the Transport Protocol Specification) in Estelle', ISO/TC97/SC6/WG4/N123, March 1986.

[SC16/N1408] ISO/TC97/SC16/WG1/FDT: 'FDT Evaluation Criteria', ISO/TC97/SC16/N1408, March 1983.

[SC16/N1707] ISO/TC97/SC16: 'Formal Specification of Session and Transport Layers', Ottawa, October 1983.

[SC21/N909] ISO/TC97/SC21/WG1/Ad Hoc Group on Conformance Testing: 'Working Draft for OSI Conformance Testing Methodology and Framework', ISO/TC97/SC21/N909, November 1985 (chapter 13 is on TTCN).

[SC21/WG1/N79] ISO/TC97/SC21/WG1/FDT/A: 'A More Precise Definition of Basic OSI Concepts', ISO/TC97/SC21/WG1/N79, August 1985.

[SC6/WG6/N187] ISO/TC97/SC21/WG6/Ad Hoc Group: 'Draft Formal Specification of the OSI Connection-Oriented Session Service in LOTOS', ISO/TC97/SC21/WG6/N187, July 1986.

[SC21/WG6/N188] ISO/TC97/SC21/WG6/Ad Hoc Group: 'Draft Formal Specification of the OSI Connection-Oriented Session Protocol in LOTOS', ISO/TC97/SC21/WG6/N188, July 1986.

[SDL]   CCITT/SGXI/WP3-1: 'SDL, Specification and Description Language', CCITT Recommendations Z.100-Z.104, 1984.

[SEDOS1] ESPRIT/SEDOS(410)/B2 Group: 'Estelle Verification/Validation', ESPRIT/SEDOS/N31-33, November 1985.

[SEDOS2] ESPRIT/SEDOS(410)/C2 Group: Verification Models for LOTOS', ESPRIT/SEDOS/N44, November 1985.

[Z.104rev] CCITT/SGX/3-1: "Draft revision of Recommendation Z.104", Stockholm, June 1986.

THE DIMENSION OF TIME IN PROTOCOL SPECIFICATION

Harry Rudin
IBM Zurich Research Laboratory
8803 Rüschlikon, Switzerland

The importance -- and even the necessity -- of formal description techniques for computer-communication protocols is now widely accepted. A dimension usually neglected in formal protocol specification is that of **time**. When the dimension of time is included, protocols can be validated, including their timeout specifications, and performance (e.g., throughput and/or response time) can be estimated, direct from the formal specification. Here, the state-of-the art is summarized and a recent result wherein the **distribution** of response time is determined direct from the formal specification is presented. An extensive bibliography is included.

I. INTRODUCTION

The CCITT effort to define SDL (CCITT84s) and the ISO efforts to define Estelle (Vissers83) and LOTOS (Brinksma84) bear witness to the energy being invested in defining techniques suited to the formal description of computer-communication protocols. The work on formal descriptions is driven by the need to provide unambiguous definitions of protocols which are to be implemented by different people in many different organizations, e.g., OSI (Open System Interconnection) protocols, (Folts83) or by different people in the same organization, e.g., IBM's SNA (System Network Architecture) (Pozefsky82).

Formal protocol description techniques not only promise the necessary precision required for compatible implementations but even more, if the formal description language used is machine readable. For the latter case, partly automated techniques can be used for

- validation and verification to remove some logical errors,
- providing early estimation of protocol performance,
- code generation for part of the implementation, and
- testing the implementation for architectural conformance.

A recent summary of the state-of-the-art in the development of these techniques is given in (Rudin85a).

While time must be acknowledged as playing an important role in protocols, it is currently only rarely included in the formal protocol specification. Frequently it is the action precipitated by a timeout which guarantees recovery from an anticipated or even from an unanticipated error condition. In such a case validation or verification of a protocol is incomplete without inclusion of time. One cannot begin to discuss many aspects of protocol performance without a notion of time.

For the analyses of many protocols, we are already at the computationally practical limit so that adding another dimension, time, makes a difficult problem even more difficult. This is the reason why research including time has appeared relatively late. But interest in the ability to handle time is growing. A recent workshop was devoted entirely to the formal modeling of time (Marsan85), although few of the papers presented there dealt specifically with protocols.

In what follows, we shall first look at one method of including time in the formal description. We shall then proceed to briefly discuss recent work dedicated to removing errors from protocol specifications: the province of validation and verification. After a review of related work in performance prediction, a new result is presented showing how the complete distribution of delay can be calculated, direct from a formal specification. We conclude with some speculation and planned further research on this last subject and with an extensive bibliography.

II. INCLUDING TIME IN THE FORMAL DESCRIPTION

It is easiest to start with an example. Figure 1 is a FSM- (Finite State Machine) based model of a communications channel or medium wherein messages experience occasional bursts of noise which cause errors and even message loss. The channel receives messages from a sender process which is not shown and forwards them on to a receiver process, not shown either. Such a model is then just one process in a several-process model of, say, a data-link-control protocol.

Shown in the model for the channel is a single state, REPEAT, with three arcs, all returning to this state. The labeling of the leftmost arc in Figure 1 symbolizes that the arrival of an event MESSAGE (from the invisible transmitter) may result in an event ERROR to be sent on (to the invisible receiver.) This corruption of MESSAGE takes place with a probability P_e and takes T units of time. The MESSAGE may also be lost, represented by the generation of a GAP with probability P_l in T units of time or MESSAGE may be faithfully repeated in the same amount of time with probability $1 - P_e - P_l$.

There are several possible representations for time; in the above, elapsed time is associated with the traversing of an arc. Elapsed time has also been associated with the lifetime of states in FSM-based descriptions. When Petri nets are used, time is associated with the lifetime of places or with the firing time of a transition. Algebraic representations are still another possibility. A number of these approaches is discussed in (Rudin85b).

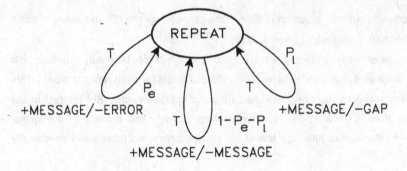

Figure 1. FSM model of a lossy, error-prone channel

III. VALIDATION AND VERIFICATION

Assuming that a protocol has a formal specification which includes time, the first order of business is clearly making as sure as possible that the specification is free from errors. Often detecting errors in protocols is divided into two parts: validation, which is a check for syntactic errors, and verification, which is a check for semantic errors, i.e., errors of function. Frequently one or the other term is used to describe both tasks.

The author has recently published a survey of techniques for examining protocol specifications which include time for errors (Rudin85b). In this paper, some recent developments will be discussed and several omissions from the earlier survey included. The discussion is organized according to the type of protocol specification used.

Petri-net-based error analysis. The earliest work on detecting errors in protocols including time specifications is based on Petri nets and appears in the papers (Merlin76) and (Symons78). Walter (Walter83) uses an approach similar to that of (Merlin76). The first automated approach is that of Menasche and Berthomieu (Menasche83) and is based on a reachability analysis.

In recent work (Sajkowski86a) analytic expressions are found which must be satisfied for a protocol to be free of various syntactic errors such as unspecified receptions, deadlocks and livelocks. Using a "projection" technique, similar to that suggested in (Lam82), Sajkowski proceeds to reduce the size of the state space required for analysis of time-dependent protocols (Sajkowski86b).

FSM-based error analysis. While much of the early work on analyzing protocols is based on Petri-net descriptions, the most widely used specification techniques are based on finite-state machines. Here, adding predicates to describe enabling conditions and actions associated with transitions, Shankar and Lam (Shankar82) use analytic, manual techniques to prove two kinds of properties for data-link-control protocols. Safety properties

properties which are always true, and liveness properties, which eventually become true, are proven. An example of the former is the received message sequence being a prefix of the transmitted message sequence — with a maximum difference of one message; an example of the latter is the length of the two sequences growing without bound in time. Shankar and Lam's method makes use of the notion of "projections" (Lam82) which reduce the size of the state space to be considered.

Bolognesi breaks the problem into two steps (Bolognesi84). First a validation is performed without taking time constraints into account; this yields message sequences which are "communications consistent." In the second step, time specifications are taken into account by converting this "time-consistency" problem into a network-flow problem for which there are known solution algorithms. This approach has been automated.

Algebraic-based error analysis. This is the class of formal description techniques most recently investigated. Using such an approach Aggarwal and Kurshan (Aggarwal83) were able to analyze a simple data-link-control protocol to check for a consistent specification of a timeout value. Using a distribution for the delay in a channel with a data-link-control protocol, Nounou and Yemini (Nounou84) were able to find an expression for the value of a timeout so that the probability of unnecessary retransmission be below some specified value.

All of these references represent first, but promising, attempts at solving aspects of the problem. Much remains to be done.

IV. PREDICTING PERFORMANCE

The traditional approach to predicting protocol performance consists of first creating a heuristic model of the protocol and then using various techniques from queueing or traffic theory to analyze said model. A different approach is favored by the present author, namely skipping the first step above and predicting performance direct from the formal specification of the protocol — and using automated techniques as far as possible (Rudin83b) and (Rudin84). Such an approach was not possible a decade ago simply because of the lack of formal protocol specifications. Currently, research in this area is proceeding at a good pace.

IV.1 Related work on predicting performance direct from formal descriptions

As above, the relevant literature is ordered in terms of the formal protocol description technique upon which it is based.

Petri-net-based performance analysis. Again, much early work was done based on Petri-net models. The earliest paper is that of Ramchandani in 1973 wherein a computation rate is calculated for a program loop (Ramchandani73). Han calculated throughput in 1978 (Han78). A faster algorithm was found later for Ramchandani's problem by Ramamoorthy and Ho (Ramamoorthy80). Part of Sajkowski's work yields equations for cycle-execution times (Sajkowski86a).

For automating the calculation, a 1980 paper by Zuberek describes a practical instantaneous description of the system of protocols — effectively, a global state including time — (Zuberek80). A convenient description of the state of the system is a prerequisite to techniques using exhaustive enumeration. Such an approach has been automated by Razouk and applied to performance calculations, including the effect of a timeout, for a simple data-link-control protocol (Razouk84).

If the assumption is made that delays follow a negative exponential distribution, than a Markov-chain analysis can be made, as did Molloy in (Molloy82) and Zuberek in a later paper (Zuberek85).

FSM-based performance analysis. In 1970, Beizer published an approach to calculating program running time by means of graphical reduction (Beizer1970). A similar approach was found for communication protocols by Gouda (Gouda78) sometime later.

In 1984, Kritzinger started work applying results from queueing theory — specifically closed, multiclass, multichain queueing networks — to the formal protocol specification expressed in finite-state-machine terms (Kritzinger84a). An example of the approach applied to an ARQ (automatic repeat request) data-link protocol may be found in (Engelbrecht85). The method can be applied to assessing the performance of multiple instances of multilayer protocols, for example protocols following Open Systems Interconnection architecture, taking into account competition and interference of the various protocol instances for a common processor resource (Kritzinger86).

When the assumption of negative exponential distribution for delays is made, a Markov-chain analysis can be made from a FSM-based definition, just as above in the case of Petri-net-based definitions. An application of this approach to analyzing the performance of mutual exclusion distributed protocols — as used for obtaining access to distributed data bases — appears in (Gravey86).

Algebraic-based performance analysis. The work by Nounou and Yemini also contains applications to estimating performance; in 1984, the response time for a simple data-link-control protocol was calculated (Nounou84). These ideas are currently being extended in a system for protocol analysis called ANALYST (Barghouti86).

After this quick review of related work, we now look at one particular algorithm for performance prediction and its recent extension.

IV.2 An automated method based on FSM descriptions

Taking into account that most formal protocol descriptions are based on FSM's (CCITT's SDL, ISO's Estelle, and IBM's FAPL) and wanting to use computer-automated techniques to the hilt, the author has been following the approach given below and summarized in Figure 2. Starting from the FSM-based formal specification, a global-state tree is generated. This same global-state tree is used for validation purposes in the "perturbation" technique (West78b), and so can be obtained as a by-product of the validation process.

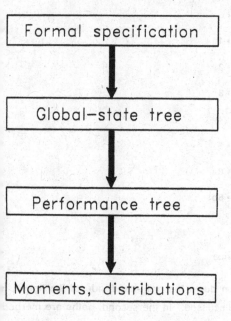

Figure 2. Outline of one method for determining performance

A global state describes the state of the overall system in terms of the state of all the component processes and lists of the messages underway between the various processes. (Often a message generated by one process has not yet been received by the destination process.) When time is added to the formal description of the protocols, it is necessary to add a notion of time, say elapsed time or age of a state, to the global-state description derived.

Typically, the aspect of performance sought requires calculation of the delay or elapsed time for some cycle of operation. The starting point of this cycle, expressed in terms of a global state, defines the root of a performance tree whose branches terminate in leaves again corresponding to the starting (the same as ending) point of the cycle. The delays and probabilities calculated for the paths from root to leaves are used to calculate the performance statistics required as shown in the last step in Figure 2.

A fragment of such a performance tree is shown in Figure 3. The initial state, number 1, is shown in the upper, left-hand corner and has a single following state, number 2. This in turn has several following states, numbers 3, 4, 5, and so on. The shortest cycle terminates after four steps and on the left-hand side of the figure, again in state number 1. The time elapsed per arc is a uniform, single unit of time in this example so that the shortest cycle ends in four time units; the remaining cycles take longer.

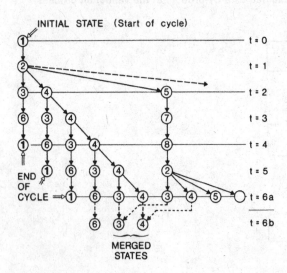

INITIAL STATE (Start of cycle)

MERGED
STATES

Figure 3. Algorithm for determining performance

Note that the sixth increment in time, shown down the right-hand side of the figure, is split into two subincrements. The first is as discussed. In the second, paths are merged, taking advantage of a Markov-like property. The algorithm keeps track of the probability of occurrence of each of the paths; when paths are merged, the new path takes a probability equal to the sum of the probabilities of the old paths. The result is narrower performance tree and sharply reduced execution times (Rudin84).

The calculation is terminated when the sum of the probability over all uncovered paths drops below a specified threshold.

Usually one is interested in some figure of merit of performance, say the mean delay around a cycle. In (Rudin84) a simple data-link-control protocol was analyzed operating over the kind of channel shown above in Figure 1. This protocol will retransmit a message if the message or its acknowledgment is corrupted or lost in the channel. Figure 4 shows the average delay — around a cycle in which two messages are successfully transmitted and acknowledged — for various values of the probability for occurrence of message error or loss.

The mean delay was calculated using the expression

$$\text{Average delay} = \sum_{i=1}^{N} \{(P_i)(D_{path\ i})\}$$

where $D_{path\ i}$ is the delay along the ith path which occurs with probability of P_i.

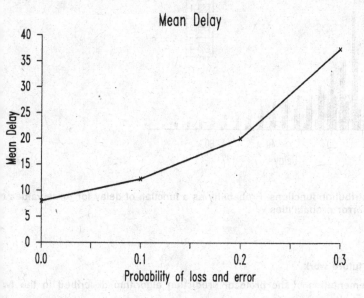

Figure 4.   Mean delay as a function of loss and error probabilities

IV.3 Determining the distribution function

An extension to this approach recently discovered is that it is possible to calculate the entire distribution function, simply by keeping track of the probabilities of various possible path delays.   For the three non-zero values of loss and error probabilities shown in Figure 4, the distributions are shown below in Figure 5. From such a distribution of delays it is possible to calculate many performance figures such as the variance of delay or the average delay in some specified, worst-case percentile.

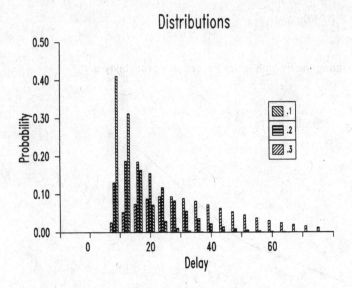

Figure 5. Delay distribution functions. Probability as a function of delay for three values of loss and error probabilities

IV.4 Speculation on future work

The current implementation of the protocol prediction algorithm described in the two preceding sections takes advantage of the assumptions of first, synchronous operation and second, uniform execution times for all transitions. The plan is to design a new package which would allow for asynchronous operation and for transitions with arbitrary execution times. In addition to complicating the bookkeeping necessary to determine delay distributions, these extensions raise a number of issues.

Equivalent global states: granularity. As soon as the assumptions above are dropped one practical question immediately arising is when are global states equivalent along the dimension of time? Being identically equivalent is too much to hope for in the general case and this raises the question of approximate equivalence. Global states which are almost but not exactly equivalent would lead to excessively wide performance trees and computation times. The plan is to introduce a notion of equivalence with a specified time granularity. States equivalent in terms of individual process states and messages queued in the communication medium and with ages equal within this granularity (or quantization in time will be considered equivalent. Equivalent states can be merged and the performance tree made correspondingly more narrow.

Transitions with time-delay distributions. The mean execution delay is a natural choice when a single-parameter characterization must be used. For many cases this will most likely be inadequate. Let us suppose that the distribution of delay which should be associated with a transition is as shown by the curve in Figure 6. The plan is to represent this distribution by a small number of discrete values: T_1 with probability P_1 , T_2 with probability P_2 , and T_3 with probability P_3 in Figure 6, as an example. Naturally the requirement

$$\sum_{i=1}^{N} P_i = 1$$

must be satisfied.

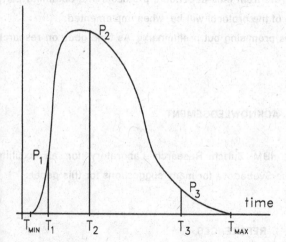

Figure 6. Distribution of delays associated with a single transition

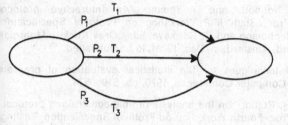

Figure 7. Multiple-arc representation of a single transition

The corresponding finite-state-machine-based representation would then be that shown below in Figure 7. Experiments will have to be carried out to determine to what extent approximations of this sort are useful.

The minimum and maximum time values, T_{MIN} and T_{MAX}, in Figure 6 would be used in a validation procedure such as that discussed in (Bolognesi84).

V. CONCLUSION

As has been seen here, activity in the area of analyzing protocols with time included is on the increase. On the one hand, adding the dimension of time increases the computational complexity substantially. On the other, it is desirable to include time, both for purposes of removing logical errors from time-dependent protocols and obtaining early indications of what the performance of the protocol will be, when implemented.

The research carried out so far is promising but preliminary. As time goes on research in this area will intensify.

VI. ACKNOWLEDGEMENT

I thank my colleagues at the IBM Zurich Research Laboratory for an exciting environment; particularly I thank Liba Svobodova for many suggestions for this paper.

REFERENCES

Aggarwal83- S. Aggarwal and R. P. Kurshan, "Modelling elapsed time in protocol specification," Proc. Workshop on Protocol Specification, Testing, and Verification, III Rüschlikon, Switzerland, May 1983, (North-Holland, Amsterdam, 1983) pp. 51 - 62.

Barghouti86- N. Barghouti, N. Nounou, and Y. Yemini, "An interactive protocol development environment," Proc. Sixth IFIP Workshop on Protocol Specification Testing, and Verification, G. Bochmann and B. Sarikaya, Eds., Gray Rocks - Montreal June 10-13, 1986, (North-Holland, Amsterdam, Dec. 1986), to be published.

Beizer70- B. Beizer, "Analytical techniques for the statistical evaluation of program running time," Proc. Fall Joint Computer Conference, 1970, pp. 519 - 524.

Bolognesi84- T. Bolognesi and H. Rudin, "On the analysis of time-constrained protocols by network flow algorithms," Proc. Fourth Workshop on Protocol Specification, Testing and Verification, Skytop, Pennsylvania, June 1984, (North-Holland, Amsterdam, 1985) pp. 491 - 513.

Brinksma84- E. Brinksma, "A specification of the OSI transport service in LOTOS," Proc. Workshop on Protocol Specification, Testing, and Verification, IV, Sky Top, Pennsylvania, June 1984, (North-Holland, Amsterdam, 1985), pp. 227 - 251.

CCITT84s- C.C.I.T.T., Red Book, Volume VI - Fascile VI.10, Functional Specification and Description Language (SDL), Recommendation Z.101- Z.104, VIIIth Plenary Assembly, Torremolinos, Oct. 8 - 19, 1984.

Engelbrecht85- J. Engelbrecht, P. Kritzinger, and H. Rudin, "Predicting protocol performance from a meta-implementation," Proc. Fifth Workshop on Protocol Specification, Testing, and Verification, M. Diaz, Ed., Moissac-Toulouse, June 10-13, 1985, (North-Holland, Amsterdam, 1985), pp. 349 - 362.

Folts83- H. C. Folts and R. desJardins, Eds., Special issue of the Proceedings of the IEEE on OSI, Vol. 71, No. 12, Dec. 1983, pp. 1331 - 1448.

Gouda78- M. G. Gouda, "Protocol machines: towards a logical theory of communication protocols," PhD. thesis, University of Waterloo, Jan. 1978 (see especially Chapter 11).

Gravey86- A. Gravey and A. Dupuis, "Performance evaluation of two mutual exclusion distributed protocols via Markovian modeling," Proc. Sixth IFIP Workshop on Protocol Specification, Testing, and Verification, G. Bochmann and B. Sarikaya, Eds., Gray Rocks - Montreal, June 10-13, 1986, (North-Holland, Amsterdam, Dec. 1986), to be published.

Han78- Y. W. Han, "Performance evaluation of a digital system using a Petri-net-like approach," Proc. National Electronics Conference, Chicago, 1978, pp. 166 - 172.

Kritzinger84a- P. Kritzinger, "Analyzing the time efficiency of a communication protocol," Proc. Fourth Workshop on Protocol Specification, Testing, and Verification, Skytop, Pennsylvania, June 1984, (North-Holland, Amsterdam, 1985), pp. 527 - 539.

Kritzinger86- P. Kritzinger, "A performance model of the OSI communication architecture," IEEE Trans. on Commun., Vol. COM-34, No. 6, June 1986, pp. 554 - 563.

Lam82- S. S. Lam and A. U. Shankar, "An illustration of protocol projections," C. A. Sunshine, Ed., Proc. Second International Workshop on Protocol Specification, Testing, and Verification, Idylwild, May 17 - 20, 1982, (North-Holland, Amsterdam, 1982), pp. 343 - 360.

Marsan85- M. A. Marsan, G. Balbo, and K. Trivedi, Eds., Proc. International Workshop on Timed Petri Nets, Torino, Italy, (IEEE Comp. Soc. Press), July 1 - 3, 1985.

Menasche83- M. Menasche and B. Berthomieu, "Time Petri nets for analyzing and verifying time dependent communication protocols", Proc. Workshop on Protocol Specification, Testing, and Verification, III, Rüschlikon, Switzerland, May 1983, (North-Holland, Amsterdam, 1983) pp. 161 - 172.

Merlin76- P. Merlin and D. J. Farber, "Recoverability of communication protocols: implications of a theoretical study," IEEE Trans. Commun., Vol. COM-24, Sept. 1976, pp. 1036-1043.

Molloy82- M. K. Molloy, "Performance analysis using stochastic Petri nets," IEEE Trans. Computers, Vol. C-31, No. 9, Sept. 1982, pp. 913 - 917.

Nounou84- N. Nounou and Y. Yemini, "Algebraic specification-based performance analysis of communication protocols", Proc. Workshop on Protocol Specification, Testing, and Verification, IV, Skytop, Pennsylvania, June 1984, (North-Holland, Amsterdam, 1985), pp. 541 - 560.

Pozefsky82- Pozefsky, D. P. and F. D. Smith, "A meta-implementation for Systems Network Architecture," IEEE Trans. Commun., Vol. COM-30, No. 6, June 1982, pp. 1348-1355.

Ramamoorthy80- C. V. Ramamoorthy and G. S. Ho, "Performance evaluation of asynchronous concurrency systems using Petri nets", IEEE Trans. Software Eng., Vol. SE-6, Sept. 1980, pp. 440 - 449.

Ramchandani73- C. Ramchandani, "Analysis of asynchronous concurrent systems by timed Petri nets", Ph.D. Thesis, M. I. T., Dept. of E. E., AD-775618, July 1973.

Razouk84- R. R. Razouk and C. V. Phelps, "Performance analysis using timed Petri nets," Proc. Fourth Workshop on Protocol Specification, Testing, and Verification, Skytop, Pennsylvania, June 1984, (North-Holland, Amsterdam, 1985), pp. 561 - 576.

Rudin83b- H. Rudin, "From formal protocol specification towards automated performance prediction," Proc. Workshop on Protocol Specification, Testing, and Verification, III, Rüschlikon, Switzerland, May 1983, (North-Holland, Amsterdam, 1983) pp. 257 - 269.

Rudin84- H. Rudin, "An improved algorithm for estimating protocol performance," Proc. Fourth Workshop on Protocol Specification, Testing, and Verification, Skytop, Pennsylvania, June 1984, (North-Holland, Amsterdam, 1985), pp. 515 - 525.

Rudin85a- H. Rudin, "An informal overview of formal protocol specification", IEEE Communications Magazine, Vol. 23, No. 3, March 1985, pp. 46 - 52.

Rudin85b- H. Rudin, "Time in formal protocol specifications", Proc. GI/NTG Conference on Communication in Distributed Systems, Karlsruhe, March 11 - 15, 1985, pp. 575 - 587.

Sajkowski86a- M. Sajkowski, "On verifying time-dependent protocols", Proc. Sixth Intl. Conference on Software Engineering for Telecommunication Switching Systems, Eindhoven, April 14 - 18, 1986, pp. 46 - 51.

Sajkowski86b- M. Sajkowski, "Protocol verification in the presence of time", Proc. Sixth IFIP Workshop on Protocol Specification, Testing, and Verification, G. Bochmann and B. Sarikaya, Eds., Gray Rocks - Montreal, June 10 - 13, 1986, (North-Holland, Amsterdam, Dec. 1986), to be published.

Shankar82- A. U. Shankar and S. S. Lam, "On time-dependent communication protocols and their projections," C. A. Sunshine, Ed., Proc. Second International Workshop on Protocol Specification, Testing, and Verification, Idylwild, May 17 - 20, 1982, (North-Holland, Amsterdam, 1982), pp. 215 - 235.

Symons78- F. J. W. Symons, "Modeling and analysis of communication protocols using numerical Petri nets", Ph.D. Thesis at University of Essex, England, May, 1978.

Vissers83- C. A. Vissers, R. L. Tenney, and G. V. Bochmann, "Formal description techniques," Special issue on OSI, Proc. IEEE, Vol. 71, No. 12, Dec. 1983, pp. 1356 - 1364.

Walter83- B. Walter, "Timed Petri-nets for modelling and analyzing protocols with real-time characteristics," Proc. Workshop on Protocol Specification, Testing, and Verification, III, Rüschlikon, Switzerland, May 1983, (North-Holland, Amsterdam, 1983) pp. 149 - 159.

West78b- C.H. West, "General technique for communications protocol validation," IBM J. Res. Develop., Vol. 22, July 1978, pp. 393 - 404.

Zuberek80- W. M. Zuberek, "Timed Petri nets and preliminary performance evaluation", Proc. 7th Annual IEEE Symposium on Computer Architecture, 1980, pp. 88 - 96.

Zuberek85- W. M. Zuberek, "Performance evaluation using extended Petri nets," in M. A. Marsan, G. Balbo, and K. Trivedi, Eds., Proc. Intl. Workshop on Timed Petri Nets, Torino, Italy, (IEEE Comp. Soc. Press), July 1 - 3, 1985, pp. 272 - 278.

From Protocol Specification to Implementation and Test

H. J. Burkhardt, H. Eckert
Gesellschaft für Mathematik und Datenverarbeitung mbH.
Institut für Systemtechnik (F2)
Darmstadt, Germany

Protocol Specification, Implementation and Test denote the three major steps to realize cooperating systems.

In this contribution, a systematic approach is outlined for modelling communication services and protocols, for specifying them formally and for deriving implementation and test specifications from protocol specifications.

This approach, called PROSIT (PROtocol Specification, Implementation and Test), has been developed within GMD during the past years, closely related with the work on Open Systems Interconnection.

1. Introduction

The objective of Open Systems Interconnection (OSI) is to enable heterogenous systems - i. e. systems differing in origin and technology - to cooperate.

More specific, OSI aims at the establishment of an interworking technology which provides the means for the transfer of data between heterogenous systems and the elementary building blocks for the construction of a wide variety of distributed applications.

Following the OSI-approach, heterogenous systems get the capability to cooperate by observing a behaviour as defined in international communication service and protocol standards. Systems having this capability are called real open systems.

To create real open systems, three major steps must be done:
- communication services and protocols have to be specified by international standardization bodies,

- to accomodate their systems with interesting cooperation
  capabilities manufacturers have to implement functional
  groupings of standardized communication protocols,
- Protocol implementations have to be tested by authorized test
  centers to assure that a real system provides for the claimed
  openess and that it can therefore cooperate with any other
  real system having passed the same tests.

Each of these steps as well as the transitions among them involve
problems.

To standardize communication services and protocols means to model
distributed systems and to describe them formally and in a strictly
implementation independent manner. The problem of standardization
is that no approved methodology exists for that purpose. Therefore,
resulting standards are not necessarily complete, error free,
consistent and unambigously understandable.

Because of these deficiencies, current communication standards form
an unsatisfactory basis for implementations. But even perfect
standards would face implementors with the problem how to interpret
the standards with respect to the reality constituted by their real
systems. In other words, the problem of implementation is that no
approved method exists to derive implementations from implementation
independent protocol specifications, but that each difference in
interpretation of a protocol standard may and usually will lead to
incompatible implementations.

Because of this problems related with standardization and implementation of communication standards and because of the absence of a
single authority which can take over the responsibility for the
functioning of the cooperation among heterogenous systems, tests
are regarded as necessary. But tests must not impose restrictions
on protocol implementations, neither with respect to the communication behaviour allowed by the related protocol specification,
nor with respect to the choices of implementations. Therefore,
tests should be derived formally from protocol specifications and
based purely on the outside visible implementation structure
following from the claimed openess.The problem of testing is that
no approved method exists which fulfills these requirements.

This "state of the art" and the belief that the OSI-objective is
worth being supported has motivated work within GMD on a systematic
approach, called PROSIT, for realizing open systems. Its aim is to
provide for all the steps (see Fig. 1) - leading from service
specifications to compatible systems - improved methods and tools.
This papers outlines the basic ideas of PROSIT for modelling
communication services and protocols and for specifying them
formally, for constructing protocols from services and verifying
them against the bordering services and for deriving implementation
specifications and test sequences from protocol specifications.

2. Refinement of the OSI-Reference Model towards models for communication services and protocols

The backbone of the PROSIT approach is formed by a refinement of the
OSI-Reference Model (OSI-RM) towards models for communication services
and protocols. In these models, the essential ingredients of the
Reference Model are put into the appropriate perspective, thus
providing an overall structure for problem oriented specifications
of OSI-communication services and protocols. Both, services and
protocols are specified with the same formal description tool, a
dedicated form of Petri-Nets (which we call PRODUCT NETS), having
individuals as tokens.
A specification of a communication service defines all interactions'
between the users of this service and its provider as sequences of
service primitives at discrete interaction points. Our specification
conforms to this procedure but is not restricted to sequences
related to a single connection. It regards all connections, thus
describing the interdependencies which are caused by the usage of
common resources among logically independent connections. Semantic
aspects of a service are expressed by modelling the change of
"consciousness" of interacting entities.

The formal specification of a communication service is a prerequisite for the development of a supporting protocol specification.
Therefore, the modelling of a service which forms the basis of its
specification, is adjusted to support the construction of protocol
specifications from service specifications. As a consequence, one
logical approach for the development of a protocol model is

to superimpose two service models at adjacent layer boundaries and
to interrelate provider-behaviour of service N with user-behaviour
of service (N-1).

The approach of specifying all interactions as concurrent, which can
concurrently occur provides the implementation independence of
service and protocol specifications which is desired for standardization purposes.
Each implementation of such a specification, therefore, appears as
a valid restriction of concurrency, i. e. as sequentialization
reasoned by the mapping onto real storages and processors.
At present, only logical relationships between interactions are
defined in our specifications, but no relationships with regard to
physical time.

2.1 Derivation and refinement of the OSI-notion "entity"

As a basis, let us consider the scenario depicted in fig. 2.1 wherein
a service provider appears as counterpart of a set of service user
instances.
A general task of the service provider is the establishment of
communication relationships among service user instances, e. g.
with respect to "connectionless" interaction or connections maintained over an extended period of time.

A communication relationship among service user instances is
always based on interactions between (1) a service user instance
acting as originator and the service provider and (2) the service
provider and a service user instance acting as recipient. All interactions between a specific service user instance and the service
provider take place at their specific interaction point by means of
service primitives.

The service provider, seen from the user's point of view as a black
box, is itself a distributed system. This system is composed of
multiple independent service provider instances. To each service
user instance, a service provider instance is dedicated, in a
static and strict one-to-one relationship. Service provider instances
cooperate following a peer-to-peer protocol in order to provide the
service requested.

Adopting a term of the OSI-Reference Model, we call an interaction point by which a service user instance can communicate with a service provider instance a "Distinct-Service-Access-Point"(SAP).

To model connection-oriented communication, it is neccessary to distinguish static preassignments, which are prerequisites to the dynamic establishment and release of connections from the connections themselves. This leads us to refine both the service user and provider instance assigned to a distinct SAP into a static instance and dynamic multiple instances respectively (see fig. 2.2).

The dynamic instances associated with one SAP are created and destroyed by a static instance which exists permanently for this purpose. Dynamic instances are created during the connection establishment phase and destroyed during the connection release phase.

Over each distinct service-access-point, exactly one service-user-static-instance communicates with one service-provider-static-instance. Distinct connection-endpoints at a service-access-point represent the resources which the static instance of the service provider allocates to its respective dynamic instances on both sides of the interface boundary during each connection establishment. All instances attached to a distinct-service-access- point reside in the same OSI-system.

In this way, it may be said, that a service-access-point associates at its global address c (locally in each system) two static instances having the same name: one static service provider instance c and its corresponding static service user instance c. Similarly, an allocated distinct-connection-endpoint (c, k) in a system associates locally two dynamic instances of the same name (c, k), i. e. a dynamic service provider instance (c, k) with its corresponding dynamic service user instance (c, k).

Consequently, there exist in a system the following interaction points between service-provider-instances and service user instances:

- for each distinct service-access-point, an interaction point
  between static instances, which is represented architecturally,

by the term service-access-point itself, and across which are
exchanged the service primites related to the establishment and
release phase of a connection
and
- for each connection, an interaction point between a dynamic
  service user instance and a dynamic service provider instance,
  which is represented architecturally, by the term connection-
  endpoint, and across which are exchanged the service primitives
  related to the data transfer phase of a connection.

To model connectionless data transmission, there is of course no
need to use dynamic instances since no connections are involved.
Consequently, connectionless data transmission has to be interpreted
as interaction among static instances. This appears to be quite
natural in recognition that a connectionless data transmission
request at one distinct service access point, possibly followed by
a connectionless data transmission indication at another distinct
service access point is of the same nature as a connect request at
one distinct service access point, possibly followed by a connect
indication at another distinct service access point forming the
first half of a connection establishment.

Keeping in mind OSI-objective of systems compatibility, we define
a relation on the set of service user and provider instances which
leads to the definition of OSI service user entities and OSI
service provider entities as classes of service user and provider
instances. These classes consist of instances which show identical
communicational behaviour.

In the course of forming classes we compose (see Fig. 2.3)

- a static service user entity consisting of static service
  user instances,
- a static service provider entity consisting of static service-
  provider instances,
- a dynamic service user entity consisting of dynamic service
  user instances
and
- a dynamic service provider entity consisting of dynamic service
  provider instances.

In consequence, the OSI-concept "service-access-point" comprises the collection of distinct-service-access-points and the OSI-concept "connection- endpoint" comprises the collection of distinct-connection-endpoints. In other words, both notions define relationships between classes of instances. These classes of instances are called entities.

2.2 The PROSIT communication service model

The PROSIT communication service model consists of a refinement of the relations introduced above. This refinement is based on the distinction between active and passive instances and leads to the notions of active and passive entities.

An active entity (service user and provider) is understood to be the collection of all calling instances (service user and provider), whereas a passive entity (service user and provider) is understood to be the collection of all called instances (service user and provider).
This refinement is necessary for making visible the boundary between OSI- Systems in addition to the service boundary. Therefore, it is possible to interpret a protocol which binds active and passive entities across the system boundary as a correlation of active and passive communication behaviour. The resulting communication service model interrelates eight entities (see Fig. 2.4):

- at one service-access-point, an active static service user entity with an active static service provider entity and an active dynamic service user entity, with an active dynamic service provider entity, and
- at the corresponding service-access-point a passive static service user entity, with a passive static service provider entity, and a passive dynamic service user entity, with a passive dynamic service provider entity.

The establishment of a connection appears within that model as the creation of the four appropriate dynamic instances, as mentioned above, and, due to interaction among four corresponding static instances, one for each static entity mentioned above.

The release of a connection appears correspondingly as deletion of the four dynamic instances previously created.

Connectionless data transmission appears within that model as interaction among four corresponding static instances, one for each static entity mentioned above.
N-way data transmission can best be interpreted as connectionless data transmission where a multicast address is used as To-Address. This has the effect that a connectionless data transmission request, issued by a static service user instance to its corresponding static service provider instance at one distinct service access point, leads to a connectionless data transmission indication at each distinct service access point, which is a member of the set of recipients named by the multicast address.

2.3 The PROSIT communication protocol model

The PROSIT communication protocol model is based on the interrelation of two communication service models at adjacent layer boundaries (see fig. 2.5).

It is therefore hierarchically structured into the following three functional levels:

- the N-service user level comprising the N-service user entities of the model for the N-service,
- the N-protocol level comprising the N-service provider entities of the model for the N-service, the (N-1)-service user entities of the model for the (N-1) service, the functionality of mapping N-service provider behaviour onto (N-1) service user behaviour and the functionality of bridging the gap between N-Service and (N-1) service ; the OSI term N-Layer corresponds to the N-protocol level.
- the (N-1) service provider level comprising the (N-1) service provider entities of the model for the (N-1)-service.

This protocol model implies that an N-Layer is to be understood as two interrelated protocol entities, namely an active N-protocol entity and a passive N-protocol entity.

Each protocol entity is composed of two N-service provider entities
and two (N-1)-service user entities.

In this refined model of the protocol entity at the Layer N, it is
apparent that:

- an N-connection can be released by destroying the related dynamic
  N-service user and provider instances without releasing the
  supporting (N-1) connection;
- an (N-1) connection can be released by destroying the related
  dynamic (N-1) service user and provider instances without releasing the supported N-connection.

It should be noted further that this refined model of the protocol
entity of layer N allows to express any mapping of connection
oriented or connectionless services at the boundary between layer
(N + 1) and layer N onto connection-oriented or connectionless
services at the boundary between layer N and layer (N - 1).

In (1) the use of the PROSIT communication service and protocol
model for structuring service and protocol specifications is
illustrated.

3. Product nets - a PROSIT-tool for the specification of
    communication services and protocols

The PROSIT communication service and protocol model establishes
logical relationships between roles. Each box in Fig. 2.4 and 2.5
represents a distinct role and the arcs between boxes stand for
the logical relationships. Each role defines communication
behaviour and is the characteristic of a set of instances. All
instances belonging to such a set behave in communication as
defined by the role. The behaviour of each instance can be defined
as relationship among productions and consumptions of data objects
representing the units of interaction with other instances.

Therefore, each box in the service and protocol models can be
interpreted - from a specification point of view - as representing
a set of relationships among productions and consumptions of data
objects wherein each relationship is associated with a single
instance. The complete models can be seen as collection of such sets.

These models are characterized by containing explicitly:

- aspects of distribution,
- concurrency of actions,
- global features of cooperating systems,
- modules, channels, interfaces, interactions,
- folding and parameterization of interactions.

Product Nets allow to express all these aspects in a unique manner.
Especially in its graphical representation, they are highly adequate
for the design of distributed systems.

Roughly speaking, a product net is a Petri net combined with arc
labels and transition inscriptions, which are as formally constructed as this is done in case of terms, predicates, atoms and formulas
in the first-order logic. In first-order logic an interpretation of
a formula is a well known concept. By introducing adequate restrictions with respect to the syntax of arc labels and transition
inscriptions, the formal semantics of nets is derived from the
interpretation of arc labels and transition inscriptions, whereby a
domain of objects is assigned to each place. A formal specification
of product-nets can be found in (2), an illustration of its use in
(1).

To illustrate some fundamental concepts of product nets, the flow
controlled transfer of data from a sending user of a transfer
service to a receiving user is modeled (see Fig. 3.1). This transfer
leads from the sending user over a service boundary to the sending
provider. The sending provider communicates over a system boundary
with the receiving service provider. At least, the receiving
service provider delivers received data to the receiving service
user.

The product net in Fig. 3.1 has the initial marking:

s1   : a set of messages   s8   : W < >-tokens
s4, s5 : one < > -token   s9, s10 : <0>

The other places are empty.

The places s3 and s4 represent the service boundary between the sending service user and the sending service provider and s5 and s6 represent the service boundary between the receiving service provider and the receiving service user. The 'ready signals' on s4 and s5 guarantee that not more than one message appears at s3 and s6. The places s7 and s8 represent the system boundary.

Receiving the messages in the order, they are sent, shows up as a consistent numbering in the sending and the receiving service provider. The flow control is enforced by 'credit tokens' on s8.

Transition T1 is activated if there exist at least one message at place S1 and if there exist a 'ready-signal' at place S4.
Transition T3 is activated if there exists at both places S3 and S8, at least one object.
In this case firing of transition T3 leads to
- the production of a transmission packet at place S7, consisting of the original message and a send-sequence number equal to the actual value of the send sequence counter at place S9.
- the removal of the message from place S3
- the incrementation of the send-sequence counter at place S9
- the removal of one credit token from place S8

Transition T4 is activated if there exist
- at place S7 a transmission packet with a sequence number equal to the actual value of the receive sequence counter at place S10,
- at place S5 a ready signal issued by the service user.

In this case firing of transition T4 leads to
- the removal of the transmission packet from place S7 and the filing of the carried message at place S6
- the removal of the ready-signal from place S5
- the incrementation of the receive sequence counter.
- the adding of one credit token at place S8.

Transition T2 is activated if place S6 carries a message.

4. The basic thoughts behind the construction of protocols from services and the verification of protocols against services

The formal specifications of commununication services adjacent in the OSI-RM form the prerequisites for the development of a protocol specification, since a protocol rules the cooperation of layer entities which have to bridge the functional difference between adjacent services.

In PROSIT, the construction of a protocol specification from service specifications is supported on one hand by the modelling approach and on the other hand by the use of elementary building blocks and rules for their composition.

The modelling approach support the construction since the PROSIT communication protocol model, which forms the basis of a protocol specification, is constituted by superimposing two service models, which form the basis for the service specifications. A protocol specification of layer N, therefore, consists of

- provider behaviour of service N and of user behaviour of service (N-1) - both already specified as part of the specification of service N and service (N-1);
- and a part which must be added to interrelate provider and user behaviour and to bridge any functional gap between the two services.

The seek for elementary building blocks and rules for their composition is motivated by the recognition that in different context always the same communication patterns occur. Such patterns are in distributed systems e. g. typical for the consistent establishment, progression and release of global context, which is just another formulation for cooperation. The flow control example shown in Fig. 3.1 is another example.

The usage of building blocks for the design of distributed systems is shown in (3) on the example of a Product Net specification of the OSI-Transport Service and in (4) on the example of a Product Net specification of a distributed application.

The use of the communication service and protocol models and of building blocks and composition rules facilitates the formal specification of services and protocols, by providing constructive aids, and increases their readability.

Beside the design, the validation of communication services and protocols and the verification of protocols against the bordering services benefit from this approach. Validation means to check services and protocols whether they have desired properties (e. g. freedom of deadlocks and lifelocks, globally consistent binding and releasing of resources).

Verification of a protocol against the bordering services means to prove that a protocol provides the service above while using correctly the service below.
To verify a protocol, we compute from each service specification the complete set of sequences of services primitives. These sequences can be derived from the product net specifications of the services as set of sequences of markings at places representing the service boundary. The same procedure has to be applied on the places representing upper and lower service boundaries in the protocol specification. This results in a set of sequences of markings at places on the upper service boundary and a set of sequences of markings at places on the lower service boundary.

Due to the constructive relationship between protocol model and service model, one can regard a protocol as verified if the sets of sequences of markings derived from the service specifications coincide with the respective sets of sequences of markings derived from the protocol specification. In (5) this verification method is demonstrated on the example of the alternating bit protocol. In (6) a checkpoint-restart-protocol is proved. In (7) a fundamental problem in the field of distributed systems with faulty communications is treated, namely the impossiblity to guarantee common termination. This again implies the necessity of static instances. Finally in (8) the PROSIT net simulation system NESSY is described.

5. The basic thoughts for deriving implementation specifications and test sequences from global protocol specifications

A protocol specification, structured according to the PROSIT communication protocol model and formally specified by means of Product Nets, defines cooperation in a distributed system. Roughly spoken, it identifies all communicating instances and describes how they interact. This view of description can be called the "designers view".

For an implementation, a specification is needed which defines - roughly spoken - a single instance and how it acts or reacts at its interaction points with the outer world. This view of description can be called the "implementors view".
For the test of a protocol implementation on conformance with a protocol specification, one needs a record of all events which might occur at boundaries accessible in a test. This view of description can be called the "observers view".

It is quite obvious that the designers view of description contains the implementators view as well as the observers view and that it must be, therefore, possible to derive an implementation speci- fication and test sequences from a protocol specification.
The global implementation independent protocol specification identifies all protocol functions and establishes their logical relationships, only. It is therefore characterized by a high degree of concurrency. In contrast, an implementation specification deals with tasks and their interrelationships.

A task is understood as a functional unit,
- which can be formally described by a finite automaton
- which is allocated as a whole to one local processor and
- which is the basic unit in an operating system being provided with system resources.

The derivation of implementation specifications from global protocol specification, therefore, comprises

- as first step the assignment of protocol functions to tasks,
- as a second step the serialization of concurrent protocol
  functions assigned to one task,
- as a third step the logical coupling of tasks.

The method for the derivation of implementation specifications from
product net specifications of protocols is described in (9).

For implementation specifications, CCITT's SDL is an accepted and
widely used language, which is supported by software development
tools of various organizations.
It seems feasible to bridge the gap between a global protocol
specification in form of product nets and an SDL-implementation
specification and to combine the benefits of product nets for
formal description of global systems and their mathematical analysis
with the SDL support for implementations.

Conformance testing is based on test sequences or more generally,
test suites which should allow comparability and wide (international)
acceptance of test results.
In this area the ISO ad-hoc group on conformance and conformance
testing has defined a set of abstract test methods and has provided
a framework for specifying conformance test suites.

The following three main categories of abstract test methods shown
in Fig. 5.1 were identified with respect to the kind of observable
and controllable interfaces within the implementation under test
(IUT) or system under test (SUT). The applicability of different
abstract test methods to a particular SUT depends on the accessi-
bility of interfaces within the SUT.

The local test methods use control and observation of the (abstract)
service primitives defined at the service boundaries directly above
and below the implementation under test.
In this case IUTs are tested in isolation from the rest of the
system which is simulated by the upper and lower tester. Test
synchronization (test coordination procedures) between activities
of the upper tester and activities of the lower tester are needed
for these methods.

The distributed test methods use control and observation of the
(N-1)-service primitives defined at the service boundary between
the lower tester and the (N-1)-service provider and control and
observation of the (N)-service primitives defined at the service
boundary between the IUT and the upper tester.
These methods also require the use of test coordination procedures
between the upper tester and the lower tester.

The remote test methods use control and observation of the (N-1)-
service primitives defined at the service boundary between the
lower tester and the (N-1)- service provider.
These methods provide the weakest tests in terms of control and
observation since no interface within the SUT is accessible.

The main difference between the local test methods and the two
other categories of test methods is that a (N-1)-service provider
is only simulated in the case of local testing whereas a real
(N-1)-service provider is involved in the case of distributed and
remote testing. Therefore, the local test methods require further
testing, since the lower tester can only be a more or less adequate
approximation of a real (N-1)-service provider.
Finally it should be mentioned that the distributed and remote test
methods are defined on the assumption that the (N-1)-service
provider behaves correctly. This assumption must be validated by
the execution of basic interconnection tests which are part of the
complete conformance test suite.

Formal protocol specification, structured according to the PROSIT
protocol model and described by means of Product Nets, offer an
enriched structuring of entities, but are in full accordance with
the ISO test methods.

Formal protocol specifications define correct actions and correct
reactions of instances. The correct reactions comprise correct
reactions on receipt of correct pdus and invalid pdus which are
either syntactically invalid our out-of-sequence pdus. Correct
actions, however, do only consist of actions which describe the
generation and sending of valid pdus. Formal protocol specifications
per se do not include definitions of incorrect actions. Therefore, those further specifications are required for the purpose of

testing (error generator) and must be added to the global protocol specification in order to get a complete test specification.

The procedure to derive test sequences from formal protocol specifications includes the following main steps:
- superposition of the protocol specification with a test configuration and identification of those interfaces within the SUT which are observable or controllable during the test (shown in Fig. 5.2 for the local test method);

- use of the formal protocol specification (specification language plus tools) in order to compute
  -- the complete state space at the considered interfaces;
  -- the complete set of state transitions at the considered interfaces;
  -- the set of allowable sequences of transitions at the considered interfaces;
- definition and selection of appropriate test strategies.

The reachability analysis produces the complete reachability graph representing the behaviour of an SUT at the considered interfaces. This graph contains all possible and allowable test sequences which can be observed or controlled at the interfaces. Therefore, this set of test sequences is considered as being complete from the theoretical viewpoint. Thus it provides a measure for the quality of a concrete set of test sequences which is selected for a particular test and which might be a subset of the full set. Test strategies are based on a superposition of the complete reachability graph with test patterns. Different test strategies are possible and so far the selection mechanism of defining an appropriate test strategy is purely intuitive and pragmatic. Test strategies are used for the following purposes:
  -- systematic selection of individual test patters;
  -- systematic structuring of test (separation into test phases, ordering of test, specification of test events, test steps, test groups and test suites);
  -- definition of quality measures for testing;

The derivation of test sequences from a global protocol specification is described in (10).

6. Conclusion

The realisation of open system means to realize distributed systems in a distributed manner.
Together with the implied systems heterogeneity, this leads to a new category of problems to which established methods, oriented to the design and programming of single systems, offer no adequate solutions.

This new category of problems is characterized by the need to distinguish clearly between various levels of abstractions and to provide for consistent transitions between them; of course, such a need had been formulated before in the context of single or homogenous systems design, but was there not of such a crucial importance.

At the highest level of abstraction, communication and cooperation in organizational structures have to be described. This description has to be based on a model deduced in an abstraction process from the existing or projected reality. This model has to express the aspects, only, relevant for cooperation; i. e. it has to emphasize which function an instance performs in its cooperation with other instances and not how it is internally conditioned to fulfill this function.
At the lowest level of abstraction, this 'how' has to be described for a distinct real open system.

The gap between these levels of abstraction must be bridged. A common understanding, what adequate modelling means, must be established and a consistent set of methods and tools for the various levels of abstraction and the transitions between them must be developed.
What has to be done in the end, is to develop the construction of distributed systems, especially the construction of distributed applications towards an engineering science.
Currently, we are far from this end, but our experience in applying PROSIT methods within industrial cooperations have given us the confidence that this end can be reached.

Reference to papers presenting PROSIT results in more details:

(1)   Burkhardt, H. J.; Eckert, H.; Prinoth, R.
Modelling of OSI-Communication Services and Protocols
using Predicate/Transition Nets
Protocol Specification, Testing, and Verification
Y. Yemini, R. Strom, and S. Yemini (ed.)
Elsevier Science Publishers B. V. (North-Holland)
C IFIP, 1985

(2)   Eckert, H.; Prinoth, R.
Produktnetze - Definition eines PROSIT-Beschreibungsmittels
Arbeitspapiere der GMD Nr. 92, 1984

(3)   Baumgarten, B.; Ochsenschläger, P.; Prinoth, R.
Building Blocks for Distributed System Design
Proceedings of the IFIP WG 6.1
Fifth International Workshop on Protocol Specification,
Testing and Verification, 1985
North Holland, 1985

(4)   Baumgarten, B.; Burkhardt, H. J.; Ochsenschläger, P.; Prinoth, R.
The Signing of a Contract - a Tree Structured Application
Modelled with Petri Net Building Blocks
Advances in Petri Nets, LNCS, 222, Springer, 1986

(5)   Eckert, H.; Prinoth, R.
A Computation-Systems Based Method for Automated
Proving of Protocols Against Services
Protocol Specification, Testing, and Verification, III
H. Ruding and C. H. West (ed.)
Elsevier Science Publishers B. V. (North-Holland)
C. IFIP, 1983

(6)   Baumgarten, B.; Ochsenschläger, P.
Modelling and Verification of a Checkpoint-Restart-Protocol
2. GI/NTG/GMR-Fachtagung: Fehlertolerierende Rechensysteme
Bonn 1984
Informatik-Fachberichte 84, pp 353-363
Springer Verlag

(7) B. Baumgarten, P. Ochsenschläger:
On Termination and Phase Changes in the Presence of
Unreliable Communication
Information Processing Letters 1985

(8) Paule, C.
Das Netzsimulationsystem NESSY
Arbeitspapiere der GMD Nr. 156, 1985

(9) Burkhardt, H. J.; Eckert, H.; Prinoth, R.
Implementing OSI Communication Protocols
- A Systematic Approach to Derive SDL Implementation
Specifications from Global Protocol Specifications
ICCC 86

(10) Burkhardt, H. J.; Eckert, H.; Giessler, A.
     Testing of Protocol Implementations
     - A Systematic Approach to Derivation of Test Sequences
     from Global Protocol Specifications -
     Proceeding of the IFIP WG 6.1
     Fifth International Workshop on Protocol Specification,
     Testing and Verification, 1985
     North Holland, 1985

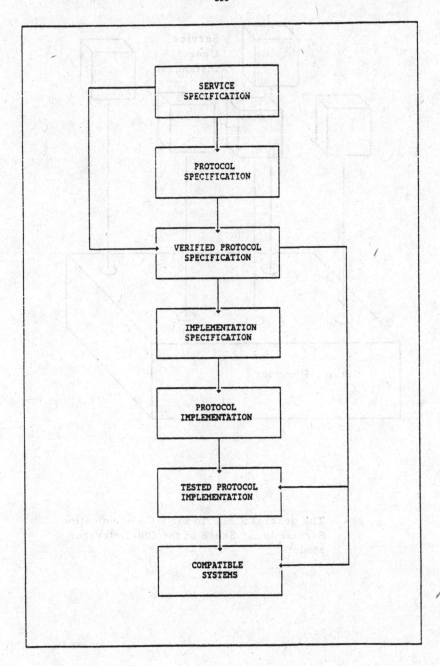

Fig. 1.1 : The intermediate steps necessary for the realization
of compatible systems from standards

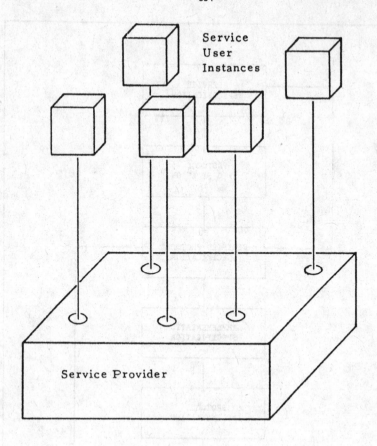

Fig. 2.1 : The general Scenario for a Communication
Service in the Sense of the OSI-Reference
Model

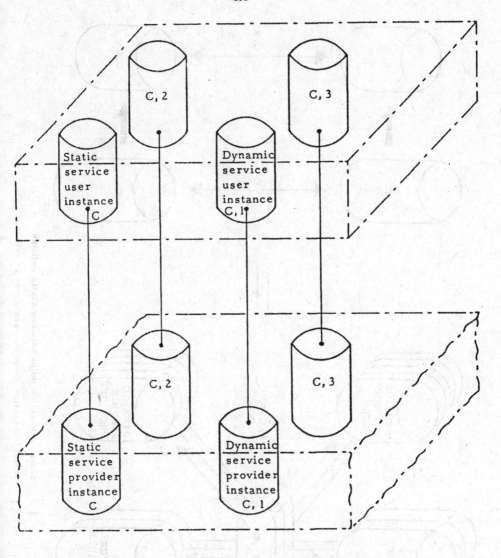

Fig. 2.2: The Refinement of both Service User and Provider
          Instances at a distinct SAP with Address C into Static
          Instances C and Dynamic Instances C,K; K = 1 to 3

396

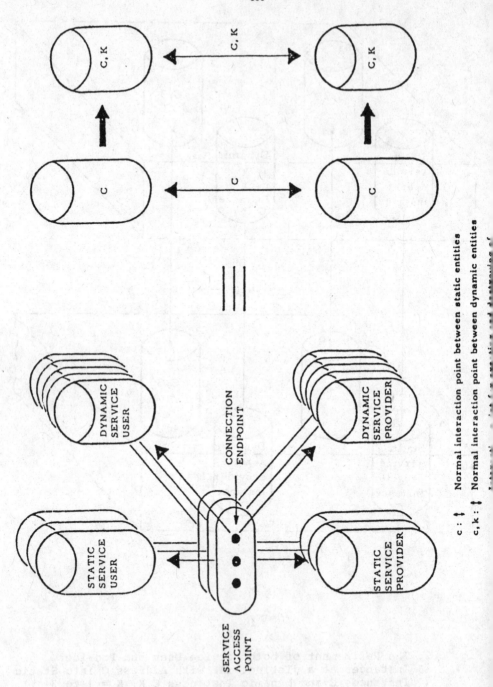

Fig. 2.3: The refined model of a Service Access Point introducin
the notions of static and dynamic entities.

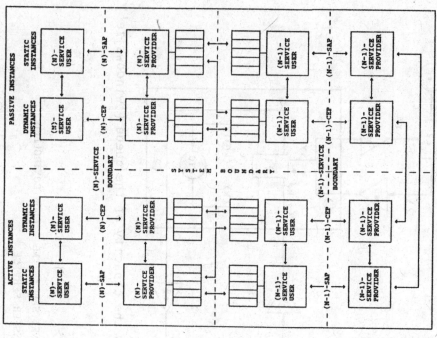

FIG. 2.5: THE PROSIT COMMUNICATION PROTOCOL MODEL

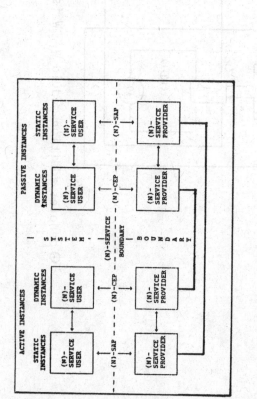

FIG. 2.4: THE PROSIT COMMUNICATION SERVICE MODEL

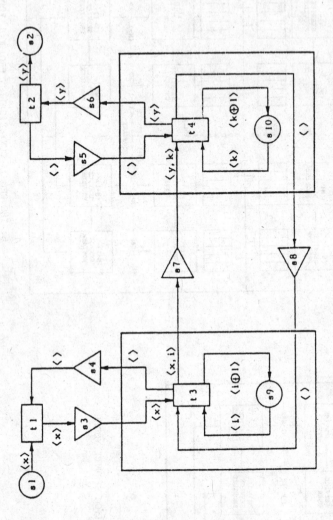

$\oplus$ denotes addition modulo W. W is the capacity of the queue ('windowsize').

FIG. 3.1: PRODUCT NET SPECIFYING A FLOW CONTROLLED
DATA TRANSFER SERVICE

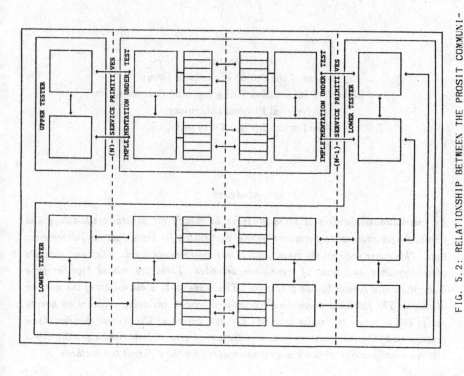

FIG. 5.2: RELATIONSHIP BETWEEN THE PROSIT COMMUNI-
CATION PROTOCOL MODEL AND THE OSI LOCAL
TEST METHOD

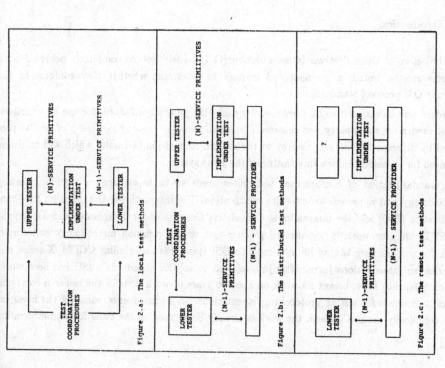

FIG. 5.1: THE OSI TEST METHODS

Methodology and Framework for OSI Conformance Testing

Dr D. Rayner
Leader of the Protocol Standards Group
Division of Information Technology and Computing
National Physical Laboratory
Teddington, Middx. TW11 0LW, UK

Abstract

An international standard is being produced to specify the general methodology and framework for protocol conformance testing, particularly for Open Systems Interconnection. This paper presents the main ideas of this emerging standard. The basic concepts of conformance and types of testing are described. These are related together by a description of a typical testing procedure. This leads on to a discussion of the analysis of results. The test suite production process is described, including guidance on how to design and structure test suites to ensure adequate coverage. The process also includes a description of the main test methods and the use of generic tests, which provide a link between corresponding abstract test suites produced for the different test methods.

0. Introduction

The objective of Open Systems Interconnection (OSI) [1] will not be completely achieved until systems can be tested in an approved manner to determine whether they conform to the relevant OSI protocol standards.

Standard test suites need to be developed for each OSI protocol standard, for use by suppliers, users, carriers or third-party test centres. This should lead to the comparability of results produced by different testers, and thereby to the mutual recognition of results which will minimise the need for repeated conformance testing of the same system.

The standardization of conformance test suites needs to be based upon a standard testing methodology and approach to test suite specification. The International Organization for Standardization (ISO) and the International Consulative Committee for Telephones and Telegraphs (CCITT) are now working together to produce just such a common methodology and framework for conformance testing [2] applicable to OSI standards and similar CCITT X-series and T-series recommendations (hereinafter just referred to as 'OSI standards'). ISO has been working on this since 1984, basing its work on about 6 years of work done in the research community on techniques for testing protocol implementations. As a consequence, many of the ideas are now technically mature. Indeed, the work is already being used as the basis for standardization of test suites for X.25 terminals and Message Handling Services (X.400 series), soon to be joined by standardization of test suites for the OSI connection-oriented Transport Protocol (ISO 8073).

1. The Meaning of Conformance in OSI

Conformance in the context of OSI is concerned only with the conformance of implementations and real systems to those OSI standards which specify requirements applicable to implementations and real systems. Those standards include protocol standards (both single-layer and multi-layer) and transfer syntax standards.

1.1. Conformance Requirements

There are three ways of looking at conformance requirements.

(1) The conformance requirements in a standard can be:-

    (a) mandatory requirements: these must be observed in all cases;

    (b) conditional requirements: these must be observed only when the conditions set out in the standard apply;

    (c) options: these can be selected to suit the implementation, so long as any requirements on which the option depends or which depend on the option are observed.

(2) The statements of conformance requirements in a standard can be:-

    (a) positive: they state what must be done;

    (b) negative (prohibitions): they state what must not be done.

(3) The requirements fall into two groups:

    (a) static conformance requirements;

    (b) dynamic conformance requirements.

    These are defined below.

1.2. Static Conformance Requirements.

Static conformance requirements are those that define the allowed minimum **capabilities** of an implementation, in order to facilitate interworking. These requirements may be at a broad level, such as the grouping of functional units and options into protocol classes, or at a detailed level, such as the range of values that must be supported for specific parameters or timers. They can come in two varieties:-

(a) those which concern the capabilities to be included in the implementation of the particular protocol;

(b) those which concern multi-layer dependencies – placing constraints on the capabilities of the underlying layers of the system.

A system can be said to conform statically if it has the capabilities specified by the static conformance requirements. If the static conformance requirements are incompletely specified then all capabilities not explicitly stated as static conformance requirements are to be regarded as optional.

1.3. Dynamic Conformance Requirements

Throughout most of the OSI protocol standard, the requirements and options define the set of allowable behaviours of an implementation or real system in instances of communication. This set defines the maximum capability that a conforming implementation or real system can have within the terms of the OSI protocol standard. These are the dynamic conformance requirements.

Thus, dynamic conformance requirements are all those requirements (and options) which determine what observable **behaviour** is permitted by the relevant OSI protocol standard(s) in instances of communication.

Also a system can be said to conform dynamically in an instance of communication if its behaviour is a member of the set of all behaviours permitted by the relevant OSI protocol standard(s).

1.4. Protocol Implementation Conformance Statement

The protocol implementation conformance statement (PICS) is a statement made by the supplier or implementor of the capabilities and options which have been implemented, and any features which have been omitted. It is needed so that the implementation can be tested for conformance against relevant requirements, and against those requirements only.

1.5. A Conforming System

A conforming system or implementation is, therefore, one which is shown to satisfy both static and dynamic requirements, consistent with the capabilities and options stated in the PICS.

2. Objectives of Conformance Testing

In principle, the objective of conformance testing is to establish whether the implementation being tested conforms to the specification in the relevant standard. Practical limitations make it impossible to be exhaustive, and economic considerations may restrict testing still further.

Therefore, four types of testing are distinguished, according to the extent to which they provide an indication of conformance:

- **basic interconnection tests**, which provide prima facie evidence that an implementation under test (IUT) conforms;
- **capability tests**, which check that the observable capabilities of the IUT are in accordance with the static conformance requirements and the capabilities claimed in the PICS;
- **behaviour tests**, which endeavour to provide as comprehensive testing as possible over the full range of dynamic conformance requirements specified by the standard, within the capabilities of the IUT;

- **conformance resolution tests**, which probe in depth the conformance of an IUT to particular requirements, to provide a definite yes/no answer and diagnostic information in relation to specific conformance issues; such tests need not be standardized.

2.1. Basic Interconnection Tests

These provide limited testing of the main features in a standard, to establish that there is sufficient conformance for interconnection to be possible, without trying to perform thorough testing.

Basic interconnection tests are appropriate:

- for potentially detecting severe cases of non-conformance;
- as a first filter before undertaking more costly tests;
- to give a prima facie indication that an implementation which has passed full conformance tests in one environment still conforms in a new environment (e.g. before testing an (N)–implementation, to check that a tested (N-1)–implementation has not undergone any severe change when linked to the (N)–implementation);
- for use by users of implementations, to determine whether the implementations are usable for communication at all with other conforming implementations, e.g. as a preliminary to data interchange.

Basic interconnection tests are inappropriate:

- as a basis for claims of conformance by the supplier of an implementation;
- as a means of arbitration to determine causes for communications failure.

Basic interconnection tests should be standardized as a very small test suite, which can be used on its own or together with a conformance test suite.

2.2. Capability Tests

These provide limited testing of each of the static conformance requirements in a standard, to ascertain what capabilities of the IUT can be observed and to check that those observable capabilities are valid with respect to the static conformance requirements and the PICS.

Capability tests are appropriate:

- to check the validity of the PICS;
- as a second filter before undertaking more in-depth and costly testing;
- to check that the capabilities of the IUT are consistent with the static conformance requirements;
- to enable efficient selection of behaviour tests to be made for a particular IUT;
- to check that the capabilities of the IUT are adequate to meet a particular user's needs.

Capability tests are inappropriate:

- on their own, as a basis for claims of conformance by the supplier of an implementation;
- as a check on the behaviour associated with each capability which has been implemented or not implemented;

- for resolution of problems experienced during live usage or where other tests indicate possible non-conformance even though the capability tests have been satisfied.

Capability tests should be standardized within a conformance test suite. They can either be separated into their own test group(s) or merged with the behaviour tests.

2.3. Behaviour Tests

Behaviour tests are intended to provide as thorough testing of an implementation as is practical, over the full range of dynamic conformance requirements specified in a standard. Since the number of possible combinations of events and timing of events is infinite, such testing cannot be exhaustive. There is a further limitation, namely that these tests are designed to be run collectively in a single test environment, so that any faults which are difficult or impossible to detect in that environment can be missed. Therefore, it is possible that a non-conforming implementation passes the conformance test suite; one aim of the test suite design is to minimise the number of times that this occurs.

It is reasonable to regard an implementation as conforming if it satisfies the conformance test suite, so long as there is no evidence to the contrary.

Behaviour tests are appropriate:

- when taken together with capability tests, as a basis for claims of conformance, so long as other tests have not revealed contrary evidence;
- as a basis for procurement.

Behaviour tests are inappropriate:

- for resolution of problems experienced during live usage or where other tests indicate possible non-conformance even though the behaviour tests have been satisfied.

Behaviour tests should be standardized as the bulk of a conformance test suite.

2.4. Conformance Resolution Tests

These provide diagnostic answers, as near to definitive as possible, to the resolution of whether an implementation satisfies particular requirements. Because of the problems of exhaustiveness noted above, the definitive answers are gained at the expense of confining tests to a narrow field.

The test architecture and test method will normally be chosen specifically for the requirements to be tested, and need not be ones that are generally useful for other requirements; they may even be ones that are regarded as being unacceptable for generally specified conformance tests, e.g. involving implementation-specific methods using, say, the diagnostic and debugging facilities of the specific operating system.

The distinction between behaviour tests and conformance resolution tests may be illustrated by the case of an event such as a Reset. The behaviour tests may include only a representative selection of conditions under which a Reset might occur, and may fail to detect incorrect behaviour in other circumstances. The conformance resolution tests would be confined to conditions under which incorrect behaviour was already suspected to occur, and would confirm whether or not the suspicions were correct.

Conformance resolution tests are appropriate:

- for providing a yes/no answer in a strictly confined and previously identified situation (e.g. during implementation development, to check whether a particular feature has been correctly implemented, or during operational use, to investigate the cause of problems);
- as a means for identifying and offering resolutions for deficiencies in a current conformance test suite.

Conformance resolution tests are inappropriate:

- as a basis for judging whether or not an implementation conforms overall;
- as a condition for procurement.

Conformance resolution tests need not be standardized.

3. Typical Test Procedures

It is important to understand how conformance testing should relate to static and dynamic conformance requirements and the PICS. There are many possible ways of interleaving various phases of testing, enabling the selection of later tests to be based on the results of earlier tests and paper studies. However, Figure 1 is used by ISO and CCITT to illustrate a typical conformance testing procedure. This illustration has the merit of separating out the various different concerns to enable the inter-relationships to be understood more clearly.

There are five main steps in this procedure. The first, a **static conformance review**, is be a paper analysis, in which the PICS accompanying the IUT will be analysed for its own consistency, and its consistency with the static conformance requirements specified in the protocol standard(s) to which the IUT is claimed to conform.

The second step, **basic interconnection testing**, is optional. If executed, it can detect severe cases of non-conformance, and thus passing this step gives confidence that it is worthwhile engaging in more comprehensive (and therefore more expensive) testing.

The third step, **capability testing**, is designed to ascertain the validity of the PICS with respect to the actual, observable capabilities of the IUT. A **second static conformance review** will combine the results of the capability tests with the results of the first review, and provide a modified PICS for use in the selection of behaviour tests.

The main step is the fourth, **behaviour testing**, which concentrates on checking the correct behaviour of the protocol implementation in a large representative sample of instances of communication. These will cover both simple to complex situations, involving both correct and erroneous behaviour on the part of the communicating partner. The results of these tests will be analysed from the dynamic conformance point of view (i.e. to check that the behaviour of the IUT was always valid). It cannot be proved by testing that an implementation conforms dynamically in all instances of communication. However, it can be shown by testing that an implementation consistently conforms dynamically in representative instances of communication.

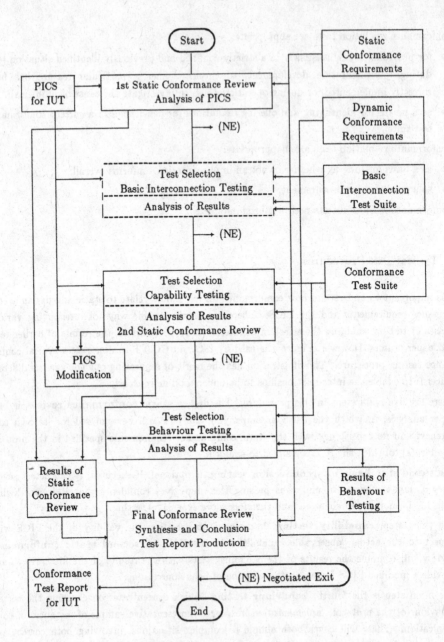

Figure 1.

The fifth step is the final conformance review, which involves a synthesis of the results of the behaviour tests with those of the second static conformance review, to see whether the behaviour tests have thrown any new light on the observed capabilities of the IUT. A conclusion on the conformance of the IUT to the requirements of the standard(s) can then be reached. This conclusion is recorded in a **Conformance Test Report**, the structure of which will be standardized to help users to compare reports produced by different organisations.

Provisions for "negotiated exits" can be seen in Figure 1. They are points where the concerned parties can decide that the results of the previous step are not good enough to justify continuing the tests.

3.1. Analysis of Results

The analysis of results is carried out by comparing the observed outcome of each test case with the required or permitted outcome; from this analysis a verdict is produced.

The *observed outcome* is the series of actual events which occur while executing a test case; it includes all input to and output from the IUT at the points of observation and control.

The *required* or *permitted outcomes* are defined by the abstract test case specification taken in conjunction with the protocol standard. For each test case there may be one required outcome or more than one permitted outcome.

The *verdict* will be *pass, fail* or *inconclusive*:

- **pass** means that the observed outcome matches one of the required or permitted outcomes;
- **fail** means that the observed outcome conflicts with all of the required or permitted outcomes, and the cause can be shown to be IUT behaviour not conforming with the standard;
- **inconclusive** means that the observed outcome neither matches any of the required or permitted outcomes, nor conflicts with them: so no conclusions can be drawn. This applies particularly to the case where the IUT behaved in a valid fashion but did not satisfy the purpose of the test case (e.g. a spontaneous disconnect occurring before the end of the test case). There are two subdivisions of inconclusive test cases, one in which the test purpose can never be achieved with this IUT, the other in which the test purpose may be achievable if the test case is repeated.

The verdict may be different in respect of static and dynamic requirements; e.g. correctly performed rejection of a connection request may produce a verdict of *pass* for dynamic requirements, but *inconclusive* for static requirements.

The verdicts made in respect of individual test cases will be synthesised into an *overall summary* for the IUT based on the complete test suite. The extent to which an overall summary of *pass* can be made for an IUT when *fail* or *inconclusive* verdicts were made for any individual test cases is debatable; practical considerations may require that a small number of *fail* or *inconclusive* verdicts for non-critical aspects may be acceptable. However, comparability of results will then be adversely affected unless common criteria are applied.

The resulting conformance test report will contain both the overall summary with respect to static and dynamic conformance plus details of which test cases were run together with their observed outcomes and verdicts. The report will also contain information to identify the implementation under test and characterise the environment in which it was tested.

4. Test Suite Production

In order to present the requirements and general guidance for test suite specification, a norma
form of the process of test suite production is assumed. Test suite designers are not required to
follow this normal form exactly, but are required to produce test suites which could have been
produced by this process. Thus, they are recommended to use a similar process involving the
same steps but possibly in a different order.

4.1. Test Suite Structure

Test suites have a hierarchical structure (see Figure 2) in which a key level is the **test case**.

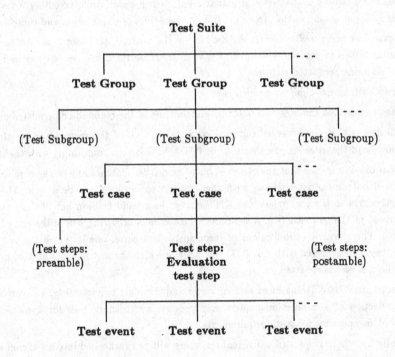

bold text: mandatory element of structure
(in brackets): optional element of structure

Figure 2. Test Suite Structure

Each test case has a narrowly defined purpose, such as that of verifying that the IUT has a certain required capability (e.g. the ability to support certain packet sizes) or exhibit a certain required behaviour (e.g. behave as required when a particular event occurs in a particular state).

Within a test suite, **test groups** and **test sub-groups** are used to provide a logical ordering of the test cases. They may be used to aid planning, development, understanding or execution of the test suite.

Each test case comprises at least one **test step**: the series of events covered by the test purpose (*evaluation test step*). It may include further test steps to put the IUT into the state required for the test (*preamble*) or to tidy up after the test (*postamble*). All test steps consist of a series of **test events** - the transfer of a single protocol data unit (PDU) or abstract service primitive (ASP) to or from the IUT.

The structure is strictly hierarchical in the sense that an item at a lower level is completely contained within a higher level item, but is not strictly hierarchical in the sense that any one item at a lower level may occur in more than one higher level item. So the same test case could occur in more than one test group and the same test step could occur in more than one test case.

4.2. Types of Test Case

Three types of test case are defined: generic, abstract and executable.

A **generic test case** is one which:

- is defined independently of the test method to be used;
- specifies the purpose of the test case;
- specifies the sequence or sequences of test events which correspond to a verdict of 'pass', using a specialised notation;
- provides some informal categorisation of sequences which correspond to verdicts of 'fail' or 'inconclusive';
- where appropriate, informally specifies a preamble.

An **abstract test case** is derived from a generic case together with the relevant protocol specification; it:

- specifies the test case in terms of a particular test method;
- adds a more precise specification for sequences of events which are only described informally in the generic test case.

An **executable test case** is derived from an abstract test case, and is in a form which allows it to be run on a real test system for testing a real implementation.

The terms generic, abstract and executable are used to describe test suites, which comprise generic, abstract and executable test cases respectively.

4.3. Main Steps of Test Suite Production Process

The main steps of the test suite production process are taken to be as follows:-

(a) study the relevant standard(s) to determine what the conformance requirements (including options) are which need to be tested, and what the implementor needs to state in the

PICS;

(b) decide on the test groups and subgroups which will be needed to achieve the appropriate coverage of the conformance requirements and refine the test groups into sets of test purposes;

(c) specify generic test cases for each test purpose, using some appropriate test notation;

(d) choose the test method(s) for which the complete abstract test cases need to be specified, and decide what restrictions need to be placed on the assumed capabilities the testers and test coordination procedures;

(e) specify the complete abstract test cases, including the test step structure to be used, in an appropriate test notation;

(f) specify the inter-relationships among the test cases and those between the test cases and the PICS, in order to determine the restrictions on the selection of test cases for execution and on the order in which they can be executed;

(g) chose the executable testers which will be used and then derive the appropriate executable test suite from the abstract test suite.

5. Determining the Conformance Requirements and PICS

Before a conformance test suite can be specified, the test suite designer shall first determine what the conformance requirements are for the relevant standard(s) and what has to be stated by the implementor in the PICS concerning the implementation of those standard(s).

ISO and CCITT have produced a checklist to give guidance to protocol defining groups on what to consider when trying to determine the conformance requirements of a specific protocol. This checklist was based on the UK input of a published conference paper [3]. In practice early OSI standards are unlikely to contain a clear specification of all the relevant conformance requirements. In particular, the static conformance requirements might be badly specified or even omitted. In such cases, the test suite designer is advised to try to progress an amendment o addendum to the relevant standard(s) to clarify the conformance requirements. In the absence of the conformance requirements being clarified, then the test suite designer is recommended to adopt as a short-term solution the assumption that everything which is not explicitly stated to be mandatory or conditional is optional. The test suite standard then needs to state clearly what the implications of this are.

The test suite designer shall state in the conformance test suite standard what information has to be supplied in the PICS. The PICS is required to contain a full list of the optional capabilities which have been implemented. In addition, the PICS is required to include specific values or ranges which are supported for each parameter and timer.

6. Abstract Test Suite Structure

6.1. Basic Requirements

An abstract conformance test suite comprises a number of test cases. The test cases are grouped into test groups and if necessary test subgroups.

6.2. The Test Group Structure

In order to ensure that the resulting conformance test suite provides adequate coverage of the relevant conformance requirements, the test suite designer is advised to design the test suite structure in terms of test groups and test subgroups in a top down manner. There are many ways of doing this; no one way is necessarily right and the best approach for one test suite might not even be appropriate for another. Nevertheless, the test suite designer should ensure that the test suite includes test cases for each of the following categories where relevant:–

(a) capability tests (for static conformance requirements);

(b) behaviour tests of valid behaviour;

(c) behaviour tests of syntactically invalid behaviour;

(d) behaviour tests of inopportune behaviour (i.e. syntactically valid events occurring at the wrong time);

(e) tests focusing on PDUs sent to the IUT;

(f) tests focusing on PDUs received from the IUT;

(g) tests focusing on interactions between what is sent and received;

(h) tests related to each protocol mandatory feature;

(i) tests related to each optional feature which is implemented;

(j) tests related to each optional feature which is not implemented;

(k) tests related to each protocol phase;

(l) variations in the test event occurring in a particular state;

(m) timing and timer variations;

(n) PDU encoding variations;

(o) variations in values of individual parameters;

(p) variations in combinations of parameter values.

This list is not necessarily exhaustive; additional categories might be needed to ensure adequate coverage of the relevant conformance requirements for a specific test suite. Furthermore, these categories overlap one another and it is the task of the test suite designer to put them into an appropriate hierarchical structure. The following is an example of a suitable structure for a single-layer test suite:–

A. Capability tests

    A.1 Mandatory features

    A.2 Optional features said by the PICS to be supported

B. Behaviour tests: response to valid behaviour by peer

    B.1 Connection establishment phase (if relevant)

B.1.1 Focus on what is sent to the IUT

    B.1.1.1 Test event variation in each state

    B.1.1.2 Timing/timer variation

    B.1.1.3 Encoding variation

    B.1.1.4 Individual parameter value variation

    B.1.1.5 Combination of parameter values

B.1.2 Focus on what is received from the IUT

    – substructured as B.1.1

B.1.3 Focus on interactions

    – substructured as B.1.1

B.2 Data transfer phase

    – substructured as B.1

B.3 Connection release phase (if relevant)

    – substructured as B.1

C. Behaviour tests: response to syntactically invalid behaviour by peer

    C.1 Connection establishment phase (if relevant)

        C.1.1 Focus on what is sent to the IUT

            C.1.1.1 Test event variation in each state

            C.1.1.2 Encoding variation of the invalid event

            C.1.1.3 Individual invalid parameter value variation

            C.1.1.4 Invalid parameter value combination variation

        C.1.2 Focus on what the IUT is requested to send

            C.1.2.1 Individual invalid parameter values

            C.1.2.2 Invalid combinations of parameter values

    C.2 Data transfer phase

        – substructured as C.1

    C.3 Connection release phase (if relevant)

        – substructured as C.1

D. Behaviour tests: response to inopportune events by peer

    D.1 Connection establishment phase (if relevant)

        D.1.1 Focus on what is sent to the IUT

            D.1.1.1 Test event variation in each state

            D.1.1.2 Timing/timer variation

            D.1.1.3 Special encoding variations

            D.1.1.4 Major individual parameter value variations

            D.1.1.5 Variation in major combination of parameter values

        D.1.2 Focus on what is requested to be sent by the IUT

            – substructured as D.1.1

    D.2 Data transfer phase

        – substructured as D.1

    D.3 Connection release phase (if relevant)

        – substructured as D.1

If the test suite is to cover more than one layer, then a single-layer test suite structure such as this could be replicated for each layer concerned. In addition, a correspondingly detailed structure could be produced for testing the capabilities and behaviour of multiple layers taken as a whole, including the interaction between the activities in adjacent layers.

6.3. Test Case Purposes

The test suite designer should provide a test purpose for each test case in a conformance test suite. It is suggested that these test purposes should be produced as the next refinement of the test suite after its structure in terms of test groups and test subgroups has been defined. The test purposes could be produced directly from studying those clauses in the relevant standard(s) which are appropriate to the test group or subgroup concerned. For some test groups or subgroups the test purposes might be derivable directly from the protocol state table; for others, they might be derivable from the PDU encoding definitions or the descriptions of particular parameters, or from text which specifies the relevant conformance requirements. Alternatively, the test suite designer could employ a formal description of the protocol(s) concerned and derive test purposes from that by means of some automated approach.

Whatever approach is used to derive the test purposes, the test suite designer should ensure that they provide an adequate coverage of the conformance requirements of the standard(s) concerned. There should be at least one test purpose related to each distinct conformance requirement.

In addition, the following example gives guidance on what 'adequate coverage' might mean:-

(a) for capability test groups:

- at least one test purpose per relevant feature;
- at least one test purpose per relevant PDU type and each major variation of each such type, using 'normal' or default values for each parameter;

(b) for test subgroups concerned with test event variation in each state:

- at least one test purpose per relevant state/event combination;

(c) for test subgroups concerned with timers and timing:

- at least one test purpose concerned with the expiry of each defined timer;
- at least one test purpose concerned with very rapid response for each relevant type of PDU;
- at least one test purpose concerned with very slow response for each relevant type of PDU;

(d) for test subgroups concerned with encoding variations:

- at least one test purpose for each relevant kind of encoding variation per relevant PDU type;

(e) for test subgroups concerned with valid individual parameter values:

- for each integer parameter, test purposes concerned with the boundary values and one randomly selected mid-range value;

- for bitwise parameters, test purposes for as many values as practical, but not less than all the 'normal' or common values;
- for other parameters, at least one test purpose concerned with a value different from what is considered 'normal' or default in other test groups;

(f) for test subgroups concerned with invalid individual parameter values:

- for each integer parameter, test purposes concerned with invalid values adjacent to the allowed boundary values plus one other randomly selected invalid value;
- for bitwise parameters, test purposes for as many invalid values as practical;
- for all other types of parameter, at least one test purpose per relevant parameter;

(g) for test subgroups concerned with combinations of parameter values:

- at least one test purpose per 'critical' value pair (representing a multi-dimensional boundary);
- at least one test purpose per pair of inter-related parameters to test a random combination of relevant values.

7. Generic Test Case Specification

The test suite designer is recommended to specify a generic test case for each test purpose. Generic test cases are defined independently of any particular abstract test method. Abstract test cases for a particular method can be derived from the protocol specification together with the generic test cases, which relate directly to each test purpose.

A generic test case is composed of 2 main parts: the preamble and the test body, defined as follows:–

(a) the preamble is a textual description of the initial state in which the test body shall start; this includes everything necessary to enable the test body to start (i.e. this is a system state not a protocol state);

(b) the test body is the minimum part of the test case necessary to be able to assign, to the outcomes, verdicts related to the test purpose. It is defined in terms of external behaviour (with respect to the IUT) without reference to any mechanisms for achieving test coordination. It should be expressed at a level of detail that:

- identifies all the 'pass' paths;
- categorizes all the 'failed' paths;
- optionally categorizes some or all of the 'inconclusive' paths;

so that all unspecified paths are 'inconclusive'.

If necessary, a textual description of the postamble can also be provided: to indicate what needs to be achieved once the test body is run in order to progress to a suitable stable state.

7.1. Relation of Generic to Abstract Test Cases

For a particular abstract test method there may be many abstract test cases that can be derived from a single generic test case.

For example, the preamble and postamble may be realized in many alternative ways depending on the degree of control and observation provided by the test method used, or depending on the variety of different possible stable states which the derived abstract test case can start from and end in. These alternative abstract test cases are simply different ways of achieving the same test purpose.

Nevertheless, the generic test suite should be used as the means of relating conformance test suites which are derived from it for different abstract test methods.

8. Abstract Test Methods

Each abstract conformance test suite will be specified in terms of the control and observation of events in accordance with one of the defined abstract test methods. The chosen test method determines the points at which control and observation is specified and the categories of events which can be used (e.g. (N-1)–ASPs, (N)–PDUs).

Test methods are defined for testing end-systems and relay systems. The two main classes of end-system test methods are **local** and **external**. Local test methods are characterised by observation and control being specified in terms of events occurring within the system under test, at the layer boundaries immediately above and below the IUT. External test methods, on the other hand, are characterised by the observation and control of PDUs taking place externally from the system under test, on the other side of the underlying service provider from the IUT. The local test methods are really only applicable to in-house testing by suppliers, whereas the external test methods are also applicable to testing carried out by users and third party test centres. They also have the advantage of being applicable to a realistic communications environment.

This paper concentrates on the test methods which are likely to be most widely used, namely the external test methods for testing end-systems. The local and relay test methods are discussed elsewhere [4].

There are three types of external test method: **distributed, coordinated** and **remote**. Each comes in three variant forms: **single-layer, multi-layer** and **embedded**. Single-layer methods are designed for testing a single-layer without reference to the layers above it. Multi-layer methods are designed for testing a multi-layer IUT as a whole. Embedded methods are designed for testing a single layer within a multi-layer IUT, using the knowledge of what protocols are implemented in the layers above the layer being tested. In fact, the preferred way of testing a multi-layer IUT is by incremental use of embedded methods, starting at the lowest layer and working up to the top.

The descriptions of the external test methods refer to a lower tester (LT) and an upper tester (UT). The **lower tester** is the abstraction of the means of providing control and observation on the other side of the underlying service provider from the IUT. The **upper tester** is the abstraction of the means of providing control and observation of the upper boundary of the IUT. The rules for cooperation between the lower and upper testers during testing are called the **test coordination procedures** (TCP). The external test methods vary according to their ability to define a **test management protocol** (TMP) to carry out the test coordination procedures, or to express the test coordination procedures only in terms of requirements.

In order to describe the test methods in a general way, independent of their variants, the IUT is taken to be an implementation of the (N-b) to (N+t)-protocols, with the (N)-protocol being the lowest which is being tested. Then for the single-layer variant t=b=0; for the multi-layer variant t≥1 and b=0; and for the embedded variant t≥1 and b≥0. Furthermore, the service primitives which are controlled and observed by the lower tester are referred to as (N-1)-ASP"s to distinguish them from the corresponding (N-1)-ASPs which occur in the system under test.

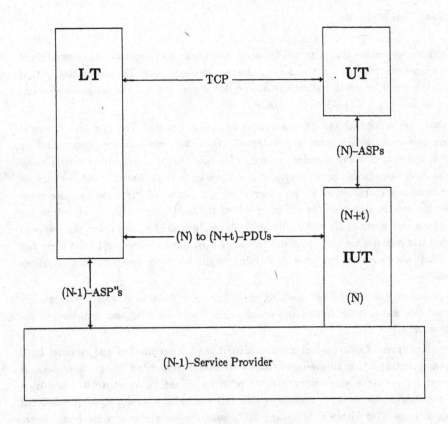

Figure 3. THE DISTRIBUTED TEST METHODS

8.1. The Distributed Test Methods

This type of test method is illustrated in Figure 3.

In these methods:-

(a) the abstract test suite is specified in terms of control and observation of (N-1)-ASP"s and (N) to (N+t)-PDUs;

(b) the abstract test suite is also specified in terms of control and observation of (N+t)-ASPs;

(c) the requirements for the test coordination procedures are specified in the abstract test suite but no assumption is made regarding their realization;

(d) the upper tester is required to achieve the effects of control and observation of the specified (N+t)-ASPs and to achieve the effects of the required test coordination procedures; no other assumptions are made.

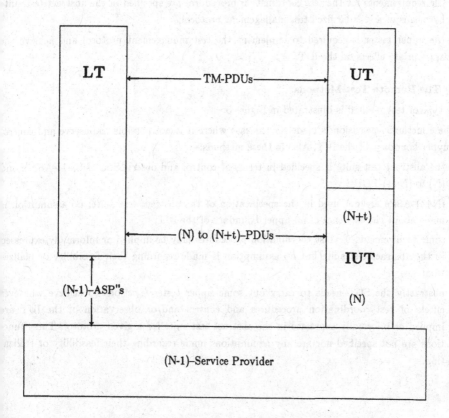

Figure 4. THE COORDINATED TEST METHODS

These methods require either direct or indirect access to the upper boundary of the IUT and a mapping between the (N+t)–ASPs and their realization within the SUT. Direct access to the upper boundary means by use of a service interface, whereas indirect access means by use of some other mapping between ASPs and real effects.

8.2. The Coordinated Test Methods

This type of test method is illustrated in Figure 4.

In these methods:–

(a) the abstract test suite is specified in terms of control and observation of (N-1)–ASP"s, (N) to (N+t)–PDUs and test management PDUs;

(b) (N+t)–ASPs need not be used in the specification of the abstract test suite; no assumption is made about the existence of an upper boundary of the IUT;

(c) the requirements for the test coordination procedures are specified in the abstract test suite by means of a standardized test management protocol;

(d) the upper tester is required to implement the test management protocol and achieve the appropriate effects on the IUT.

8.3. The Remote Test Methods

This type of test method is illustrated in Figure 5.

In these methods, provision is made for the case where it is not possible to observe and control the upper boundary of the IUT. Also in these methods:–

(a) the abstract test suite is specified in terms of control and observation of (N-1)–ASP"s and (N) to (N+t)–PDUs;

(b) (N+t)–ASPs are not used in the specification of the abstract test suite; no assumption is made about the existence of an upper boundary of the IUT;

(c) some requirements for test coordination procedures may be implied or informally expressed in the abstract test suite but no assumption is made regarding their feasibility or realization;

(d) abstractly the SUT needs to carry out some upper tester functions to achieve whatever effects of test coordination procedures and control and/or observation of the IUT are implied or informally expressed in the abstract test suite for a given protocol. These functions are not specified nor are any assumptions made regarding their feasibility or realization.

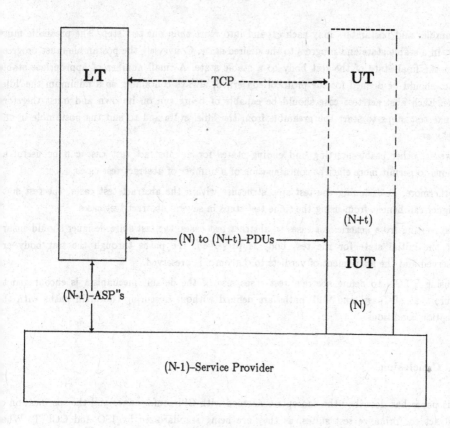

Figure 5. THE REMOTE TEST METHODS

9. Abstract Test Case Specification

The abstract test suite specifier should select an appropriate notation in which to define the abstract test cases. ISO and CCITT have defined an informal for this purpose: the Tree and Tabular Combined Notation (TTCN) [5]. It is recommended that if the test suite specifier is not already committed to another notation, then TTCN should be chosen.

Once the test notation and test method have been chosen, then the generic test cases can be expanded into abstract test cases. There are two main kinds of change required to convert a generic test case into an abstract test case. The first is to express the test body in terms of control and observation required by the test method and if relevant express the synchronization needed between upper and lower testers. The second kind of change is to fill in the detail of the preamble and postamble.

Preambles and postambles may each expand into more than one test step. The preamble must start in a stable state and progress to the desired state. Conversely the postamble must progress from the final state of the test body to a stable state. A small number of appropriate stable states should be defined for the protocol concerned, always containing as a minimum the 'idle' state. Each abstract test case should be capable of being run on its own and must therefore include test steps to start the preamble from the 'idle' state and to end the postamble in the 'idle' state.

However, other stable starting and ending states for an abstract test case can be useful as options to permit more efficient concatenation of a number of abstract test cases.

Furthermore, in designing the test step structure within the abstract test cases, the test suite designer can benefit from using the same test steps in several abstract test cases.

In converting from generic test cases to abstract test cases, the test suite designer should ensure that the initial state for the test body is preserved, the paths through the test body are preserved and the assignment of verdicts to outcomes is preserved.

In using TTCN to define abstract test cases, use of the default mechanism is encouraged to ensure that all 'pass' and 'fail' paths are defined without obscuring the main paths with the exception conditions.

10. Conclusion

This paper has described the concepts concerned with OSI conformance and the specification of abstract conformance test suites, as they are being standardized by ISO and CCITT. When completed, the standard will contain 6 parts covering:–

   (i)   general concepts;

   (ii) abstract test suite specification;

   (iii) executable test suite derivation;

   (iv) requirements for protocol implementors;

   (v) test execution;

   (vi) interpretation of test reports.

This paper draws on the work done on the first two of these parts, which are well advanced [2]. The other four, however, are only at a very early stage of development. When they are complete there will be a specific part for each of the different rôles that people can play in relation to testing: test suite designers; test realizers; protocol implementors; test operators; and users of tested products.

This standard will lead the way to the production of standard conformance test suites for OSI protocols and to the harmonization of testing activities carried out by different test centres, suppliers and users. This in turn should enable OSI to succeed in its objective of widespread inter working of open systems.

Acknowledgements

The author, as ISO rapporteur for OSI conformance testing, is in a privileged position for producing this paper. He wishes to acknowledge the help of both the ISO and CCITT rapporteur groups, without which this paper could not have been written. He also wishes to acknowledge the advice and encouragement of his colleagues at NPL and in the NCC COMMS-AID team at the National Computing Centre, Manchester, UK.

References

1. ISO 7498, Information processing systems – Open Systems Interconnection – Basic Reference Model, International Organization for Standardization, 1984.

2. D. Rayner (rapporteur), OSI conformance testing methodology and framework, Parts 1-6, San Diego, April 1986 (unpublished).

3. D. Rayner, Towards an objective understanding of conformance, in: Protocol Specification, Testing and Verification III, proc. 3rd international workshop on this subject, held in Zurich on 31 May – 2 June 1983, edited by H. Rudin and C.H. West, North Holland, 1983, 477-492.

4. D. Rayner, Towards standardized OSI conformance tests, in: Protocol Specification, Testing and Verification V, proc. 5th international workshop on this subject, held in Toulouse-Moissac on 10-13 June 1985, edited by M. Diaz, North Holland, 1985, 441-460.

5. J.R. Pavel (editor), The tree and tabular combined notation, Annex D of Part 2 of [2], San Diego, April 1986 (unpublished).

EUROPEAN HARMONISATION OF OSI USAGE AND CONFORMANCE TESTING SERVICES

Hubert ZIMMERMANN
Centre National d'Etude des Télécommunications
Issy les Moulineaux - FRANCE

Abstact

Actual usage of OSI layer standards in products implies choices of options and parameter values, as well as test tools to check conformity of products to protocol specifications. In order to promote usage of OSI standards, and avoid inconsistent implementations, the European Community has established a European-wide harmonisation program in this area. This program includes the definition of OSI functional standards and the setting up of OSI conformance testing services. This paper presents an overview of this harmonisation program.

1. INTRODUCTION

As time goes on, the structure of data processing systems will have to match closer and closer the structure of human organisations which use them. Just as any human or any group of humans need to communicate and interwork with a variety of other humans and groups of humans, all computers will have to be capable of interworking with a variety of other computers. In other words, computers will have to be able to communicate regardless of their origin and of their size, i.e. interwork openly, as required by the organisations which use them.

The need for open networking was recognised before the end of the seventies, and the OSI standardisation program was launched in 1978 with the objective of providing gradually the international standards required the support progressive development of open networking.

Two major factors have influenced the organisation of the OSI standardisation program:

(1) The inherent complexity of networking standards imposed a top-down approach, in which the overall organisation of networking functions (the OSI Reference Model) is defined first and serves as a basis for specifying in details the functions to be performed (the Layer Services) and the Protocols to be used to this effect.

(2) The dynamics of open networking requirements which grow from simple at the beginning to more and more sophisticated as time goes on, imposes a stepwise approach in which basic OSI standards are developed first to meet the most urgent requirements, while extensions are planned and progressivily developed.

Today, in 1986, the basic set of OSI standard is available. It contains:

a) Basic standards for layers 1 through 3, covering packet switching, circuit switching and local area networks,

b) Basic standards for layer 4 covering a variety of application requirements on a variety of telecommunication networks,

(c) Basic standards for layer 5 covering dialogue control requirements of most known applications,

(d) Basic standards for layers 6 and 7 corresponding to the most demanded open networking applications, i.e. Messaging, File Transfer and Document Transfer.

The availability of this initial set of OSI standards represents a major step in the development effort towards open networking, and most computer manufacturers have announced products to support these standards.

However, it turns out that these standards alone are not sufficient to guarantee actual open networking between products which conform to them. Indeed, OSI standards have left open options and parameter values which, if chosen inconsistently, may lead to incompatibilities among products.

It is therefore essential to move one step forward in the standardisation program, and define precisely which options are to be retained and which parameter values are to be choosen in each specific interworking case.

In order to avoid incompatible application of OSI standards by European nations, the European Community took the initiative to set up a program for harmonised application of OSI standards throughout Europe. This initiative is described in section 3.

Furthermore, actual usage of standards implies that conformity of products to standards can be checked by adequate test tools. However, there is a risk that these tools introduce incompatibilities, for instance by imposing different interpretations of the standards.

Here again, in order to avoid the setting up of inconsistent conformance test services, the European Community has launched a program for the setting up of conformance testing services harmonised at European level. This program is described in section 4.

These European actions for harmonised application of OSI standards are part of a general policy on Information Technology adopted by the European Community. This background policy is outlined in Section 2.

2. HARMONISED APPLICATION OF IT STANDARDS IN EUROPE

2.1. Objectives

During the past few years, it became clearer and clearer that Information Technology was going to be a key factor in the development of the European Economic Community, and that the problem could no longer be tackled in a piecemeal fashion by each European nation.

A consensus emerged on the fact that further development of Information Technology should be based on a voluntary standardisation policy with four major objectives, outlined below:

(a) Helping to establish a genuine Community information technology and telecommunications market.

The rapid implementation of a Community-wide coordinated strategy for standards is essential in view of the urgency and importance of this issue for the future of information technology and telecommunications in the Community.

The adoption of common standards is vital to:

- the establishment of a more transparent and competitive Community information technology and telecommucations market,

- the ability of Community industry to gain maximum advantage from the Community dimension and to satisfy the pressure of demand for fast-developing products and systems that are more and more frequently interconnected.

(b) Improving conditions of competition

The Community market is today divided between the customers of the different firms. Such a commercial and industrial stategy requires that firms obtain a large enough market share, either on the basis of their own products or on the basis of licensing or subcontracting agreements enabling them to supply all the products and services required.

This will be possible only if the harmonised application of IT standards do open the market by removing the barriers among contries as well as among proprietary IT architectures.

(c) The efficiency of information exchanges in the Community

Information exchanges are increasingly dependent on conventions established between data-processing systems connected by networks. Standardization provides an essential basis for these exchanges and their efficiency is directly dependent upon it. It must however, be effected in such a way that IT standards and common technical specifications which apply to the telecommunications sector contribute to the efficiency of these exchanges throughout the Community.

(d) Taking user requirement into account

Users are faced with incompatibility problems that can only be solved by costly conversions and that reduce the reliability of their systems. User requirements, especially as regards the compatible working of systems, must be taken into account so that they can assemble their equipment as a function of the work profiles required.

2.2. European policy for IT standards

The importance of this European policy for IT standards was recognized officially in June 84 by the Ministers of Industry of the European Community Member States. The decision was made to set up a European IT standardisation program as a chain of linked elements:

- the choice of reference international standards corresponding to the needs expressed;

- the polishing of standards to make them directly useable or the development of functional standards to provide services based on the chaining of several reference standards;

- conformance testing without which applications would inevitably differ;

In order to implement this harmonisation policy, a joint association was set up between CEN, CENELLEC and CEPT.

In the following of this paper, we will focus on the OSI part of this standardisation policy.

The initial choice of reference international standards was agreed at the beginning of 1985. It is described in the CEN/CENELEC Harmonisation Document number 1 (HD 0001).

The development of functional standards was immediatly started (see section 3 below) and a program for the setting up of conformance testing service was launched by the EEC (see section 4 below).

3. OSI FUNCTIONAL STANDARDS

3.1. The concept of OSI functional standard

The OSI standards currently published by ISO and CCITT emphasise the generality of layer standards which are able to support a variety functions and network configurations. Depending on the actual configuration used and the application supported, specific options and parameter values have to be chosen within each layer of the OSI Reference Model.

Products designed to have only a limited capability will not need to implement many of the options defined in some of the relevant protocol standards. For example, where a wide range of values is allowed (but not mandatory) in the protocol standard for a particular parameter, it may be appropiate to support only a small subset of values. Different users and different suppliers will tend to be interested in different subsets.

Anarchy would be the result of uncontrolled selection fo subsets, options and parameters, with no openness possible as envisaged by the OSI Model. Where a number of ways exist to combining protocols from adjacent layers to perform a particular function, it is certain that each valid one (as well as some invalid ones) will be chosen by at least on implementer or user, unless some constraints on this freedom can be agreed and maintained.

It is to control this aspect of applying standards to products that the concept of Functional Standards has been introduced for the OSI environment.

More precisely, a Functional Standard specifies the application of one or more OSI standards in support of a specific requirement for communication between computer systems.

A Functional Standard does not alter the standards to which it refers, but makes explicit the relationships among a set of standards used together (relationships which are implicit in the definitions of the standards themselves) and may also specify particular details of each standard being used.

It follows that a Functional Standard:

- does not require any change to the structure defined by the Basic Reference Model for OSI;

- does not alter the nature of the conformance requirements for fully open systems of the standards to which it refers (though it may specify other requirements which are relevant to their use in the particular environment for which it is intended);

- does not define the total OSI interworking functionality of a system.

Each Functional Standard is a document which comprises:

(a) a simple definition of the function;

(b) an illustration of the scenario within which the function is applicable;

(c) a single working set of standards, including precise references to the actual texts of the standards being used, and any other relevant source documents;

(d) specifications of the application of each referenced standard, covering recommendations on the choices of classes or subsets, and the selection of options, ranges of parameter values, etc.

(e) where necessary, recommendations on the resolution of ambiguities in the working set of standards and on the correction of errors within them. (But these are not the primary purpose of a Functional Standard, and would be relevant only until the appropriate amendments were applied to the standard concerned);

(f) a statement defining the requirements which must be observed by products claiming conformity, including any remaining permitted options of the referenced standards, which thus become options of the Functional Standard.

As far as end systems are concerned, two major types of Functional Standard are defined:

(a) Application Functions (A-Functions) governing the relationship between an end system and another end system independently of the telecommunications facilities being used. The Functional standards for A-Functions define the precise usage of OSI standards in layers 5 through 7 for a given application (e.g. Electronic Mail);

(b) Telecommunication Functions (T-Functions) governing the use of telecommunications facilities by end systems. The Functional standards for T-Functions define the precise usage of OSI standards in layers 1 through 4 for a given networking configuration (e.g. on a public packet switching network).

3.3. Initial set of Functional Standards

the initial set of Functional Standards available around mid 1986 included in particular:

(a) T-Functions

- Transport over an X25 Public Packet Switching Data Network
- Transport over an X21 Public Circuit Switching Data Network
- Transport over a Local Area Network;

(b) A-Functions

- Document Transfer via Teletex
- Acces to Public Message Handling Service
- Acces to Private Message Handling Service
- File Transfer.

A complete program has been set up to cover progressively the most urgent needs for Functional Standards.

4. OSI CONFORMANCE TESTING SERVICES (CTS)

4.1. Background

It is generally agreed that it is practically impossible to ensure harmonized application of standards of extreme complexity, as in the case of OSI Functional Standards, if testing services are not available to verify the conformance of products to such standards. Furthermore conformance testing services are essential to ensure the credibility of standards for the end users.

Moreover, it is essential that the various conformance test services intalled in different countries be consistent with each other, i.e. conduct equivalent tests providing equivalent results on identical products.

To this effect, the Commission of the European Community has launched at the beginning of 1986, a set of contracts for development and installation of harmonised OSI conformance testing services. (In parallel with this technical program, the procedures for certification and mutual recognition of certificates are being discussed at European level with the objective of setting up rapidly a European certification scheme).

The OSI-CTS program was divided into two parts:

(a) WAN-CTS for the testing of A-Functions and T-Functions over Wide Area Networks;

(b) LAN-CTS for the testing of T-Functions over Local Area Networks.

In the following, we will concentrate on the WAN-CTS part of the program, which the author of this paper is more familiar with.

4.2. Participants in the WAN-CTS program

In order to ensure optimum harmonisation it was felt important to involve directly in the project those countries which had launched or were about to launch the development of a national OSI conformance testing service.

The main contractors involved in the WAN-CTS project are:

- British Telecom (BT) (UK)
- Centre National d'Etudes des Télécommunications (CNET) (F)
- Centre Studi e Laboratori Telecommunicazioni (CSELT) (I)
- Fernmelde Technisches Zentralamt (FTZ) (D)
- National Computing Center (NCC) (UK)
- Communication Laboratory Post og Telegrafvaesenet (PTT) (DK)
- Compania Telefonica Nacional Espana (CTNE) (ES).

The WAN-CTS program also involves also a number of subcontractors from research laboratories and from the industry.

4.3. Organisation of the WAN-CTS project

The WAN-CTS project is divided into six technical area:

(a) Methodology

This area covers all aspects of harmonisation of tests and test procedures among the other technical areas. This part of the work is conducted in close relationship with the ISO work on Conformance Testing Methodology and Framework.

(b) Networks

This area covers tests for layers 1 through 3 of the OSI Reference Model.

(c) Transport and Session

(d) Teletex

(e) MHS

(f) FTAM

4.4. Test services to be offered

Indeed a comprehensive range of services is needed to meet the requirements of the various types of products to be tested.

For example a 7-layer product may need all layers tested, layers 1-7 tested, layers 4-7 tested, layers 5-7 tested or layers 6-7 to be tested.. Moreover, within these categories the product can be either layered (with accessible interfaces) or monolithic (with no accessible interfaces).

Each test service offered to customers is characterised by:

- The protocol layers to be tested in the product under test.

- The lower layer services needed in the product in order to access the test service.

- The need (if any) to access inter-layer boundaries in the product under test.

The precise definition of the test services to be offered by the WAN-CTS project is still under discussion. However, the following table gives an indication of the structure of test services currently envisaged.

TEST SERVICE	PROTOCOL LAYERS IN PRODUCT TO BE TESTED						
	1	2	3	4	5	6	7
--NET--							
(a) X.21 bis	├──┤------------------------------------						
(b) X.21	├────────────┤----------------------						
(c) X.25	......├──────────┤-------------------						
(d) X.75	......├───┤--------------------------						
--Transport & Session--							
(e) Layer 4 only	...................├────┤[UT]						
(f) Layer 5 only	.........................├────┤[UT]						
--Teletex--							
(g) X.25 terminals	...................├────────────────						
(h) X.21 terminals	...................├────────────────						
--MHS--							
(i) Layers 4-7	...................├────────────────						
(j) Layers 6-7	...............................├────						
--FTAM--							
(k) Layers 6-7	...............................├────						

Notes

├──────┤	=	Protocol layers to be tested in the product under test.
....	=	Lower layer services in the product.
-------	=	Upper layer services in the product.
[UT]	=	Upper Tester used to control an accessible interface in the product.

4.5. Examples of Products to be Tested

Products submitted for testing will make use of one or more of the test services identified in the previous section. Three examples follow:

Example 1

An FTAM product for use on a X.25 network allowing access to both the Transport and Session interfaces will require the following services:

(a) or (b) for layer 1
(c) for layers 2 and 3
(e) for layer 4
(f) for layer 5
(k) for layers 6 and 7.

Example 2

A Teletex product for use on a X.21 circuit and with no access to internal inter-faces will require the following services:

(a) or (b) for layer 1
(d) for layer 2
(h) for layers 4 - 7.

Example 3

An MHS product for use on an X.25 network and with no access to internal interfaces will require the following services:

(a) or (b) for layer 1
(c) for layers 2 and 3
(i) for layers 4 - 7.

5. CONCLUSION

The benefits expected from OSI standards will be obtained only if beyound the production of layer standards by ISO and CCITT the additional effort is done to put
them consistently in practice throughout the world.

To this effect, the European Community has taken the initiative of setting up a
program for consistent application of OSI standards throughout Europe. This program
includes the definition of functional standards and the setting up of conformance
testing services.

Other similar initiatives follow in the United States and in Japan.

There is hopefully a declared willingness to coordinate among those initiatives so
as to ensure world-wide consistency of application of OSI standards.

ENC - IBM European Networking Center

Priv.-Doz. Dr. Günter Müller
IBM European Networking Center
Tiergartenstr. 15
D - 6900 Heidelberg
EARN/BITNET: MUE at DHDIBM1

What is ENC

ENC, the IBM European Networking Center, was established in Heidelberg, Germany on July 16, 1985. It is IBM Europe's research and competence center for advanced projects in networking and telecommunications.

Objectives of ENC

ENC seeks to be a focal point for research and experimentation in telecommunications and advanced networking systems. In this role ENC builds a variety of prototype networks, including OSI (Open Systems Interconnection).

The center conducts research and development in the area of communication software. Its emphasis is on Open Systems Interconnection, advanced applications in computer networks, software engineering tools for rapid prototyping, and operating systems support for advanced networks.

ENC is open to cooperation with leading European research partners from the scientific and industrial community. The Center is currently cooperating closely with 16 European Universities, several major public research institutes, and with the German Research Network (DFN), as well as EARN.

ENC shares its results with the research community. ENC scientists publish and contribute to international journals and conferences. Some of them also lecture at universities throughout Europe, and are active members of international professional societies.

ENC participates in several international standards committees, e.g. ISO and ECMA and participates in the technical definition of new communication standards.

Research Areas

Currently, the main research activities of ENC are the development of prototypes in Network Operating Systems, Distributed Applications, Conformance Testing, new concepts for European Research Networks, and an investigation of Integrated Services Digital Networks (ISDN).

Network Operating Systems

The Network Operating Systems Research activities aim at easing the development of distributed applications by end users by providing generic programming support services and environments.

The emphasis of this project is to provide transparent access to network resources. A user can access remote facilities, e.g. file servers, as if they were locally available. Instead of building a new operating system from scratch, the local operating systems in the heterogeneous networks are augmented by network oriented functions. The ENC has designed and developed together with the University of Karlsruhe a global transport system and a remote service call facility for this purpose. These extensions can be ported into all network nodes without affecting the standard use of the local operating systems.

The ENC prototype of this network operating system currently includes three IBM mainframes under VM/CMS, a number of IBM Personal Computers, and VAX machines under VMS. The lower layer network services are provided by the IBM token ring and an Ethernet bus.

The local area network expertise of the ENC is also being exploited in a joint project with CERN, the European Laboratory for Particle Physics in Geneva. Here the ENC is helping to set up experimental control facilities using the IBM Token-Ring Local Area Network for interconnecting nine buildings distributed along the 27km circumference particle accelerator ring.

Distributed Applications

The number of applications available to the end user is currently limited in the practical OSI world of today. Indeed only the Teletex and Message Handling Services (MHS) services are currently available. Hence acceptance of OSI will depend largely upon the development of new applications for the end user. These new applications are expected to arise in both the office and manufacturing environments.

One such application is access to remote databases in a network. ENC is participating in ECMA Technical Committee 22 to develop an international standard for Remote Data Base Access (RDA). This facility would allow end user applications to access databases in an OSI network without knowing the operating environment of the data base system. Based on the experience with these protocols a distributed database environment is under study and will be developed in a joint project with the University of Frankfurt. Here a novel database programming language allows easy developments of distributed database applications on workstations.

In order to facilitate the development of OSI software a specification technique called PASS (Parallel Activity Specification Scheme) was developed and used for session and transport layer implementation at ENC. PASS consists of an abstract specification language and a generator for automatic code production.

Conformance Testing

A basic requirement for the interconnection of Open Systems is total conformance of the participating systems with the communication standards. For this purpose the ENC is developing a generalized tool to allow rigorous conformance testing of all higher OSI layers. The tool consists of a test driver, test responder, and a language to describe test scenarios. Unlike some existing test tools the ENC tool can be tailored to incorporate new layers and applications that are not yet standardized.

European Research Networks

In Europe many countries are developing their own networks for their research community. The danger of incompatibility is great and hence, in order to preserve and foster compatibility in European Research Networks, close cooperation across the European borders is required.

The ENC activities in this area have resulted in two prototypes:

- The X.400 gateway between EARN (European Academic and Research Network) and any upcoming OSI network providing message handling services

- A native X.400 message handling system using PROFS as a user agent

Both prototypes were designed and developed jointly with GMD (Gesellschaft fuer Mathematik und Datenverarbeitung). They are based upon standards developed by ISO and CCITT. Under the guidance of DFN and GMD this experimental code was demonstrated at the Hannover fairs in 1985 and 1986 with international multi-vendor participation.

This project also assists the EARN community in migrating to OSI protocols. The ENC has provided its prototype software to conduct a first international experiment for OSI networks on a European scale. A later version of the software is to be used within the German Research Network. These experiments will allow the ENC to study OSI network management issues and performance of OSI in general.

Current research is being concentrated on directory services for message handling and computer conferencing.

Integrated Services Digital Network (ISDN)

In a new project the ENC will investigate broadband communication and its effects on present day communication architectures, e.g. OSI standards, and applications in ISDN networks. The environment to study high speed links, gateways to local area networks, and network management issues will be provided by the German PTT in the BERKOM project in Berlin. BERKOM provides ISDN services over fiber optics links, and studies open protocol questions and potential applications for high speed networks.

List of abbreviations

ISO	International Standards Organization
CCITT	Comité Consultatif International de Télégraphique et Téléphonique
DFN	Deutsches Forschungsnetz (German Research Network)
GMD	Gesellschaft fuer Mathematik und Datenverarbeitung
VMS	Virtual Memory System. An operating system on DEC machines
VM/CMS	Virtual Machine/Communication Monitor System. An IBM /370 operating system
MHS	Message Handling System
EARN	European Academic and Research Network
VAX	Virtual address extension. Name of a DEC Mini.
BERKOM	Berliner Kommunikationssystem

Vol. 219: Advances in Cryptology – EUROCRYPT '85. Proceedings, 1985. Edited by F. Pichler. IX, 281 pages. 1986.

Vol. 220: RIMS Symposia on Software Science and Engineering II. Proceedings, 1983 and 1984. Edited by E. Goto, K. Araki and T. Yuasa. XI, 323 pages. 1986.

Vol. 221: Logic Programming '85. Proceedings, 1985. Edited by E. Wada. IX, 311 pages. 1986.

Vol. 222: Advances in Petri Nets 1985. Edited by G. Rozenberg. VI, 498 pages. 1986.

Vol. 223: Structure in Complexity Theory. Proceedings, 1986. Edited by A. L. Selman. VI, 401 pages. 1986.

Vol. 224: Current Trends in Concurrency. Overviews and Tutorials. Edited by J. W. de Bakker, W.-P. de Roever and G. Rozenberg. XII, 716 pages. 1986.

Vol. 225: Third International Conference on Logic Programming. Proceedings, 1986. Edited by E. Shapiro. IX, 720 pages. 1986.

Vol. 226: Automata, Languages and Programming. Proceedings, 1986. Edited by L. Kott. IX, 474 pages. 1986.

Vol. 227: VLSI Algorithms and Architectures – AWOC 86. Proceedings, 1986. Edited by F. Makedon, K. Mehlhorn, T. Papatheodorou and P. Spirakis. VIII, 328 pages. 1986.

Vol. 228: Applied Algebra, Algorithmics and Error-Correcting Codes. AAECC-2. Proceedings, 1984. Edited by A. Poli. VI, 265 pages. 1986.

Vol. 229: Algebraic Algorithms and Error-Correcting Codes. AAECC-3. Proceedings, 1985. Edited by J. Calmet. VII, 416 pages. 1986.

Vol. 230: 8th International Conference on Automated Deduction. Proceedings, 1986. Edited by J. H. Siekmann. X, 708 pages. 1986.

Vol. 231: R. Hausser, NEWCAT: Parsing Natural Language Using Left-Associative Grammar. II, 540 pages. 1986.

Vol. 232: Fundamentals of Artificial Intelligence. Edited by W. Bibel and Ph. Jorrand. VII, 313 pages. 1986.

Vol. 233: Mathematical Foundations of Computer Science 1986. Proceedings, 1986. Edited by J. Gruska, B. Rovan and J. Wiedermann. IX, 650 pages. 1986.

Vol. 234: Concepts in User Interfaces: A Reference Model for Command and Response Languages. By Members of IFIP Working Group 2.7. Edited by D. Beech. X, 116 pages. 1986.

Vol. 235: Accurate Scientific Computations. Proceedings, 1985. Edited by W. L. Miranker and R. A. Toupin. XIII, 205 pages. 1986.

Vol. 236: TEX for Scientific Documentation. Proceedings, 1986. Edited by J. Désarménien. VI, 204 pages. 1986.

Vol. 237: CONPAR 86. Proceedings, 1986. Edited by W. Händler, D. Haupt, R. Jeltsch, W. Juling and O. Lange. X, 418 pages. 1986.

Vol. 238: L. Naish, Negation and Control in Prolog. IX, 119 pages. 1986.

Vol. 239: Mathematical Foundations of Programming Semantics. Proceedings, 1985. Edited by A. Melton. VI, 395 pages. 1986.

Vol. 240: Category Theory and Computer Programming. Proceedings, 1985. Edited by D. Pitt, S. Abramsky, A. Poigné and D. Rydeheard. VII, 519 pages. 1986.

Vol. 241: Foundations of Software Technology and Theoretical Computer Science. Proceedings, 1986. Edited by K. V. Nori. XII, 519 pages. 1986.

Vol. 242: Combinators and Functional Programming Languages. Proceedings, 1985. Edited by G. Cousineau, P.-L. Curien and B. Robinet. V, 208 pages. 1986.

Vol. 243: ICDT '86. Proceedings, 1986. Edited by G. Ausiello and P. Atzeni. VI, 444 pages. 1986.

Vol. 244: Advanced Programming Environments. Proceedings, 1986. Edited by R. Conradi, T. M. Didriksen and D. H. Wanvik. VII, 604 pages. 1986

Vol. 245: H. F. de Groote, Lectures on the Complexity of Bilinear Problems. V, 135 pages. 1987.

Vol. 246: Graph-Theoretic Concepts in Computer Science. Proceedings, 1986. Edited by G. Tinhofer and G. Schmidt. VII, 307 pages. 1987.

Vol. 247: STACS 87. Proceedings, 1987. Edited by F. J. Brandenburg, G. Vidal-Naquet and M. Wirsing. X, 484 pages. 1987.

Vol. 248: Networking in Open Systems. Proceedings, 1986. Edited by G. Müller and R. P. Blanc. VI, 441 pages. 1987.